Verlagsgesellschaft mbH, D-6940 Weinheim (Federal Republic of Germany), 1990

>ution

Verlagsgesellschaft, P. O. Box 10 1161, D-6940 Weinheim (Federal Republic of Germany)

rland: VCH Verlags-AG, P. O. Box, CH-4020 Basel (Switzerland)

d Kingdom and Ireland: VCH Publishers (UK) Ltd., 8 Wellington Court, Wellington Street,
Cambridge CB1 1HZ (England)

ind Canada: VCH Publishers, Suite 909, 220 East 23rd Street, New York, NY 10010-4606 (USA)

527-27984-9 (VCH Verlagsgesellschaft)          ISBN 0-89573-937-2 (VCH Publishers)

**En**
**in**

edit
Wo

© VC

Dist
VC
Swi
Uni

USA

ISBN 3

# Enzymes in Industry

## Production and Applications

Edited by Wolfgang Gerhartz

Dr. Wolfgang Gerhartz
Degussa AG
ZN Wolfgang
P. O. Box 1345
D-6450 Hanau 1
Federal Republic of Germany

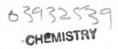

The *cover illustration* shows the Stuart-Briegleb model of lysozyme. It was produced by E. E. POLYMEROPOULOS of ASTA Pharma AG, Frankfurt, Federal Republic of Germany. The data were taken from the Cambridge Crystallographic Data System (F. H. Allen, O. Kennard and R. Taylor, *Acc. Chem. Res.* **16,** 146–153 (1983)). The picture was generated on a Silicon Graphics workstation using the molecular modelling program MOLCAD of the Darmstadt Institute of Technology, Darmstadt, Federal Republic of Germany.

Published jointly by
VCH Verlagsgesellschaft, Weinheim (Federal Republic of Germany)
VCH Publishers, New York, NY (USA)


Editorial Director: Dr. Hans-Joachim Kraus
Production Manager: Myriam Nothacker

Library of Congress Card No.: 90-12099

British Library Cataloguing-in-Publication Data:
Enzymes in industry.
  1. Industrial chemicals: Enzymes
  I. Gerhartz, Wolfgang
  661.8

  ISBN 0-89573-937-2 U.S
  ISBN 3-527-27984-9 W. Germany


Deutsche Bibliothek Cataloguing-in-Publication Data:
**Enzymes in industry** : production and applications / ed. by Wolfgang Gerhartz. – Weinheim ; Basel (Switzerland) ; Cambridge ; New York, NY : VCH, 1990
  ISBN 3-527-27984-9 (Weinheim ...)
  ISBN 0-89573-937-2 (New York)
NE: Gerhartz, Wolfgang [Hrsg.]

Composition, printing and bookbinding: Graphischer Betrieb Konrad Triltsch, D-8700 Würzburg
Printed in the Federal Republic of Germany

# Foreword

"Enzymes" was the last major manuscript I edited for "*Ullmann's Encyclopedia of Industrial Chemistry*" – and it was the most rewarding article in my ten-year period as executive editor of Ullmann's.

Its careful planning was headed by Professor Günter Schmidt-Kastner at Bayer AG in Wuppertal, Federal Republic of Germany, who invited two other German enzyme specialists to a small symposium with the sole purpose of determining the contents of the article and proposing potential authors. The two other specialists were Professor Maria-Regina Kula of Düsseldorf University and Dr. Georg-Burkhard Kreße of Boehringer Mannheim, Tutzing. The help and advice given by these three experts are gratefully acknowledged.

The response of the suggested authors was very enthusiastic. The list of authors (page XIII) reads like an international "Who is Who" of industrial enzymology.

I accepted the publisher's invitation to make a monograph from the Ullmann article with a great deal of pleasure and cooperation with the authors has once again been perfect. All chapters have been updated, and the literature is covered until late 1989. Even the latest nomenclature of DNA modification methyltransferases has been included.

The book is recommended to all those non-enzymologists who want to know what is really happening as regards enzymes in industry. The term "industry" is used in its broadest sense: it includes the production of enzymes as well as their use in the production of bulk products, such as detergents, glucose, or fructose; in fine chemicals synthesis; in food processing and food analysis; in clinical diagnosis and therapy; and – last but obviously not least – in genetic engineering.

I hope that the reader will derive as much pleasure from reading this book as I did from producing it.

The Editor

# Contents

# Authors

RICHARD N. PERHAM, University of Cambridge, Cambridge, England
(Chap. 1: Fermentation)

MARIANNE GRASSL, GERHARD MICHAL, BERNHARD REXER, Boehringer Mannheim
GmbH, Penzberg and Tutzing, Federal Republic of Germany (Chap. 2: Catalytic
Activity of Enzymes)

ANDREAS HERMAN TERWISSCHA VAN SCHELTINGA, Gist-brocades N.V., Delft,
The Netherlands (Section 3.1: Fermentation)

CHRISTIAN GÖLKER, Bayer AG, Wuppertal, Federal Republic of Germany
(Section 3.2: Isolation and Purification)

SABURO FUKUI, ATSUO TANAKA, Department of Industrial Chemistry,
Faculty of Engineering, Kyoto University, Kyoto, Japan
(Section 3.3: Immobilization)

HELMUT UHLIG, Röhm GmbH, Darmstadt, Federal Republic of Germany
(Sections 4.1, 4.5, 4.7, and 4.8: Survey of Industrial Enzymes; Enzymes in Meat
Processing; Fruit, Vegetables and Wine Processing; and Hydrolysis of Protein,
Fat, and Cellulose)

WALTER E. GOLDSTEIN, ESCA Genetics Corporation, San Carlos, California 94070,
United States (Section 4.2 and Chap. 8: Enzymes in Starch Processing and
Baking; Safety and Environmental Aspects)

HANS A. HAGEN, SVEN PEDERSEN, Novo-Nordisk a/s, Bagsværd, Denmark
(Section 4.3: Glucose Isomerization)

BOB POLDERMANS, Gist-brocades N.V., Delft, The Netherlands (Section 4.4:
Proteolytic Enzymes)

ERNST H. REIMERDES, Meggle AG, Günzburg, Federal Republic of Germany
(Section 4.6: Dairy Products)

WOLFGANG LEUCHTENBERGER, ULF PLÖCKER, Degussa AG, Hanau-Wolfgang,
Federal Republic of Germany (Section 4.9: Amino Acids and Hydroxycarboxylic
Acids)

HERBERT WALDMANN, SVEN GRABOWSKI, ETHAN S. SIMON,
GEORGE M. WHITESIDES, Department of Chemistry, Harvard University,
Cambridge, Massachusetts 02138, United States (Section 4.10: Enzymes in
Organic Synthesis)

GEORG-BURKHARD KRESSE, Boehringer Mannheim GmbH, Penzberg, Federal Republic of Germany (Section 5.1: Enzymes in Analysis and Medicine, Survey)

KARL WULFF, Boehringer Mannheim GmbH, Mannheim, Federal Republic of Germany (Section 5.2: Enzymes in Diagnosis)

GÜNTHER HENNIGER, Boehringer Mannheim GmbH, Penzberg, Federal Republic of Germany (Section 5.3: Enzymes for Food Analysis)

LEOPOLD FLOHÉ, WOLFGANG A. GÜNZLER, Grünenthal GmbH, Aachen, Federal Republic of Germany (Section 5.4: Enzymes in Therapy)

CHRISTOPH KESSLER, Boehringer Mannheim GmbH, Penzberg, Federal Republic of Germany (Chap. 6: Enzymes in Genetic Engineering)

KNUD AUNSTRUP, Novo-Nordisk a/s, Bagsværd, Denmark (Chap. 7: Economic Aspects)

# Abbreviations

| | |
|---|---|
| A | adenosine |
| ACA | acetamidocinnamic acid |
| ACL | α-amino-ε-aprolactam |
| ADH | alcohol dehydrogenase |
| ADI | acceptable daily intake |
| ADP | adenosine 5′-diphosphate |
| Ala | alanine |
| Arg | Arginine |
| AMP | adenosine 5′-monophosphate |
| ATC | D,L-2-amino-$\Delta^2$-thiazoline-4-carboxylic acid |
| ATP | adenosine 5′-triphosphate |
| C | cytidine |
| cDNA | copy DNA |
| CL | citrate lyase |
| CMP | cytidine 5′-monophosphate |
| CoA | coenzyme A |
| CS | citrate synthetase |
| CTP | cytidine 5′-triphosphate |
| d | deoxy |
| *dam* | gene locus for *E. coli* DNA adenine methylase ($N^6$-methyladenine) |
| *dcm*I | gene locus for *E. coli* DNA cytosine methylase (5-methylcytosine) |
| dd | dideoxy |
| ddNTP | dideoxynucleoside 5′-triphosphate |
| DEAE | diethylaminoethyl |
| DNA | deoxyribonucleic acid |
| DNase | deoxyribonuclease |
| dNTP | deoxynucleoside 5′-triphosphate |
| DOPA | 3-(3,4-dihydroxyphenylalanine)[3-hydroxy-L-tyrosine] |
| dpm | decays per minute |
| ds | double-stranded |
| E.C. | Enzyme Commission |
| F6P | fructose 6-phosphate |
| fMet | *N*-formylmethionine |
| FMN | flavin mononucleotide |
| $FMNH_2$ | flavin mononucleotide, reduced |
| G | quanosine |
| GDP | guanosine 5′-diphosphate |

| | |
|---|---|
| Glu | glutamic acid |
| Gly | glycine |
| GMP | guanosine 5′-monophosphate |
| GOD | glucose oxidase |
| GOT | glutamate – oxaloacetate transaminase |
| G6P | glucose 6-phosphate |
| GPT | glutamate – pyruvate transaminase |
| GTP | guanosine 5′-triphosphate |
| 3-HBDH | 3-hydroxybutyrate dehydrogenase |
| HFCS | high-fructose corn syrup |
| *hsd*M | *E. coli* gene locus for methylation |
| *hsd*R | *E. coli* gene locus for restriction |
| *hsd*S | *E. coli* gene locus for sequence specificity |
| IDP | inosine 5′-diphosphate |
| Ile | isoleucine |
| INT | iodonitrotetrazolium chloride |
| ITP | inosine 5′-triphosphate |
| LDH | lactate dehydrogenase |
| Lys | lysine |
| $^m$(superscript) | methylated |
| MDH | malate dehydrogenase |
| Met | methionine |
| M6P | mannose 6-phosphate |
| mRNA | messenger RNA |
| MTT | 3-(4,5-dimethylthiazolyl-2)-2,5-diphenyltetrazolium bromide |
| N | any nucleotide |
| NAD | nicotinamide – adenine dinucleotide |
| NADH | nicotinamide – adenine dinucleotide, reduced |
| NADP | nicotinamide – adenine dinucleotide phosphate |
| NADPH | nicotinamide – adenine dinucleotide phosphate, reduced |
| NMN | nicotinamide mononucleotide |
| NTP | nucleoside 5′-triphosphate |
| p | phosphate groups |
| $^{32}$P | phosphate groups containing $^{32}$P phosphorus atoms |
| $p_i$ | inorganic phosphate |
| PEP | phosphoenolpyruvate |
| 6-PGDH | 6-phosphogluconate dehydrogenase |
| Phe | phenylalanine |
| PMS | 5-methylphenazinium methyl sulfate |
| poly(dA) | poly(deoxyadenosine 5′-monophosphate) |
| $pp_i$ | inorganic pyrophosphate |
| Pro | proline |
| PRPP | phosphoribosyl pyrophosphate |
| Pu | purine |
| Py | pyrimidine |

| | |
|---|---|
| r | ribo |
| RNA | ribonucleic acid |
| RNase | ribonuclease |
| SAM | S-adenosylmethionine |
| SMHT | serine hydroxymethyltransferase |
| ss | single-stranded |
| T | thymidine |
| TMP | thymidine 5′-monophosphate |
| tRNA | transfer RNA |
| TTP | thymidine 5′-triphosphate |
| U | uridine |
| UMP | uridine 5′-monophosphate |
| UTP | uridine 5′-triphosphate |
| Val | valine |

**Bacteriophages:**

fd
ghl
M13
N4
PBS1
PBS2
SPO1
SP6
SP15
T3
T4
T5
T7
XP12
$\lambda$
$\lambda$gt11
$\Phi$SM11
$\Phi$X174

**Plasmids:**

pBR322
pBR328
pSM1
pSP64
pSP65
pSPT18, pSPT19
pT7-1, pT7-2
pUC 18, pUC 19
pUR222

**Eukaryotic viruses:**

Ad2
SV40

# 1. Introduction

Enzymes are the catalysts of biological processes. Like any other catalyst, an enzyme brings the reaction catalyzed to its equilibrium position more quickly than would occur otherwise; an enzyme cannot bring about a reaction with an unfavorable change in free energy unless that reaction can be coupled to one whose free energy change is more favorable. This situation is not uncommon in biological systems, but the true role of the enzymes involved should not be mistaken.

The activities of enzymes have been recognized for thousands of years; the fermentation of sugar to alcohol by yeast is among the earliest examples of a biotechnological process. However, only recently have the properties of enzymes been understood properly. Indeed, research on enzymes has now entered a new phase with the fusion of ideas from protein chemistry, molecular biophysics, and molecular biology. Full accounts of the chemistry of enzymes, their structure, kinetics, and technological potential can be found in many books and series devoted to these topics [1.1]–[1.5]. This chapter reviews some aspects of the history of enzymes, their nomenclature, their structure, and their relationship to recent developments in molecular biology.

## 1.1. History

Detailed histories of the study of enzymes can be found in the literature [1.6], [1.7].

**Early Concepts of Enzymes.** The term "enzyme" (literally "in yeast") was coined by KÜHNE in 1876. Yeast, because of the acknowledged importance of fermentation, was a favorite subject of research. A major controversy at that time, associated most memorably with LIEBIG and PASTEUR, was whether or not the process of fermentation was separable from the living cell. No belief in the necessity of vital forces, however, survived the demonstration by BUCHNER (1897) that alcoholic fermentation could by carried out by a cell-free yeast extract. The existence of extracellular enzymes had, for reasons of experimental accessibility, already been recognized. For example, as early as 1783, SPALLANZANI had demonstrated that gastric juice could digest meat in vitro, and SCHWANN (1836) called the active substance pepsin. KÜHNE himself appears to have given trypsin its present name, although its existence in the intestine had been suspected since the early 19th century.

**Enzymes as Proteins.** By the beginning of the 20th century, the protein nature of enzymes had been recognized. Knowledge of the chemistry of proteins drew heavily on the improving techniques and concepts of organic chemistry in the second half of the 19th century; it culminated in the peptide theory of protein structure, usually credited to FISCHER und HOFMEISTER. However, methods that had permitted the separation and synthesis of small peptides were unequal to the task of purifying enzymes. Indeed, there was no consensus that enzymes were proteins. Then, in 1926, SUMNER crystallized urease from jack bean meal and announced it to be a simple protein. Against this, WILLSTÄTTER argued that enzymes were not proteins but "colloidal carriers" with "active prosthetic groups." However, with the conclusive work by NORTHROP and his colleagues who isolated a series of crystalline proteolytic enzymes, beginning with pepsin in 1930, the protein nature of enzymes was established.

The isolation and characterization of intracellular enzymes was naturally more complicated and, once again, significant improvements were necessary in the separation techniques applicable to proteins before, in the late 1940s, any such enzyme became available in reasonable quantities. Because of the large amounts of accessible starting material and the historical importance of fermentation experiments, most of the first pure intracellular enzymes came from yeast and skeletal muscle. However, as purification methods were improved, the number of enzymes obtained in pure form increased tremendously and still continues to grow. Methods of protein purification are so sophisticated today that, with sufficient effort, any desired enzyme can probably be purified completely, even though very small amounts will be obtained if the source is poor.

**Primary Structure.** After the protein nature of enzymes had been accepted, the way was clear for more precise analysis of their composition and structure. Most amino acids had been identified by the early 20th century. The methods of amino acid analysis then available, such as gravimetric analysis or microbiological assay, were quite accurate but very slow and required large amounts of material. The breakthrough came with the work of MOORE and STEIN on ion-exchange chromatography of amino acids, which culminated in 1958 in the introduction of the first automated amino acid analyzer [1.8]. Modern machines have lowered the time required for an analysis to less than 1 h and the amount of protein required to $< 1$ µg [1.9].

The more complex question – the arrangement of the constituent amino acids in a given protein, generally referred to as its primary structure – was solved in the late 1940 s. The determination in 1951 of the amino acid sequence of the $\beta$-chain of insulin by SANGER and TUPPY [1.10] demonstrated for the first time that a given protein does indeed have a unique primary structure. The genetic implications of this were enormous. The introduction by EDMAN of the phenyl isothiocyanate degradation of proteins stepwise from the N-terminus, in manual form in 1950 and subsequently automated in 1967 [1.11], provided the principal chemical method for determining the amino acid sequences of proteins. The primary structures of pancreatic ribonuclease [1.12] and egg-white lysozyme [1.13] were published in 1963. Both of these enzymes, simple extracellular proteins, contain about 120 amino acids. The first intracellular enzyme to have its primary structure determined was glyceraldehyde 3-phosphate dehydrogenase [1.14], which has an amino acid sequence of 330 residues and represents a size (250–400 residues) typical of many enzymes. The longest sequence analyzed to date is $\beta$-galactosidase from *Escherichia coli,* which contains 1021 amino acids [1.15]. The methods of protein sequence analysis are now so well developed that no real practical deterrent exists, other than time or expense, to determination of the amino acid sequence of any polypeptide chain [1.9].

**Active Site.** The fact that enzymes are very specific in their choice of substrate and, in general, are much larger than the substrates on which they act, quickly became apparent. The earliest kinetic analysis of enzymatic reactions pointed to the transient formation of an enzyme–substrate complex. These observations could be explained easily if the conversion of substrate to product was assumed to occur at a restricted site on an enzyme molecule. This site soon became known as the "active center" or, as is more common today, the "active site."

Particular compounds were found to react with specific amino acid side chains and lead to inhibition of particular enzymes. This suggested that such side chains might be taking part in the catalytic mechanisms of these enzymes. An early example was the inhibition of glycolysis or fermentation by iodoacetic acid, which was later recognized as resulting from reaction with a unique cysteine residue of glyceraldehyde 3-phosphate dehydrogenase, which normally carries the substrate in thioester linkage [1.16].

Many such group-specific reagents have now been identified as inhibitors of individual enzymes; often they are effective because of the hyper-reactivity of a functionally important side chain in the enzyme's active site. However, a more sophisticated approach to the design of enzyme inhibitors became possible when the reactive group was attached to a substrate; in this way, the specificity of the target enzyme was utilized to achieve selective inhibition of the enzyme [1.17]. Such active-site-directed inhibitors have acquired major importance not only academically in the study of enzyme mechanisms but also commercially in the search for a rational approach to selective toxicity or chemotherapy.

**Three-Dimensional Structure.** Chemical studies showed that the active site of an enzyme consists of a constellation of amino acid side chains brought together spatially from different parts of the polypeptide chain. If this three-dimensional structure was disrupted by denaturation, i.e., without any covalent bonds being broken, the biological activity of the enzyme was destroyed. In addition, it was found that all the information required for a protein to fold up spontaneously in solution and reproduce its native shape was contained in its primary structure. This was part of the original "central dogma" of molecular biology.

The X-ray crystallography of proteins [1.18] demonstrated unequivocally that a given protein has a unique three-dimensional structure. Among the basic design principles was the tendency of hydrophobic amino acid side chains to be associated with the hydrophobic interior of the folded molecule, whereas charged side chains

**Figure 1.** A molecular model of the enzyme lysozyme: the arrow points to the cleft that accepts the polysaccharide substrate (Reproduced by courtesy of J. A. RUPLEY)

were almost exclusively situated on the hydrophilic exterior or surface. The first high-resolution crystallographic analysis of an enzyme, egg-white lysozyme, confirmed these principles and led to the proposal of a detailed mechanism [1.19]. The active site was located in a cleft in the structure (Fig. 1; see also book cover), which has subsequently proved to be a common feature of active sites. According to this, the enzymatic reaction takes place in a hydrophobic environment, and the successive chemical events involving substrate and protein side chains are not constrained by the ambient conditions of aqueous solution and neutral pH.

## 1.2. Enzyme Nomenclature

Strict specificity is a distinguishing feature of enzymes, as opposed to other known catalysts. Enzymes occur in myriad forms and catalyze an enormous range of reactions. By the late 1950s the number of known enzymes had increased so rapidly that their nomenclature was becoming confused or, worse still, misleading because the same enzyme was often known to different workers by different names; in addition, the name frequently conveyed little or nothing about the nature of the reaction catalyzed.

To bring order to this chaotic situation, an International Commission on Enzymes was established in 1956 under the auspices of the International Union of Biochemistry (IUB). Its terms of reference were as follows: "To consider the classification and nomenclature of enzymes and coenzymes, their units of activity and standard methods of assay, together with the symbols used in the description of enzyme kinetics." The Commission's recommendations have formed the basis of enzyme nomenclature since its first report in 1961 [1.1].

Responsibility for enzyme nomenclature passed to the Nomenclature Committee of IUB in 1977, which has subsequently published two reports, the latest in 1984 being a new edition [1.20] and one supplement [1.21]; it is expected that further supplements will be published from time to time in the *European Journal of Biochemistry*. The growth in scale can be appreciated from the fact that the *Report of the Enzyme Commission* (1961) listed 712 enzymes whereas the present version of *Enzyme Nomenclature* (1984) lists 2477.

## 1.2.1. General Principles of Nomenclature

The accepted system for classification and nomenclature of enzymes embodies three general principles.

The first is that enzyme names, especially those ending in *-ase,* should be used only for single enzymes, i.e., single catalytic entities. They should not be applied to systems containing more than one enzyme.

The second general principle is that an enzyme is named and classified according to the reaction it catalyzes. This refers only to the observed chemical change produced by the enzyme, as expressed in the chemical equation. The mechanism of action is ignored, and intermediate cofactors or prosthetic groups are not normally included in the name. Thus, an enzyme cannot be named systematically until the reaction it catalyzes has been identified properly.

The third general principle is that enzymes are named and classified according to the type of reaction catalyzed, which enables Enzyme Commission (E.C.) code numbers to be assigned to enzymes to facilitate subsequent unambiguous identification. For the purpose of systematic nomenclature, all enzymes in a particular class are considered to catalyze reactions that take place in a given direction, although only the reverse direction may have been demonstrated experimentally. However, the recommended name for the enzyme may well be based on the presumed direction of the reaction in vivo.

Thus, a given enzyme often has two names, one systematic and the other recommended or trivial. The latter is generally the name in current usage, shorter and more readily applied. After an enzyme has been identified by its systematic name and E.C. code number, the recommended name can be used without fear of ambiguity. This practice is now generally followed in the literature.

## 1.2.2. Classification and Numbering of Enzymes

According to the report of the first Enzyme Commission in 1961, enzymes are divided into six main classes according to the type of reaction catalyzed. They are assigned code numbers, prefixed by E.C., which contain four elements separated by points and have the following meaning:

1) the first number indicates to which of the six classes the enzyme belongs,

2) the second indicates the subclass,
3) the third number indicates the sub-subclass, and
4) the fourth is the serial number of the enzyme in its sub-subclass.

The six classes are distinguished in the following manner:

### 1. Oxidoreductases

This class encompasses all enzymes that catalyze redox reactions. The recommended name is *dehydrogenase* whenever possible, but *reductase* can also be used. *Oxidase* is used only when $O_2$ is the acceptor for reduction. The systematic name is formed according to *donor:acceptor oxidoreductase*.

### 2. Transferases

Transferases catalyze the transfer of a specific group, such as methyl, acyl, amino, glycosyl, or phosphate, from one substance to another. The recommended name is normally *acceptor grouptransferase* or *donor grouptransferase*. The systematic name is formed according to *donor:acceptor grouptransferase*.

### 3. Hydrolases

Hydrolases catalyze the hydrolytic cleavage of $C-O$, $C-N$, $C-C$, and some other bonds. The recommended name often consists simply of the substrate name with the suffix *-ase*. The systematic name always includes *hydrolase*.

### 4. Lyases

Lyases catalyze the cleavage of $C-C$, $C-O$, $C-N$, and other bonds by elimination. The recommended name is, for example, *decarboxylase, aldolase, dehydratase* (elimination of $CO_2$, aldehyde, and water, respectively). The systematic name is formed according to *substrate group-lyase*.

### 5. Isomerases

Isomerases catalyze geometric or structural rearrangements within a molecule. The different types of isomerism lead to the names *racemase, epimerase, isomerase, tautomerase, mutase,* or *cycloisomerase*.

### 6. Ligases

Ligases catalyze the joining of two molecules, coupled with the hydrolysis of a pyrophosphate bond in ATP or another nucleoside triphosphate. Until 1983, the recommended name often included *synthetase,* but the current recommendation is that names of the type *X−Y ligase* be used instead, to avoid confusion with the name *synthase* (which is not confined to enzymes of class 6). The systematic name is formed according to *X:Y ligase* (ADP-forming).

A few examples will serve to illustrate how this system works. (The full list can be found in *Enzyme Nomenclature* 1984 [1.20].)

The enzyme alcohol dehydrogenase (recommended name) catalyzes the reaction

Alcohol + $NAD^+$ $\rightleftharpoons$ Aldehyde or Ketone + $NADH + H^+$

The enzyme has been assigned E.C. number 1.1.1.1. It may also be called aldehyde reductase, but its systematic name is alcohol: $NAD^+$ oxidoreductase.

Similarly, the enzyme hexokinase (recommended name), which catalyzes the reaction

ATP + D-Hexose $\rightleftharpoons$ ADP + D-Hexose 6-phosphate

has been given the E.C. number 2.7.1.1. It has such other names as glucokinase and hexokinase type IV, and its systematic name is ATP: D-hexose 6-phosphotransferase.

# 1.3. Structure of Enzymes

Enzymes are proteins (for an exception, see Section 1.3.4) and, as such, are amenable to structural analysis by the methods of protein chemistry, molecular biology, and molecular biophysics.

## 1.3.1. Primary Structure

The primary structure of enzymes can be determined by direct chemical methods which, in sensitivity and automation, have reached very high levels of sophistication [1.9], [1.22]. However, for many proteins, particularly those with long polypeptide chains, direct sequence analysis would be very time-consuming; others may be available only in very small amounts. In these cases, a more profitable approach is to clone the relevant structural gene and determine its DNA sequence [1.9], [1.23], [1.24]. From this, the amino acid sequence can be inferred. Whenever possible, this sequence should be checked, e.g., for genetic reading frame, against whatever amino acid sequence information is available from direct methods. The recombinant DNA approach is so quick and so powerful, however, that amino acid sequence information about enzymes is growing much more rapidly from this source than from direct chemical analysis [1.25], [1.26]. Indeed, the information now available is so large in total that computer data banks are required to store it and make it available for systematic access [1.27].

## 1.3.2. Three-Dimensional Structure

The three-dimensional structure of an enzyme can only be obtained at high resolution by the application of X-ray crystallography [1.28]. By this means, the detailed structures of many enzymes have been determined, and a broad understanding of the principles of protein structure has resulted [1.29], [1.30]. Proteins are generally well-ordered; their interiors are well-packed (comparable to other crystalline organic molecules) to produce a hydrophobic core with a di-electric constant similar to that of a hydrocarbon. Proteins vary in the amount of regular secondary structure ($\alpha$-helix and $\beta$-sheet) they contain and can be grouped into four classes according to the combination and packing of these structural features [1.31].

Despite their close-packed and generally well-ordered structure, enzymes are usually not entirely rigid molecules, and some conformational flexibility in solution is widely observed, particularly by NMR spectroscopy [1.32], [1.33]. These conformational changes may be limited to a molecular "breathing" or flexing of the structure, they may involve various "hinge-bending" motions, or they may extend to more substantial conformational mobility in parts of the polypeptide chain. All such motions contribute to the mechanisms of enzyme catalysis [1.2], [1.34].

## 1.3.3. Quaternary Structure, Folding, and Domains

Many enzymes consist of more than one polypeptide chain (or subunit), and these must form an aggregate, usually with relatively simple symmetry, before full (or even any) biological activity is conferred (Table 1). The subunits within an

**Table 1.** Quaternary structures of some typical enzymes

| Enzyme | E.C. number [CAS registry number] | Source | Number of subunits | Point symmetry | |
|---|---|---|---|---|---|
| | | | | Crystallographic symbol | Schönflies symbol |
| Alcohol dehydrogenase | 1.1.1.1 [9031-72-5] | horse liver | 2 | 2 | $C_2$ |
| Glutathione reductase | 1.6.4.2 [9001-48-3] | human red blood cells | 2 | 2 | $C_2$ |
| Triose phosphate isomerase | 5.3.1.1 [9023-78-3] | chicken muscle | 2 | 2 | $C_2$ |
| Lactate dehydrogenase | 1.1.1.27 [9001-60-9] | dogfish muscle | 4 | 222 | $D_2$ |
| Glyceraldehyde 3-phosphate dehydrogenase | 1.2.1.12 [9001-50-7] | *Bacillus stearothermophilus* | 4 | 222 | $D_2$ |
| Pyruvate kinase | 2.7.1.40 [9001-59-6] | cat muscle | 4 | 222 | $D_2$ |
| Aspartate carbamoyl-transferase | 2.1.3.2 [9012-49-1] | *Escherichia coli* | 6 + 6 | 32 | $D_3$ |
| Dihydrolipoamide acetyltransferase | 2.3.1.12 [9032-29-5] | *Escherichia coli* | 24 | 432 | 0 |
| Dihydrolipoamide acetyltransferase | 2.3.1.12 [9032-29-5] | *Bacillus stearothermophilus* | 60 | 532 | Y |

oligomer or multimer are often identical or at least limited to a few different types. Aggregation is generally some form of self-assembly dictated by coherent binding patterns between the subunits, which provide the necessary recognition sites in sorting out the subunits required for assembly [1.29], [1.35].

The complexity of this sorting process in a cell becomes evident from the fact that many intracellular enzymes are dimers or tetramers. Increasingly more complicated structures are being recognized and their design principles analyzed. These range from enzymes with simple cyclic symmetry up to those with the most elaborate cubic point group symmetry, e.g., octahedral and icosahedral [1.35].

The folding of polypeptide chains, along with their aggregation into ordered structures, is a spontaneous process in solution, which implies that it is exergonic [1.36]. Calculation of the time required for a protein to explore all possible structures during the folding process indicates, however, that the search for the "right" structure cannot be entirely random. Thus, even for a small protein such as bovine pancreatic ribonuclease (124 amino acid residues), such a search might take around $10^{95}$ years, whereas the experimentally determined time in vivo is a few milliseconds. This dramatic discrepancy led to the concept of *kinetic pathways* during folding. Such pathways have been experimentally explored and intermediates identified for various proteins. The stable structure of a protein in solution is, therefore, identified as the lowest free energy form of the kinetically accessible structures [1.29], [1.30], [1.35], [1.36].

A typical enzyme is not an entity completely folded as a whole, as is evident from the growing catalogue of three-dimensional protein structures determined by X-ray crystallography. On the contrary, enzymes frequently consist of apparently autonomous or semiautonomous folding units, called *domains* (Fig. 2). Sometimes, these may be identified as products of limited proteolysis, i.e., regions of the polypeptide chain that can be excised from the chain with retention of their biological properties. Indeed, this has proved in many instances to be a valuable guide to the actual activity contributed by that part of the enzyme. Classical examples of such functional domains can be found in the study of muscle contraction and antibody–antigen recognition [1.29], [1.30].

In other cases, domains are not readily released as biologically active entities, and their existence must be inferred from the three-dimensional structure of the enzyme. Most globular proteins can in fact be subdivided into such regions, which generally have molecular masses of 20 000 or less [1.29]. The active site of an enzyme is often located at the interface between two such domains as, for example, in the well-known cleft of lysozyme (Fig. 1) or in glutathione reductase. Other domains appear to represent favored folding patterns in the assembly of proteins, but biological activity associated with them can often be inferred from comparison of the structures of related proteins: a typical example is the NAD-binding domain present in dehydrogenases.

Structural domains may be regions of the polypeptide chain that fold independently of each other. Functional domains, as defined above, do indeed fold independently; and individual subunits of oligomeric enzymes appear to fold before association [1.29], [1.30], [1.35], [1.36].

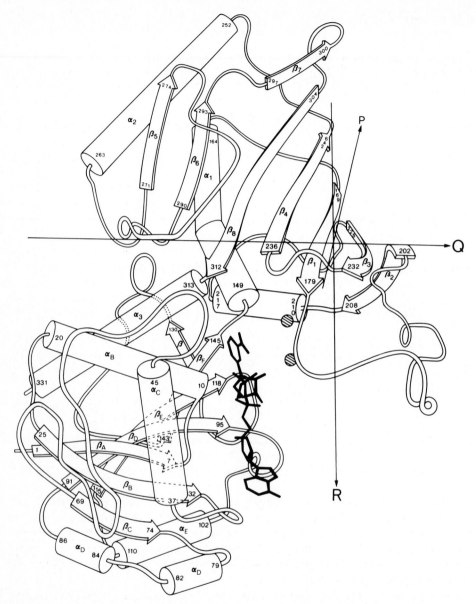

**Figure 2.** The domains in glyceraldehyde 3-phosphate dehydrogenase from *Bacillus stearothermo-philus* [1.44] Reproduced with permission

### 1.3.4. The Ribozyme

Enzymes are proteins, but the specific involvement of RNA molecules in certain reactions concerned with RNA processing in vivo is worth noting. Thus, in *Escherichia coli,* tRNA precursors are cleaved by ribonuclease P to generate the correct 5′-ends of the mature tRNA molecules, and the enzyme contains an essential RNA moiety that can function in the absence of the protein. In fact, this RNA moiety fulfills all the criteria of an enzyme [1.37]. Similarly, the ribosomal RNA of *Tetrahymena thermophila* undergoes self-splicing, performing a highly specific intramolecular catalysis in the removal of an intervening sequence. A truncated version of the intervening sequence, lacking the first 19 nucleotides of the original excised RNA, can then behave as an enzyme in vitro, capable of acting as an RNA polymerase and a sequence-specific ribonuclease under appropriate conditions [1.38].

Such reactions are, in a sense, exciting oddities in the broader world of conventional enzymology, but they have important implications for the evolution of enzymatic activity and may point to interesting developments for the future.

# 1.4. Enzymes and Molecular Biology

## 1.4.1. Biosynthesis of Enzymes

Enzymes are synthesized in cells by the normal machinery of protein synthesis. The structure of any given enzyme is encoded by a structural gene, whose DNA base sequence is transcribed into a messenger RNA, and the mRNA is translated from its triplet code into the amino acid sequence of the desired protein by the ribosomes and associated factors [1.39], [1.40]. The enzyme then folds spontaneously to its active conformation. Posttranslational modifications may be required to target an enzyme to its ultimate intracellular or extracellular location.

## 1.4.2. Enzymes and DNA

For many years, the chemical manipulation of DNA lagged behind that of proteins. The chemical complexity and variety of proteins, with up to 20 different naturally occurring amino acids, served to make them more amenable to increasingly sophisticated methods of analysis. On the other hand, DNA, composed of only four different nucleotides, appeared dauntingly large, with few structural features to make it yield to available methodology.

Paradoxically, this very lack of variety in the nature of the constituent nucleotides of DNA has permitted the recent exciting revolution in genetic engineering – innovations in which the enzymology of DNA [1.41] has played a prominent part. For example, the discovery and purification of restriction enzymes (Chap. 6) enabled DNA to be cleaved selectively into defined fragments; phosphatases and ligases permit the fragments to be rejoined selectively; and DNA polymerases allow DNA to be synthesized and sequenced at astonishing speed, all in vitro [1.23], [1.39]–[1.41].

### 1.4.3. Protein Engineering

The combination of these important advances in the enzymology of DNA with impressive new techniques for the chemical synthesis of oligodeoxyribonucleotides has put an entirely new technology, *protein engineering,* in the hands of the enzymologist. A cloned gene can now be overexpressed to generate large quantities of a desired enzyme and the gene (and thus the amino acid sequence) can be mutated selectively to produce a novel form of the enzyme with changed specificity or kinetic parameters; the synthesis even of an entirely new enzyme, either by combination of gene fragments encoding domains from other previously studied enzymes or by de novo synthesis of an entire gene, can be contemplated.

Detailed reviews of this new field of enzymology, which has obvious and important implications for developments in biotechnology, can be found elsewhere [1.2], [1.42], [1.43], [1.45].

# 2. Catalytic Activity of Enzymes

The theory of enzyme-catalyzed reactions proposed by MICHAELIS and MENTEN in 1913 [2.4] is based on the assumption that the enzyme (the catalyst, E) and the substrate (the reactant, S) form a complex (ES). This reaction is reversible and can be subjected to the law of mass action (see also [2.5]). The reaction proceeds as follows:

$$S + E \underset{k_{-1}}{\overset{k_1}{\longrightarrow}} ES \tag{2.1}$$

and

$$\frac{c_E \cdot c_S}{c_{ES}} = K_s = \frac{k_{-1}}{k_1} \tag{2.2}$$

where $K_s$ is the equilibrium constant.

The conversion of substrate to product (P) proceeds via the enzyme–substrate complex:

$$ES \overset{k_2}{\longrightarrow} P + E \tag{2.3}$$

In most cases this reaction determines the decomposition rate of the enzyme–substrate complex, and equals the overall reaction rate:

$$v = k_2 \cdot c_{ES} \tag{2.4}$$

Therefore, $c_{ES}$ is the parameter that determines the reaction rate.

According to Equation (2.2), the concentration of the enzyme–substrate complex depends on the concentration of the components E and S. If, for example, $c_E$ is constant, an increase of $c_S$ yields an increased reaction rate ($v$) until all enzyme is bound as enzyme–substrate complex. Under these conditions, the reaction rate reaches its maximum ($V$). The enzyme is said to be saturated with substrate. The typical course of such a dependence of reaction rate on substrate concentration is shown in Figure 3.

**Michaelis Constant.** The Michaelis constant $K_m$ is defined as the substrate concentration at half the maximal reaction rate. The value $K_m$ can be obtained by plotting the experimentally measured reaction rate against the various substrate concentrations (Fig. 3); for further discussion, see [2.5]. The Michaelis constant approaches the dissociation constant $K_s$ of the enzyme–substrate complex and is therefore valuable for estimating individual reaction kinetics.

Michaelis constants for enzymes usually range from $10^{-2}$ to $10^{-5}$ mol/L; a low $K_m$ indicates a high affinity between enzyme and substrate.

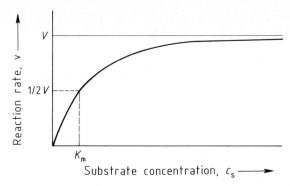

**Figure 3.** Reaction rate as a function of substrate concentration (enzyme at constant concentration) For explanation of symbols, see text

**Molar or Molecular Activity** (Turnover Number). The efficiency of an enzyme-catalyzed reaction is indicated by the molar activity, formerly called turnover number, which is defined as the number of substrate molecules converted in 1 min by one enzyme molecule under standardized conditions. This can be calculated from the specific activity of a particular enzyme if its molecular mass is known (cf. Section 2.3.2). The average molar activity ranges from $10^3$ to $10^4$; peak values have been measured for acetylcholinesterase (E.C. 3.1.1.7) [*9000-81-1*] at $1 \times 10^6$ and for catalase (E.C. 1.11.1.6) [*9001-05-2*] at $5 \times 10^6$.

# 2.1. Factors Governing Catalytic Activity [2.6]

## 2.1.1. Temperature

The temperature dependence of enzyme-catalyzed reactions exhibits an optimum because the thermodynamic increase of reaction rate (1 in Fig. 4) is followed by a steep drop caused by thermal degradation of the enzyme (2 in Fig. 4). The optimum is generally between 40 and 60 °C. Some temperature-insensitive enzymes may exhibit an optimum at almost 100 °C. Data on various frequently used enzymes are given in [2.7]. Figure 4 illustrates temperature dependence [2.3a].

## 2.1.2. Value of pH

All enzymes have an optimum pH range for activity. The optimum depends not only on pH but also on ionic strength and type of buffer. It may also be influenced by temperature, substrate, and coenzyme concentrations. For most enzymes, the pH optimum lies in the range from 5 to 7. Extreme values of 1.5 and 10.5 have been

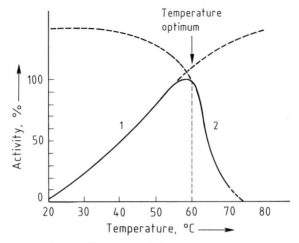

**Figure 4.** Temperature optimum of enzyme activity

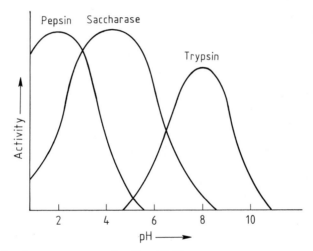

**Figure 5.** Activity of various enzymes as a function of pH

found for pepsin (E.C. 3.4.23.1) [*9004-07-3*] and for alkaline phosphatase (E.C. 3.1.3.1) [*9001-78-9*], respectively. Figure 5 shows some examples [2.3a].

## 2.1.3. Activation

Many chemical effectors activate or inhibit the catalytic activity of enzymes. In addition to substrates and coenzymes many enzymes require nonprotein or, in some cases, protein compounds to be fully active. Enzyme activation by many inorganic ions has been adequately described. The activating ion may be involved directly in the reaction by complexing the coenzyme or cosubstrate (e.g., Fe ions bound to flavin or the ATP–Mg complex). In other cases, the ion is part of the enzyme and either acts as a stabilizer for the active conformation (e.g., Zn ions in alkaline phosphatase) or participates directly at the active site (e.g., Mn ions in isocitrate dehydrogenase (E.C. 1.1.1.42) [*9028-48-2*] and Zn or Co ions in carboxypeptidases).

## 2.1.4. Inhibition [2.8]

In vivo and in vitro inhibition studies of enzymatic reactions contributed important knowledge to various fields of biochemistry. For example, the mechanism of action of many toxic substances and antidotes has been found to affect enzymes directly. In many cases, the importance of these enzymes for metabolism has been revealed. On the other hand, discovery of end-product inhibition elucidated many metabolic pathways. Enzyme inhibition can be either reversible or irreversible. Depending on the type of inhibitory effect, the following mechanisms of enzyme inhibition may be distinguished.

**Irreversible Inhibition.** An irreversible inhibitor frequently forms a stable compound with the enzyme by covalent bonding with an amino acid residue at the active site. For example, diisopropyl fluorophosphate (DIFP) reacts with a serine residue at the active site of acetylcholinesterase to form an inactive diisopropylphosphoryl enzyme. Alkylating reagents, such as iodoacetamide, inactivate enzymes with mercapto groups at their active sites by modifying cysteine.

**Reversible Inhibition.** Reversible inhibition, in contrast, is characterized by an equilibrium between enzyme and inhibitor. Several main groups of reversible inhibitory mechanisms can be differentiated.

*Competitive Inhibition.* The inhibitor competes with the substrate or coenzyme for the binding site on the active center by forming an enzyme-inhibitor complex EI. In most cases, the chemical structure of the inhibitor resembles that of the substrate. Inhibition can be made ineffective by excess substrate, as is the case for inhibition of succinate dehydrogenase (E.C. 1.3.99.1) [*9002-02-2*] by malonate.

*Noncompetitive Inhibition.* The inhibitor decreases the catalytic activity of an enzyme without influencing the binding relationship between substrate and enzyme. This means that inhibitor and substrate can bind simultaneously to an enzyme molecule to form ES, EI, or ESI complexes. Noncompetitive inhibition is dependent solely on the inhibitor concentration and is not overcome by high substrate concentration. An example is the blocking of an essential cysteine residue by such heavy metals as copper or mercury.

*Uncompetitive Inhibition.* The inhibitor reacts only with the intermediary enzyme–substrate complex. An example is the reaction of azide with the oxidized form of cytochrome oxidase (E.C. 1.9.3.1) *[9001-16-5]*. Lineweaver–Burk plots of the

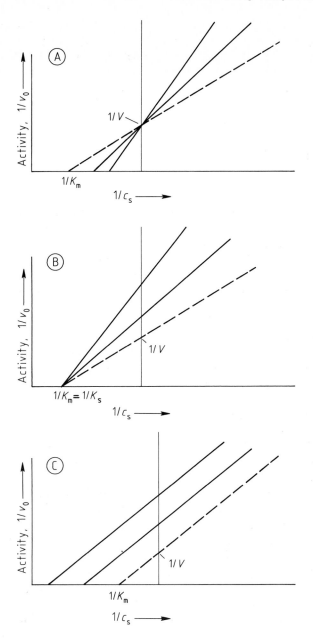

**Figure 6.** Lineweaver–Burk graphs of reversibly inhibited enzyme reactions
A: Competitive inhibition
B: Noncompetitive inhibition
C: Uncompetitive inhibition
– – –: Uninhibited reaction

reciprocal initial reaction rate $1/v_0$ versus the reciprocal substrate concentration for these three modes of reversible inhibition are shown in Figures 6A–6C [2.1].

*Substrate Inhibition.* High concentration of substrate (or coenzyme) may decrease the catalytic activity of an enzyme. Examples are the action of ATP on phosphofructokinase (E.C. 2.7.1.11) [*9001-80-3*] or of urea on urease (E.C. 3.5.1.5) [*9002-13-5*].

*End-Product Inhibition.* In many multi-enzyme systems, the end product of the reaction sequence may act as a specific inhibitor of an enzyme at or near the beginning of the sequence. The result is that the rate of the entire sequence of reactions is determined by the steady-state concentration of the end product (Fig. 7). This type of inhibition is also called *feedback inhibition* or *retroinhibition*.

## 2.1.5. Allostery [2.9]

Cosubstrates with a central role in metabolism, such as acetyl-CoA, ATP, or AMP, may also influence the rate of reaction sequences by allosteric regulation. For example, phosphofructokinase, the first enzyme in the energy-supplying Embden–Meyerhof–Parnass pathway, is inhibited by a high concentration of ATP (i.e., positive energy balance). A high concentration of AMP, on the other hand, (i.e., energy deficiency) terminates this inhibition. Allosterically regulated enzymes have a quaternary structure and are composed of two or more structurally similar or

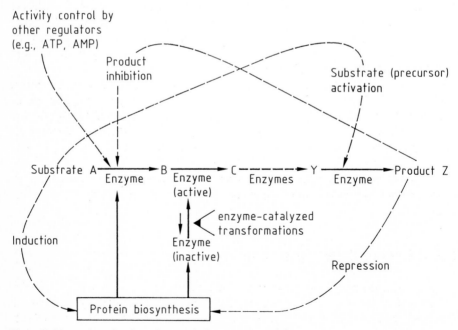

**Figure 7.** Natural mechanisms for regulating enzyme activity

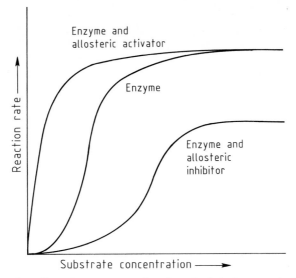

**Figure 8.** Activity of an allosteric enzyme as a function of substrate concentration in absence and presence of an allosteric activator or inhibitor

identical subunits (protomers), each with a binding site for the substrate and another independent binding site for the allosteric effector. The binding of the effector modifies the conformation of the subunit and its active center, which then affects the conformation and hence the catalytic activity of the entire molecule. The cooperation of substrate and effector regulates the overall catalytic activity of the enzyme depending on the concentration of metabolite.

Allosteric enzymes usually do not show the classical Michaelis–Menten kinetic relationship of $c_s$, $V$, and $K_m$. Many allosteric enzymes give a sigmoid plot of initial rate vs. substrate concentration rather than the hyperbolic plots predicted by the Michaelis–Menten equation. The shape of the curve is characteristically changed by an allosteric activator (positive cooperativity) or an allosteric inhibitor (negative cooperativity), as shown in Figure 8.

This sigmoidal curve implies that within a certain range of substrate concentration, the enzyme is able to respond to small concentration changes by great activity changes.

## 2.1.6. Biogenic Regulation of Activity

In principle, enzyme activity may also be controlled by regulating the amount of enzyme in the cell. This can be accomplished by regulating the biosynthesis of individual enzymes or of several functionally related enzymes by induction or repression, or by specific attack of proteolytic enzymes (Fig. 7).

Chymotrypsin (E.C. 3.4.21.1) [*9004-07-3*], for example, is synthesized as inactive zymogen and converted to the active enzyme by trypsin (E.C. 3.4.21.4) [*9002-07-7/*], a typical protease. A large number of natural protease inhibitors have been isolated and characterized. They act as protease antagonists and are capable of selectively affecting many proteolytic reactions. These proteinase inhibitors are proteins with molecular masses of 5000–25 000; they inhibit proteases by specific complex formation [2.10].

## 2.2. Enzyme Assays

### 2.2.1. Reaction Rate as a Measure of Catalytic Activity

As biological catalysts, enzymes increase the rate of a reaction or permit it to proceed. Therefore, the conversion rate of substrates, $v$, is measured to determine the catalytic activity. In principle, the laws of classical reaction kinetics can be applied [2.5]. Because of intermediate formation of an enzyme–substrate complex during the reaction, a special case for many enzyme-catalyzed reactions according to the Michaelis–Menten equation results

$$v = k \frac{c_E \cdot c_S}{c_S + K_m} \tag{2.5}$$

Although these reactions are mainly of first and second order, a zero-order reaction can often be achieved by carefully selecting the reaction conditions, e.g., an excess of substrate ($c_s \gg K_m$). In this case,

$$v = k \cdot c_E = V$$

Consequently, catalytic activity is linearly dependent on the amount of enzyme used.

### 2.2.2. Definition of Units

Originally, units were defined by the investigator who first discovered and described an enzyme. Therefore, in the older literature, enzyme activity was expressed in arbitrary units, e.g., changes in absorbance, increase of reducing groups, amount of converted substrate expressed in milligrams or micromoles. These parameters were related to various time units, such as 1 min, 30 min, or 1 h.

To obtain standardized values for each enzyme, in 1961 the Enzyme Commission of the International Union of Biochemistry defined the *International Unit* U as the activity of an enzyme which, under optimized standard conditions, catalyzes the conversion of 1 µmol of substrate per minute. With respect to basic SI units, the Expert Panel on Quantities and Units (EPQU) of the International Federation of Clinical Chemists (IFCC) and the Commission on Quantities and Units in Clinical Chemistry (CQUCC) of IUPAC defined the base unit *katal* as the catalytic amount of any enzyme that catalyzes a reaction (conversion) rate of 1 mol of substrate per second in an assay system [2.11]. This unit, however, is not in common use.

The temperature must be stated for each assay. As a general rule, the rate of enzymatically catalyzed reactions is about doubled by a temperature increase of 10 °C in the range of 0–40 °C. The temperature must be controlled precisely and kept constant to achieve reproducible results [2.6].

However, for practical reasons, many enzyme reactions cannot be monitored by measuring the stoichiometric amount of substrate consumed or product formed. Therefore, the catalytic activity of a particular enzyme may be impossible to express in International Units.

The definition of units for most enzymes used in molecular biology is rather arbitrary. For example, the unit of a restriction endonuclease (E.C. 3.1.21.3–5), is defined as the catalytic activity of the enzyme that yields a typical cleavage pattern, detectable after electrophoresis, with a precise amount (usually 1 µg) of a particular DNA under defined incubation conditions [2.12], [2.13]. Other parameters that define the activity of such enzymes are the degradation of a nucleic acid or the incorporation of nucleotides into a nucleic acid, expressed in micrograms, nanomoles, absorbance units, or number of base pairs. For practical reasons, different time periods such as 1, 10, 30, or 60 min, may be chosen as references.

## 2.2.3. Absorption Photometry [2.14]

**Basic Considerations.** Because of its simple technique and reliable, reasonably priced instruments, photometry is today one of the preferred methods of enzyme assay. It can be carried out most quickly and conveniently when the substrate or the product is colored or absorbs light in the ultraviolet region because the rate of appearance or disappearance of a light-absorbing product or substrate can be followed with a spectrophotometer.

According to the Bouguer–Lambert–Beer law, which is valid for very dilute solutions, the following relationship exists between absorbance $A$ and concentration:

$$A = \log \frac{I_0}{I} = \varepsilon \cdot c \cdot d \tag{2.6}$$

and

$$c = \frac{A}{\varepsilon \cdot d} \tag{2.7}$$

where $c$ is the concentration in millimoles per liter, $\varepsilon$ the millimolar absorption coefficient in liters per mole and per millimeter, and $d$ the path length in millimeters.

The catalytic activity $z$ then corresponds to the absorbance change per minute.

$$z = \frac{\Delta A \cdot V \cdot 1000}{\varepsilon \cdot d \cdot \Delta t} \tag{2.8}$$

where $V$ is the assay volume in liters and $t$ the time in minutes.

The unit of $z$ is then micromoles per minute and corresponds to the definition of the International Unit U given in the previous section.

*Example 1: Assay of Lactate Dehydrogenase (LDH).* In the reaction catalyzed by LDH (E.C. 1.1.1.27) [9001-60-9], hydrogen is transferred from NADH to pyruvate, to yield L-lactate and NAD [2.15]:

$$\text{Pyruvate} + \text{NADH} + \text{H}^+ \rightleftharpoons \text{L-Lactate} + \text{NAD}^+ \tag{2.9}$$

The reduced coenzyme NADH absorbs at 340 nm, whereas the oxidized form NAD, lactate, and pyruvate do not. Thus the progress of Reaction (2.9) can be followed by measuring the decrease in light absorption at 340 nm with a mercury line photometer emitting at 334 or 365 nm. The enzyme can also be measured by monitoring the reverse reaction under slightly alkaline conditions. However, the reverse reaction is much slower than the reaction starting with pyruvate [2.16].

The principle of the optical assay may also be used to follow an enzymatic reaction in which neither the substrate nor the product has any characteristic light absorption maxima. In that case, the reaction is coupled to some other enzymatic reaction which can be followed easily by photometry. The activity of the nonabsorbing enzyme system can then be measured if the enzyme considered is made the rate-determining component by appropriate choice of assay conditions.

*Example 2: Assay of Pyruvate Kinase (PK).* The reaction between phosphoenolpyruvate and ADP yields pyruvate and ATP by transfer of a phosphate group; it is catalyzed by pyruvate kinase (E.C. 2.7.1.40) [9001-59-6]:

$$\text{Phosphoenolpyruvate} + \text{ADP} \longrightarrow \text{Pyruvate} + \text{ATP} \tag{2.10}$$

This reaction is easily measured when a large excess of LDH and NADH is added to the system, which couples Reactions (2.10) and (2.9) [2.17]. The formation of pyruvate in Reaction (2.10) is followed by the very rapid reduction of pyruvate to lactate in Reaction (2.9). For each molecule of pyruvate formed and reduced, one molecule of NADH is oxidized to NAD, causing a decrease in light absorption at 340 nm.

The catalytic activity of enzymes such as phosphatases [2.18], whose natural substrates do not have suitable spectral properties, can be determined by using a colorless synthetic substrate, which is split enzymatically to yield a colored product.

*Example 3: Assay of Phosphatase.* For the assay of phosphatases, glycosidases, and several other hydrolases, colorless 4-nitrophenyl compounds are incubated, and 4-nitrophenolate is formed under alkaline conditions with a characteristic maximum between 400 and 420 nm. In this way, the catalytic activity of such enzymes can be measured conveniently at this wavelength:

$$\text{4-Nitrophenyl phosphate} + \text{H}_2\text{O} \longrightarrow \text{4-Nitrophenol} + \text{Phosphate} \tag{2.11}$$

$$\text{4-Nitrophenol} \xrightarrow{\text{OH}^-} \text{4-Nitrophenolate} \tag{2.12}$$

Another principle involves the absorption of a colored metabolite formed directly or indirectly, by the action of the enzyme being analyzed. Again, the activity of the enzyme can be determined only if it is made the rate-determining component by proper choice of conditions.

*Example 4: Assay of Glycerol Phosphate Oxidase.* Among the class of oxidoreductases utilizing $O_2$ as the sole final electron acceptor, many enzymes exist that generate $H_2O_2$ during oxidation of their individual substrates [2.19]. Highly sensitive and accurate colorimetric assays of $H_2O_2$ have been developed during the last two decades [2.20]; in one system, phenol gives a purple color with 4-aminoantipyrine in the presence of oxidizing agents. Therefore, this color reaction can be used for colorimetric determination of $H_2O_2$-generating enzyme reactions with peroxidase as oxidation catalyst and indicator enzyme.

Although this principle is used predominantly to determine metabolites in body fluids, it is often employed in the assay of such enzymes as glycerol phosphate oxidase (E.C. 1.1.3.21) or glucose oxidase (E.C. 1.1.3.4) *[9001-37-0]*.

The catalytic activity of the first is determined according to the following reaction sequence:

$$2 \text{ L-}\alpha\text{-Glycerol phosphate} + 2 \text{ O}_2 \longrightarrow 2 \text{ Dihydroxyacetone phosphate} + 2 \text{ H}_2\text{O}_2 \qquad (2.13)$$

$$2 \text{ H}_2\text{O}_2 + 4\text{-Aminoantipyrine} + \text{Phenol} \longrightarrow \text{Quinonimine dye} + 4 \text{ H}_2\text{O} \qquad (2.14)$$

The broad absorption maximum of the quinonimine dye is centered around 500 nm, with an absorption coefficient of about $13 \times 10^2 \text{ L mol}^{-1} \text{ mm}^{-1}$. The increase in absorbance per minute at 500 nm is measured to determine the activity of glycerol phosphate oxidase.

## 2.2.4. Fluorometry [2.21]

The fluorometric method is rarely used for determining the catalytic activity of raw or purified enzyme preparations. Because of its high sensitivity, it permits the assay of small amounts of enzymes in organs or tissue sections [2.22]. For example, systems that depend on NAD and NADP can be measured by fluorometry; the reduced pyridine coenzymes exhibit a fluorescence of low intensity. To enhance sensitivity, the oxidized form is treated with alkali, to yield strongly fluorescing compounds; in addition, selective filters must be used. The overall sensitivity of this method is a thousand times that of absorption photometry.

## 2.2.5. Luminometry [2.23]

Luminometry uses fluorescence, phosphorescence, and chemiluminescence as detector systems. Chemiluminescence observed in living organisms is termed bioluminescence. Bioluminescence is catalyzed by enzymes called luciferases, whose substrates, known as luciferins, are converted to light-emitting products.

In luminometry, the number of photons emitted by the reaction system per unit time is measured with specially designed instruments called luminometers; they are based on single photon counting detectors, usually photomultiplier tubes.

An example is the reaction catalyzed by the luciferase from *Photinus pyralis, Photinus*-luciferin 4-monooxygenase (ATP-hydrolyzing) (E.C. 1.13.12.7) *[61970-00-1]*:

$$\text{ATP} + \text{D-Luciferin} + \text{O}_2 \longrightarrow \text{Oxyluciferin} + \text{AMP} + \text{pp} + \text{CO}_2 + 0.9 \, h\nu \qquad (2.15)$$

In this reaction, ATP is consumed as a substrate, and photons at a wavelength of 562 nm are emitted. The quantum yield is 0.9 einstein per mole of luciferin, i.e., for one ATP molecule consumed, approximately one photon is emitted. This reaction is therefore suitable for the assay of ATP and, hence, of enzymes that catalyze ATP-consuming or ATP-producing reactions.

To monitor enzyme activity with the aid of firefly  bioluminescence, the light intensity must increase linearly for several minutes and must be strictly proportional to the catalytic activity of the enzyme. Such measuring conditions have been realized for determination of the catalytical activity of creatine kinase (E.C. 2.7.3.2) [*9001-15-4*]:

$$\text{Creatine phosphate} + \text{ADP} \longrightarrow \text{Creatine} + \text{ATP} \tag{2.16}$$

Other reactions that depend on NAD(P) can be followed by using the bioluminescence from lucibacteria.

## 2.2.6. Radiometry

When radioactively labeled substrates are used, the activities of some enzymes can be determined with high sensitivity. This technique is widely employed in the field of molecular biology to monitor (1) the incorporation of radioactively labeled nucleotides into acid-insoluble nucleic acids or polynucleotides (DNA and RNA polymerases), (2) the decomposition of radioactively labeled DNA (exonuclease III), (3) the transfer of a radioactively labeled phosphate group from $\gamma$-$^{32}$P-ATP [*2964-07-0*] to the 5'-hydroxyl end of a polynucleotide (polynucleotide kinase), or (4) the exchange of radioactively labeled pyrophosphate on a carrier matrix (T$_4$ DNA ligase). The most common isotopes for labeling are $^{32}$P, $^{14}$C, $^{3}$H, and to a minor degree, $^{35}$S.

In experiments with radioactively labeled compounds, special safety and legal regulations must be observed. In some determinations, however, the amounts of radioactive material needed are below the limits regulated by law.

## 2.2.7. Potentiometry [2.24]

A pH-sensitive glass electrode can be used to measure reactions in which protons are produced or consumed. For this purpose, the pH is kept constant by counter-titration, and the consumption of acid or base required to do this is measured. The electrode controls an automatic titrator, and this concept  is the *pH-stat* technique described by BÜCHER [2.25]. A typical example is the determination of the catalytic activity of lipase (E.C. 3.1.1.3) [*9001-62-1*] [2.26]. A fat (triglyceride) is hydrolyzed by this enzyme, and the fatty acid formed is neutralized by countertitration with NaOH in a pH-stat mode:

$$\text{Triglyceride} + \text{H}_2\text{O} \longrightarrow \text{Diglyceride} + \text{Fatty acid} \tag{2.17}$$

The substrate, olive oil, is incubated with the diluted sample containing lipase, and the mixture is titrated at constant pH. A recorder plots the consumption of

NaOH vs. time, and the resulting slope correlates with catalytic activity. Other examples are the assay of papain (E.C. 3.4.22.2) [*9001-73-4*] and glucose oxidase (E.C. 1.1.3.4.) [*9001-37-0*].

## 2.2.8. Conductometry

In principle, all enzymatic reactions that lead to a change in overall ionic mobility can be measured by conductometry. In this way, the elastolytic activity of elastase (E.C. 3.4.21.36) by using unmodified elastin as substrate has been determined [2.27]. Other applications have also been described [2.28]. In this reaction, protons are liberated by cleavage of peptide bonds.

## 2.2.9. Calorimetry

Many enzymatic reactions evolve heat; therefore, some interest in calorimetric (enthalpimetric) methods has developed. In a suitable experimental arrangement, a temperature sensor serves as a device for measuring the catalytic activity of the enzyme. A new area of analytical chemistry has developed from this approach. Previously called microcalorimetry, it is now commonly known as enthalpimetry [2.29]. The method is used mainly in research.

## 2.2.10. Polarimetry

Polarimetry is rarely used, partly because of the inconvenience involved. However, it is required in determining the catalytic activity of mutarotase (E.C. 5.1.3.3.) [*9031-76-9*], which catalyzes the equilibrium between $\alpha$- and $\beta$-glucose [2.30].

## 2.2.11. Manometry

Manometry is one of the classical methods in biochemistry. It is no longer used for routine assay of enzymes. Formerly, the catalytic activity of glucose oxidase [2.31], arginase (E.C. 3.5.3.1) [*9000-96-8*] [2.32], and other enzymes was determined by this method.

## 2.2.12. Viscosimetry

Although viscosimetry has been virtually abandoned for enzyme assays, it is still very important in determining cellulase activity (E.C. 3.2.1.4) [*9012-54-8*]. The change in viscosity per unit time indicates the catalytic activity [2.33].

## 2.2.13. Immobilized Enzymes [2.34]

Immobilized enzymes are used in analytical chemistry and as catalysts for the production of chemicals, pharmaceuticals, and food (Section 3.3). They also serve as simple and well-defined models for studying membrane-bound enzymes.

Because of their particular structure, immobilized enzymes require specific assays. In addition to requiring optimal conditions different from those of soluble enzymes, particle size, particle-size distribution, mechanical and chemical structure, stability and structure of the matrix, and the catalytic activity used for immobilization must be considered.

At least two different assay procedures are used for insoluble enzymes: one employs a stirred suspension in a vessel; the other, a packed bed or a column reactor. The conditions of such assay are very close to those used in industrial applications. Enzyme activity can be assayed continuously or batchwise.

Conductometry, potentiometry, and polarimetry are better suited to detection than photometry, because they allow the reaction to be followed directly, without the need for additional indicator enzymes, coenzymes, or second substrates. An example is the assay of immobilized penicillin amidase (E.C. 3.5.1.11) [*9014-06-6*] [2.35]:

Penicillin G + $H_2O \longrightarrow$  6-Aminopenicillanate + Phenyl acetate + $H^+$ (2.18)

## 2.2.14. Electrophoresis

Electrophoresis is an indispensable tool for determining the catalytic activity of nucleases, especially restriction endonucleases. It is also used for other important enzymes in genetic engineering, e.g., DNA methylases. Restriction enzymes catalyze the specific cleavage of DNA, e.g., that of *Escherichia coli* phage $\lambda$ DNA, which is a typical substrate in genetic engineering (see Chap. 6). The DNA is split into smaller fragments of defined lengths. Because the negative charge per base pair is the same for all fragments, these can be separated according to length. The products of cleavage can be separated from intact or incompletely split molecules by electrophoresis under conditions that are carefully optimized for each specific enzyme–substrate reaction. Electrophoresis is usually performed with high-quality agarose gels as carrier material. Each resolution problem requires a specific agarose concentration, e.g., a relatively low concentration (0.5 g per 100 mL) for large fragments and a relatively high concentration (1.6–2 g per 100 mL) for small fragments. When very small fragments must be separated, polyacrylamide gels may be used as an alternative carrier material. Separated DNA fragments are visualized by staining with the fluorescent dye ethidium bromide [*1239-45-8*].

The fragments can be labeled by staining the gel in a separate tank after completion of electrophoresis or, more easily, by electrophoresis in the presence of the dye in the gel and buffer (e.g., at 1 µg/mL). When the gel is illuminated with long-wavelength UV light (e.g., at 366 nm), the separated fragments become visible and can be photographed.

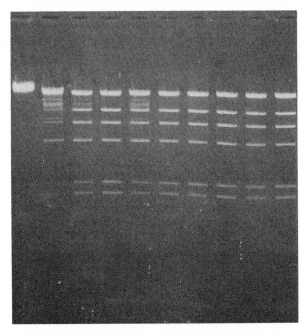

**Figure 9.** Determination of the catalytic activity of *Hin*dIII on λ DNA
a) λ DNA without enzyme; b–d) 1, 2, and 3 μL, respectively, 1:30 dilution; e–g) 1, 2, and 3 μL, respectively, 1:20 dilution; h–j) 1, 2, and 3 μL, respectively, 1:10 dilution

To calculate enzyme activity, the minimum amount of enzyme must be estimated which converts a given substrate completely to the fragment pattern typical for that enzyme. For practical reasons, this amount is divided by the volume of the enzyme solution.

As an example, electrophoretic assay of the restriction enzyme *Hin*dIII is described briefly (Fig. 9) [2.12].

*Definition of unit:* one unit is the catalytic activity of *Hin*dIII, which completely splits 1 μg of λ DNA in a total volume of 0.025 mL. The reaction is terminated after 60 min incubation at 37 °C in 0.025 mL of a defined buffer mixture.

*Assay:* different volumes (1–3 μL) of several enzyme dilutions (1:10, 1:20, 1:30) are incubated with 1 μg of λ DNA in a total volume of 0.025 mL. The reaction is terminated after 60 min by cooling with ice and adding 0.015 mL of a stop solution; the mixture (0.02 mL) is then placed in the slots of an agarose gel.

*Agarose gel:* the gel consists of 1 g of agarose per 100 mL and 1 mg of ethidium bromide per liter. The gel dimensions are 200 × 200 mm; total volume is 250 mL; and slots of 1 × 7 mm are prepared with a comb.

*Electrophoresis:* the apparatus is designed for submarine electrophoresis (2 h at 100 V). The buffer system is tris(hydroxymethyl)aminomethane–acetate, 40 mmol/L, and disodium ethylene-diamine-tetraacetate, 2 mmol/L; pH 8.2. The buffer contains 1 mg of ethidium bromide per liter.

*Detection:* the gel is illuminated directly after electrophoresis with long-wavelength UV light (366 nm) and photographed (Polaroid CU 5/film type 107); the amount of enzyme is estimated at which the complete typical fragment pattern is obtained [e.g., slot (c) in Fig. 9]. The activity is calculated according to

$$\text{Activity} = \frac{\text{Dilution}}{\text{Sample volume, mL}} \times \frac{\text{Units}}{\text{Volume, mL}}$$

(of original enzyme solution)

In the example, complete digestion was obtained with a minimum of 0.003 mL at 1:30 dilution of the original enzyme solution, resulting in an activity of 10 000 U/mL.

# 2.3. Quality Evaluation of Enzyme Preparations

## 2.3.1. Quality Criteria

The quality of enzyme preparations is characterized by activity, purity, stability, formulation, and packaging. These parameters depend on each other, but the formulation and packaging are easy to control and keep constant. The other parameters influence each other in such a way that quality is considered to be a function of activity, purity, and stability [2.36].

## 2.3.2. Specific Activity

One of the most important quality criteria of an enzyme preparation is its *specific activity,* i.e., the catalytic activity related to its protein content. Specific activity is usually expressed as units per milligram or, for less purified products, units per gram (see Section 2.2.2).

Specific activity data can be evaluated correctly only if the specific activity can be compared with that of a highly purified enzyme of the same origin. For this purpose, catalytic activities must be measured under identical conditions including the determination of protein.

## 2.3.3. Protein Determination

Since protein content is the most important reference point for determination of the specific activity of an enzyme preparation, several methods of protein determination are briefly mentioned in the following paragraphs [2.37]. All of these procedures are based on different principles and depend on the amino acid composition of the enzyme proteins. They will, therefore, yield different values.

**Ultraviolet Absorption.** Because of their content of aromatic amino acids, proteins exhibit an absorption maximum at 270–280 nm. For many pure proteins, reference values have been established for the 280-nm absorbance of a solution containing 10 mg/mL ($A_{280}^{1\%}$). WARBURG and CHRISTIAN found a formula which accounts for the nucleic acid content [2.38]. For greater precision, absorbance is also measured at shorter wavelengths, e.g., 235 nm [2.39].

**Biuret Method** [2.40]. The reaction of peptide bonds with copper ions in an alkaline solution yields a purple complex which can be determined photometrically. The intensity is a linear function of protein concentration.

**Lowry Method** [2.41]. The *Lowry* method combines the biuret reaction of proteins with reduction of the *Folin-Ciocalteu* phenol reagent (phosphomolybdic–phosphotungstic acid) by tyrosine and tryptophan residues. The reduction is promoted by the copper–protein complex. This method is very sensitive, but it is affected by many other compounds. The method has been modified to overcome these problems and to obtain a linear relationship between absorbance and protein content.

**Protein-Dye Binding.** Attempts have been made to determine protein concentration by using dyes. The method published by BRADFORD now predominates [2.42]. It is based on the shift of the absorption maximum of *Coomassie* Brilliant Blue G 250 [*6104-58-1*] from 465 to 595 nm, which occurs when the dye binds to the protein.

**Kjeldahl Analysis.** Before colorimetric procedures were established, protein concentration was calculated from nitrogen content by using an empirical factor [2.40].

## 2.3.4. Contaminating Activities

The content of contaminating activities, that is, the presence of other enzymes in the original material, is an important quality criterion for enzymes. This is usually related to the activity of the main enzyme and expressed in percent. Since the absolute amount is often very small, it cannot be expressed as protein mass (in milligrams) and, thus, does not influence the overall specific activity of the enzyme preparation. For example, lactate dehydrogenase from rabbit muscle has a specific activity of 500 U/mg and contains 0.001 % pyruvate kinase. Even if the content of pyruvate kinase increased tenfold to 0.01 %, the corresponding change in specific activity could not be measured. However, this contaminating activity is so high that such an enzyme preparation is useless for the determination of pyruvate kinase in blood.

Depending on the special applications of an enzyme, different impurities must be determined. For example, enzymes used in genetic engineering all act on a common substrate, nucleic acid; therefore, they must be free from impurities that also act on that substrate, such as specific or unspecific nucleases or phosphatases. Traces of unspecific endodeoxyribonucleases, for example, are detected routinely by incubating 10–100 U of an enzyme for a prolonged period (e.g., 16 h) with a susceptible substrate such as the supercoiled form of a plasmid DNA, e.g., pBR322. A minimum of 50 units should not influence the structure of that substrate, whereas another application may require that only 10 units do not change substrate structure. Such impurities are usually detected by electrophoresis (see Section 2.2.14) or by radiometry (see Section 2.2.6).

### 2.3.5. Electrophoretic Purity

Electrophoresis is of minor importance in the evaluation of purity because of its low sensitivity (detection of less than 0.5% contaminating protein is impossible). It is far inferior to the determination of contaminating activities. Furthermore, enzymatically inactive proteins usually do not interfere with enzymatic analyses. Electrophoresis gained importance when isoenzymes had to be analyzed which could not be distinguished by their catalytic function but only by physical properties such as electric charge. It is an indispensable tool for the identification of lactate dehydrogenase isoenzymes.

Electrophoresis is also used in the isolation of various enzymes of RNA, DNA, and protein biosynthesis. For this purpose, the introduction of disc electrophoresis on polyacrylamide gels and the use of dodecyl sulfate for the separation of enzyme complexes became valuable tools [2.43]. The system developed by LAEMMLI [2.44] is widely used and exhibits very high sensitivity, especially if gradient gels are employed [2.45]. The sensitivity of the method has been enhanced considerably by silver staining [2.46].

### 2.3.6. Performance Test

For many applications, partially purified enzyme preparations can be used, provided they do not contain any interfering contaminating activities. They are less costly and, therefore, preferred to highly purified products. However, they may contain unknown byproducts that can interfere with enzymatic analyses, for example. To avoid such problems, a performance test should be carried out. Examples are the determination of glucose with glucose oxidase and peroxidase (E.C. 1.11.1.7) [9003-99-0] or the determination of glycerol with glycerokinase (E.C. 2.7.1.30) [9030-66-4].

For some enzymes used in molecular biology, the determination of activity is not directly correlated to their application. In such cases, even highly purified enzymes must be analyzed for proper function in a typical performance experiment. An example is $T_4$ DNA ligase (E.C. 6.5.1.1) [9015-85-4] which functions properly by joining together fragments created by the action of a restriction endonuclease; in addition, the joined fragments can be recleaved by the same restriction endonuclease.

### 2.3.7. Stability [2.36]

A very important factor in the application of enzymes is their stability in concentrated or dilute form and after mixing with other substances. This applies to the manufacture of products for pharmaceutical purposes, food chemistry, or enzymatic analysis. Some enzymes can be stabilized by adding glycerol (50 vol%), ammonium sulfate (ca. 3.2 mol/L), or sodium chloride (3 mol/L) to their aqueous solution. Furthermore, many enzymes can be kept in lyophilized form for a long period of time in the presence of stabilizers such as salts, preservatives, inert proteins (predominantly bovine serum albumin), or carbohydrates.

Most enzymes used in analysis are stored at ca. 4 °C; solutions of restriction endonucleases must be kept at −20 °C or lower to maintain catalytic activity. To avoid degradation by moisture, the chilled enzyme preparation must be warmed before opening. Freezing and thawing may, in some cases, impair the activity of enzymes. Contamination by heavy metals or oxidants often inactivates enzymes, for example, by blocking mercapto groups. The activity of metalloenzymes or metal-dependent enzymes may be reduced by complexing agents such as ethylenediaminetetraacetate.

## 2.3.8. Formulation of Enzyme Preparations

An enzyme preparation should be formulated according to its application. For analytical purposes, it should be easy to pipette and – if possible – free of stabilizers and preservatives that might impair its function. For example, glutamate dehydrogenase (E.C. 1.4.1.3) [9029-12-3] must not contain any traces of ammonia if it is to be used in the enzymatic determination of urea or ammonium.

In reagent kits employed for enzymatic analysis in clinical laboratories or for food analysis, the enzyme may be used preferably in lyophilized form. Compared to enzyme solutions, the solid material is in some cases easier to mix with other solid components and stable for a longer period of time, even at slightly elevated temperature. When immobilized enzymes are to be used in columns, their particle size must ensure fast flow.

**Packaging.** Careful selection of packaging materials is very important for handling enzymes. Bottles and stoppers used for lyophilized enzymes must be absolutely tight to prevent access of moisture. Glass or plastic bottles as well as stoppers (rubber or plastic) should not release any traces of heavy metals or other enzyme-inactivating substances into the enzyme solution or suspension. In some cases, enzymes must be protected from light and packaged in brown glass bottles.

# 3. General Production Methods

## 3.1. Fermentation

Use of an aerobic submerged culture in a stirred tank reactor is the typical industrial process for enzyme production involving a microorganism that produces an extracellular enzyme. An example is given in Figure 13 (p. 40). This section concentrates on the typical process, with only short references to variations. Figure 10 shows the unit operations of the production process, and basic information can be found in handbooks on microbiological principles and methods [3.1]–[3.3]. This section is concerned with fermentation itself, and three elements – the organism, the

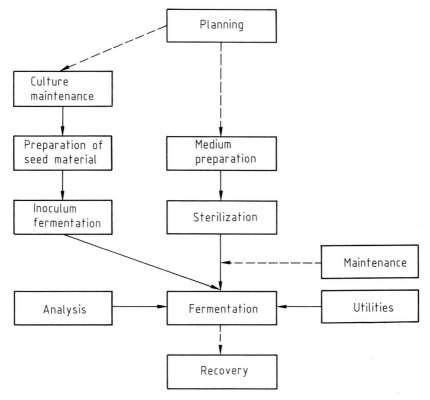

**Figure 10.** Unit operations of a fermentation process

equipment, and the protocol for fermentation – are discussed. These elements must be arranged in the most effective way possible to realize economically optimal results in industrial processes. Optimization of biotechnological processes is a multidisciplinary effort; for further information, see [3.4]–[3.6].

In the past, biotechnological process development may have looked somewhat like an art because adequate process control was lacking. This chapter should show that, based on the present state of the art, a production process can be highly controlled. To achieve this level of control in fermentation, the organism must be genetically adapted, and the process protocol must match the physiological possibilities of the organism and the limitations of the apparatus.

## 3.1.1. Organism and Enzyme Synthesis

An organism can be viewed as a metabolic system converting substrates into cell mass and byproducts. Enzymes function in this system as catalysts for the different reactions. Each cell is equipped with mechanisms that regulate the synthesis and activity of the enzymes to enable the cell to respond adequately to environmental changes. In its elementary form, therefore, an organism can be described as a set of metabolic components with a mechanism for enzyme synthesis and a regulatory apparatus. The kinetics of the process are determined by structural components of the organism and by various physical and chemical factors, as shown in Figure 11. The key role that enzymes play in biological processes has led to an extensive study

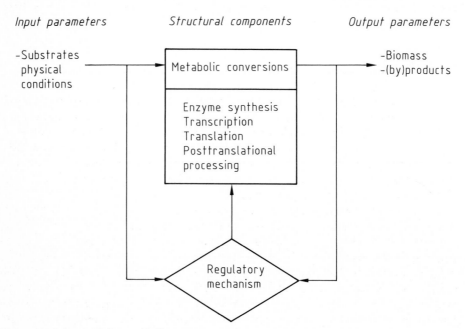

**Figure 11.** Scheme of the metabolic process

of both the mechanism and the regulation of their synthesis [3.7], [3.8], and some of the results are related to the problem of process development.

The basic mechanisms of enzyme synthesis, including transcription, translation, and posttranslational processing, seem to be highly conservative [3.9]. However, several differences exist between various classes of organisms, as well as some fundamental differences between prokaryotic and eukaryotic organisms. The enzymes themselves differ enormously in molecular mass, number of polypeptide chains, isoelectric point, and degree of glycosylation. In addition, a variety of enzyme-producing species exist [3.10]. Although all the differences may influence the characteristics of synthetic patterns, the basic mechanisms underlying enzyme synthesis are similar enough to allow a general treatment of the microbiological production process. However, the differences in production kinetics among various species are large enough to make individual optimization programs necessary.

Different organisms may also differ in their suitability for fermentation; such process characteristics as viscosity or recoverability, legal clearance of the organism, and knowledge available about the selected organism, must be considered.

Because of the action of the regulatory mechanism, enzyme synthesis rates range from no synthesis to maximum synthesis allowed by the synthetic apparatus, as in a normal control loop. The complexity of mechanisms ranges from relatively simple and well-understood induction and repression systems, to very complex global regulatory networks [3.11].

Process development must deal with the complexity of the enzyme synthesis system either by changing the structural characteristics, including structural elements of the regulatory systems (strain improvement, Section 3.1.2), or by selecting optimal environmental conditions (physiological optimization, Section 3.1.3).

## 3.1.2. Strain Improvement

Strain improvement for production purposes requires changing the genetic structure of the microorganism. The techniques for obtaining such genetic changes can be divided into mutation and selection, in vivo recombination, and genetic engineering [3.4]. Recombination techniques require the use of DNA exchange methods, either naturally existing ones such as conjugation, transformation, and sexual cycle or the recently developed technique of protoplast fusion. These processes and techniques are described in the literature [3.12].

Strain improvement programs can be very successful, as examples of antibiotic-producing strains show. However, data on industrial enzyme production are scarce, although 100- to 1000-fold increases in enzyme activity are reported [3.10]. Such data may be misleading when they are based upon low, inaccurate starting values and incompletely optimized production processes. The amount of protein produced and the time required for its production should be stated in a complete report on productivity level. Production levels can be as high as 10 g of protein per liter, in less than 100 h fermentation time, by combining different genetic techniques for strain improvement and optimal physiological conditions for production [3.13]. The result of strain improvement is generally expressed as an increase in the slope of the enzyme activity line, as shown in Figure 13 (p. 40). Strain improvement aims not only at an increased rate of synthesis but also at other benefits. EVELEIGH includes minimal byproduct formation, lack of formation of toxic metabolites, and ready recoverability in the list of desired qualities [3.14]. WHITE mentions the possibility of incorporating genes into a yeast strain to allow its growth on cheap, complex substrates [3.15]. Improvement of such characteristics can be obtained basically by the techniques described in the following paragraphs.

**Mutation and Selection.** The frequency of desired mutations and the chance of combining desired characteristics in a recombination process are low because of the random nature of mutation and selection. Therefore, many cells must be subjected to a mutation or recombination procedure and then tested for the desired combination of characteristics by selection. The success of strain improvement programs often depends on development of an effective selection method for finding one mutant among 10 000 – 100 000 cells. Methods range from plate selection to the continuous culture technique [3.16], [3.17].

Mutation changes the protein structure and most probably results in a deterioration of function. Changes in structural components by mutation are, therefore, rarely improvements unless the specific loss of function is required for production purposes, e.g., when a loss of regulatory function results in enhanced enzyme production.

Mutation and selection are directed primarily toward higher overall productivity rather than mutation of a specific function, but a loss of regulatory function is highly probable. However, some studies describe screening for a mutation in a specific function [3.14].

**In Vivo Recombination.** Based on the complex interdependency of pathways and the competition for substrates in the organism, random mutation may also affect the rate of enzyme synthesis. However, only minor improvements should be expected, although a series of mutations can still result in an interesting degree of improvement. In vivo recombination techniques have been used to complement mutation techniques by bringing together mutations in different cell lines and cleaning strains of undesired deleterious mutations [3.4]

**Genetic Engineering.** If the absence of regulation is assumed, considerable improvement can be realized by increasing the turnover rate of the limiting step in synthesis. Techniques of genetic engineering can be used to increase the rate of mRNA synthesis by constructing plasmids with the desired gene (gene multiplication). The number of plasmids per cell can be very high. The rate of mRNA synthesis should be related linearly to the number of gene copies, but the rate of enzyme synthesis may be limited again by the next step in the process. These techniques also allow replacement of the promotor by a more effective one [3.18]. The same methods allow replacement of the leader sequence of a gene, resulting in the excretion of a formerly intracellular enzyme [3.19].

Genetic engineering also uses microorganisms to produce enzymes of higher organisms by placement of the corresponding gene into the microorganism. The presence of introns may then prevent proper expression of the gene, but techniques have been developed to overcome this difficulty [3.15]. Chymosin (E.C. 3.4.23.4) [9001-98-3], calf rennet, has been cloned by several groups either in prokaryotes or in yeast and will probably be the first cloned mammalian enzyme to be produced industrially by microorganisms [3.20].

## 3.1.3. Physiological Optimization

Figure 11 shows that not only the structural characteristics of the system, but also the input parameters, determine its metabolic activity. Once a suitable organism has been found, either genetically improved or not, the next task of process development is to define optimal input parameters.

The rate of enzyme synthesis $r_e$ is defined by

$$r_e = q_e \cdot c_x$$

where $r_e$ is the enzyme synthesis rate, units per liter per hour (for definition of Unit, see p. 356); $q_e$ is the specific enzyme synthesis rate, units per gram of biomass per hour; and $c_x$ is biomass concentration, grams per liter.

In kinetic terms, optimization of the enzyme synthesis rate means finding the highest specific synthesis rate for a given amount of biomass. Because enzyme synthesis depends so much on primary metabolism, the complete cellular machinery must function at all times. Therefore, conditions favoring enzyme synthesis also favor growth, as suggested in Figure 11. The search for optimal conditions for enzyme synthesis can then be reduced to a search for (1) the conditions influencing growth rate and (2) the relation between specific enzyme synthesis rate and growth rate.

Growth can be expressed by a specific growth rate $\mu$ as the amount of biomass synthesized per unit biomass and unit time:

$$\mu = \frac{1}{c_x} \cdot \frac{dc_x}{dt}$$

where $c_x$ is the biomass concentration in grams per liter and $t$ the time in hours.

Specific enzyme synthesis rate and specific growth rate express the metabolic activities of the cell. These are abbreviated as synthesis rate ($q_e$) and growth rate ($\mu$). MONOD [3.21] formulated the relationship between growth rate and substrate concentration as

$$\mu = \mu_{max} \cdot \frac{c_s}{K_s + c_s}$$

where $\mu_{max}$ is the maximum obtainable growth rate per hour; $K_s$ is the saturation constant and $c_s$ the concentration of substrate, both expressed as moles per liter.

According to this relation actual growth rates at low substrate concentration are lower than the maximum possible value determined by the structural properties of the organism. When $c_s \gg K_s$, the growth rate reaches its maximum value. If all substrates are present in excess ($c_{s_i} \gg K_{s_i}$ for any i), the growth rate is limited by internal structure. The parameters $\mu_{max}$ and $K_s$ are dependent on pH, temperature, osmotic pressure, and such factors as medium type (minimal vs. nutrient medium). When all substrates are present in excess except one ($\mu < \mu_{max}$), that substrate is called the *limiting factor;* in this case, an external growth limitation exists. Nutrient-limited growth is considered the natural state for microorganisms [3.22].

The kinetics of enzyme synthesis cannot be expressed in one equation for all enzymes. Instead, a number of patterns of synthesis rate $q_e$ vs. growth rate $\mu$ are shown in Figure 12. The curves are somewhat idealized, but many examples can be found in the literature [3.23–3.26].

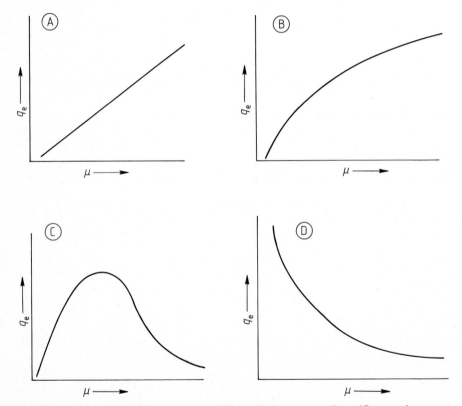

**Figure 12.** Types of relationship between specific synthesis rate $q_e$ and specific growth rate $\mu$
A: Growth-coupled synthesis [3.25]
B: Saturated synthesis [3.26]
C: Saturated synthesis with catabolic repression [3.27]
D: Repressed synthesis [3.28]

The importance of these physiological considerations is that a limiting factor can be used to control growth rate and, thereby, synthesis rate. If the $q_e$ vs. $\mu$ relationship is known, the value of $\mu$ that results in a maximum $q_e$ corresponds to the optimal physiological condition for synthesis.

Complete physiological characterization includes knowledge of the relationship of $q_e$ to $\mu$, the yield on different substrates, the effect of the limiting factor, the physicochemical conditions, and the possible role of regulatory agents. Because these factors are interdependent, establishment of the complete physiological characteristics of an organism is a very extensive task. In industrial practice, experience, general knowledge, and use of well-known organisms or techniques are generally employed to find shortcuts.

The previous discussion emphasizes the importance of the limiting factor for control of growth and product synthesis. Control of growth is possible if the substrate concentration $c_s$ is in the range of the saturation concentration $K_s$. The $K_s$ values for most substrates ($\pm 0.1$ mmol/L) are rather low compared to the consumption rate ($\pm 10$ mmol $L^{-1} h^{-1}$ for glucose). Continuous feed is, therefore, necessary to control growth rate by limiting substrate concentration, as applied in fed batch and continuous cultures. In batch cultures, all substrates are present at the start of fermentation. The growth rate will be maximal until the substrate is nearly consumed. Therefore, direct control of growth is not possible in a batch culture.

## 3.1.4. The Fermentor and its Limitations

For enzyme production, economy of scale leads to the use of fermentors with a volume of 20–200 m$^3$. The higher energy yield from aerobic combustion results in the use of aerobic processes which require continuous transfer of poorly soluble oxygen into the culture broth. The concomitant problems of mass and heat transfer are usually neglected in small fermentors and at low cell densities. However, in industrial microbiology, with the above-mentioned fermentor volumes and the economic necessity of using the highest possible cell densities, transport processes must be considered. Such processes can limit metabolic rates; e.g., the oxygen supply may become limiting and the microorganism may respond by changes in its physiological pattern. Under those conditions, the desired control of microbial metabolism may be lost. In controlled operation of an industrial process, metabolic rates must be limited to a level just below the transport capacity of the fermentor. Therefore, the highest possible productivity in a fermentor is obtained at maximal transport capacity. This is reflected in low concentrations of dissolved oxygen during fermentation, as shown in Figure 13.

Maximizing the transport processes is chiefly a problem of fermentor design and is generally treated in handbooks on bioengineering [3.3], [3.5], [3.27]. In the common stirred tank reactor, agitation and injection of compressed air are used to mix the contents of the reactor and to transfer gases. The oxygen transport rate $n$, for example, can be expressed as

$$n = k_1 a \, [c_0 \, (\mathrm{g}) - c_0 \, (\mathrm{l})]$$

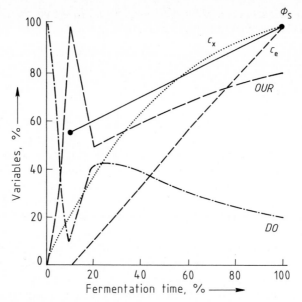

**Figure 13.** Schematic representation of a fermentation
This illustrates the production of Maxamyl by Gist-brocades N.V. After an inital exponential growth period, the process is controlled by an increasing feed rate which results in the oxygen uptake rate and biomass profile shown. The final result is an almost constant rate of enzyme synthesis during fermentation
Variables (expressed as percent of maximum value during fermentation):
   $c_x$, biomass concentration
   $c_e$, enzyme activity
   *DO*, dissolved oxygen
   *OUR*, oxygen uptake rate
   $\Phi_s$, carbon feed rate

where $k_1 a$ is the transfer coefficient per hour; $c_0$ (g) is the equilibrium oxygen concentration in the gas phase and $c_0$ (l) the equilibrium oxygen concentration in the medium, expressed both in millimoles per liter.

The transfer coefficient $k_1 a$ shows that the main resistance to oxygen transport is the gas–liquid interface, $k_1$ being the resistance coefficient in meters per hour and *a* the total gas–liquid interface area in square meters per cubic meter. The transfer capacity of a fermentor can be expressed as the $k_1 a$ value. The $k_1 a$ value is determined primarily by the power input of the agitator and the air jet system [3.28]. Bioengineering, therefore, aims at the design of a fermentor with the highest $k_1 a$ value at the lowest power input.

The $k_1 a$ value is also influenced by factors such as viscosity, ionic strength, and presence of surfactants. The dependence of $k_1 a$ on viscosity explains the preference for using pellet growth to cultivate mycelium-forming microorganisms [3.29].

The complex relationships of factors determining transport processes make the scaleup of biotechnological processes a very difficult task [3.4].

## 3.1.5. Process Design

Process design is the complicated task of choosing the optimal conditions for maximal process outcome. The number of interdependent factors is high, and the available physiological knowledge is seldom complete. The relationship between synthesis rate and growth rate may be very complex and depend on the presence of inductors or the absence of repressors. The use of genetically developed strains may ease process design considerably. The designed process is usually first tested on a pilot-plant scale and optimized in a number of fermentation runs; it is then scaled up to production size.

The total synthesis rate $r_e$ depends not only on growth rate but also on biomass concentration. However, when both a high biomass concentration and an optimal growth rate are maintained, the limit of transfer capacity is soon reached because of the increasing metabolic activity. Therefore, the prime goal of process design is to find an optimal compromise between these conflicting demands. This is realized by selecting a particular feed rate profile like the one shown in Figure 13.

The *enzyme synthesis rate* is not the only criterion for process optimization. In industrial practice, the profits of the entire production process must be maximized. The *recovery costs* may be a substantial part of the production costs of the enzyme; they are at least partly proportional to the concentration of enzyme in the mash at harvesting. The costs of fermentation also depend on the *enzyme yield* per amount of substrate used. A number of criteria, therefore, do not necessarily coincide.

**Culture Type.** The primary decision in process design concerns culture type. In *batch culture,* the growth rate cannot be controlled by dosed feeding, because all substrates are added at the beginning. Batch processes are rarely used today.

In *fed-batch processes,* a low initial biomass concentration should be chosen to maintain the desired growth rate for a certain period, without exceeding the transport capacity of the equipment. In addition, fed-batch processes can be designed so that the enzyme concentration at harvesting is higher than in batch or continuous cultures, and the productivity of fed-batch processes can be increased severalfold compared to batch processes. Fed-batch processes probably constitute the most frequently used process type [3.31]

*Continuous cultures* are ideally suited for high productivity because the excess biomass is continuously withdrawn, and both synthesis rate and biomass concentration can be optimal. Therefore, in principle, continuous culture is preferred for biotechnological production [3.30]. However, enzyme concentrations are lower than those reached in fed-batch cultures. In addition, the use of continuous cultures is limited by technical reasons such as the higher contamination risk and the problem of strain degeneration.

**Reactor Design.** The common *stirred tank* reactor can be used for all culture types. Continuous culturing can also be done in *plug flow reactors* or *multiphase reactors* [3.32]. In both types, the cells are transferred from one physiological phase to another. The method is proposed for enzyme production when regulatory mechanisms do not allow synthesis even at low growth rates.

## 3.1.6. Modeling and Optimization

In the previous sections, cells are envisaged as systems that are directly related to the environment, and the feed rate is treated as an independent variable with which to manipulate the desired output activities of the origanism. This approach allows the development of kinetic models of growth and production that can be used to optimize the process. If productivity is the main criterion of optimal fermentation, and the synthesis rate $r_e = q_e \cdot c_x$ at any time (Section 3.1.3), the productivity can be defined as

$$P_e = \frac{1}{t_{tot}} \cdot \int_0^{t_e} q_e(t) \cdot c_x(t) \cdot dt$$

where $q_e$ and $c_x$ are as defined in Section 3.1.3; $t_{tot}$ is the total fermentation time including turn-around time and $t_e$ is the fermentation time, both expressed in hours; and $P_e$ is the productivity in units per liter per hour.

To handle this equation, $q_e$ and $c_x$ must be expressed as functions of the feed rate, and the feed rate can be described as a function of the fermentation time. For continuous cultures the equation can be solved easily at constant feed rate, if steady-state conditions are assumed and the start-up phase is ignored. For fed-batch processes, the task is to define an optimal trajectory of values for the feed rate so that the value at any moment contributes optimally to final productivity. Therefore, the maximum value of $P_e$ must be chosen from all possible feed trajectories, and mathematical methods are available for this task [3.33].

This approach can be a powerful tool in process design and can reduce the amount of experimentation necessary. The method can be extended by incorporating other physiological factors and using other goal functions based on different criteria. In Section 3.1.3, the Monod equation is used as a kinetic model for biomass growth, but this equation cannot describe cellular growth under all conditions. One problem is that it does not express the change in cellular composition with growth rate. More complicated growth models, based on a description of the cell in different compartments (structured models) [3.34], can be used to express the biomass concentration as a direct or indirect function of feed rate.

Modeling the enzyme production rate is more complicated. As shown in Figure 12, the enzyme synthesis patterns cannot be described by a single function. Models of enzyme production must express (1) the dependence of regulatory functions on the concentration of signal molecules and (2) the differences in production rates, depending on growth rates [3.35].

Not only are models used to calculate optimal feed trajectories, they also reduce the complexity of physiological responses and can be used heuristically to detect all factors of relevance to a process. Industrially, modelling, optimization, and experimental work are combined to reach satisfactory production levels as quickly as possible. The process is then optimized experimentally in more detail.

### 3.1.7. Instrumentation and Control

When an optimal trajectory of the feed rate has been found and desired values for physicochemical variables have been established, the process is ready for a production scale. Elementary physical variables, pH, and feed rate are maintained by proper control. If process development has been of reasonable quality, the resulting output will be somewhere in the expected range. In practice, however, deviations between actual and expected results may arise for several reasons: (1) the model and the experimental knowledge are never complete; (2) errors of measurement and set point always exist, along with deviations in the variables regulated; (3) raw materials may vary in composition; and (4) transport conditions can cause deviations in response.

If deviations from the expected results are minimal, the process can be operated with a minimum level of instrumentation and control. If the deviations cannot be ignored, output variables may be measured to correct input variables; for example, the carbon dioxide production rate can be used for feedback regulation of the feed rate. More advanced control can be realized when a model of the process is used to relate different input and output variables. The measured input and output variables of the process can be compared with the model, and some type of regulatory action can be derived from this comparison [3.36].

Fermentation processes are complicated, and in principle, control of the process can be improved by measuring as many variables as possible and using them for computer-based process control and optimization. A general limitation of such control systems is lack of adequate (cheap and sterilizable) sensors for measuring the chemical and biological variables [3.37]. Another problem is the lack of qualified people to develop the software and operate such systems [3.38].

## 3.2. Isolation and Purification [3.39]–[3.54]

The degree of purity of commercial enzymes ranges from raw enzymes to highly purified forms and depends on the application. Raw materials for the isolation of enzymes are animal organs, plant material, and microorganisms.

Enzymes are universally present in living organisms; each cell synthesizes a large number of different enzymes to maintain its metabolic reactions. The choice of procedures for enzyme purification depends on their location. Isolation of intracellular enzymes often involves the separation of complex biological mixtures. On the other hand, extracellular enzymes are generally released into the medium with only

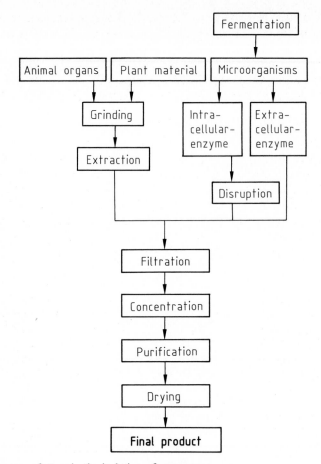

**Figure 14.** Sequence of steps in the isolation of enzymes

a few other components. Enzymes are very complex proteins, and their high degree of specificity as catalysts is manifest only in their native state. The native conformation is attained under specific conditions of pH, temperature, and ionic strength. Hence, only mild and specific methods can be used for enzyme isolation. Figure 14 shows the sequence of steps involved in the recovery of enzymes.

## 3.2.1. Preparation of Biological Starting Materials

**Animal Organs.** Animal organs must be transported and stored at low temperature to retain enzymatic activity. The organs should be freed of fat and connective tissue before freezing. Frozen organs can be minced with machines generally used in the meat industry, and the enzymes can be extracted with a buffer solution. Besides mechanical grinding, enzymatic digestion can also be employed [3.55]. Fat attached to the organs interferes with subsequent purification steps and can be removed with organic solvents. However, enzymatic activity might be influenced negatively by this procedure.

**Plant Material.** Plant material can be ground with various crushers or grinders, and the desired enzymes can be extracted with buffer solutions. The cells can also be disrupted by previous treatment with lytic enzymes.

**Microorganisms** are a significant source of enzymes. New techniques, summarized under genetic and protein engineering, have much to offer the enzyme industry. A gene can be transferred into a microorganism to make that organism produce a protein it did not make naturally. Alternatively, modification of the genome of a microorganism can change the properties of proteins so that they may be isolated and purified more easily. Such modifications might, for example, cause the release of intracellular enzymes into the medium; change the net charge and, therefore, the chromatographic properties of proteins; or lead to the formation of fused proteins [3.56].

Most enzymes used commercially are *extracellular enzymes,* and the first step in their isolation is separation of the cells from the solution. For *intracellular enzymes,* which are being isolated today in increasing amounts, the first step involves grinding to rupture the cells. A number of methods for the disruption of cells (Table 2) are known, corresponding to the different types of cells and the problems involved in isolating intracellular enzymes. However, only a few of these methods are used on an industrial scale.

**Table 2.** Methods for disruption of cells

| Mechanical methods | Nonmechanical methods | |
|---|---|---|
| High pressure (Manton–Gaulin, French-press) | Drying (freeze-drying, organic solvents) | |
| Grinding (ball mill) | Lysis | |
| Ultrasound | physical: | freezing, osmotic shock |
| | chemical: | detergents, antibiotics |
| | enzymatic: | enzymes (e.g., lysozyme) |

### 3.2.1.1. Cell Disruption by Mechanical Methods

*High-pressure homogenization* is the most common method of cell disruption. The cell suspension is pressed through a valve and hits an impact ring (e.g., Manton–Gaulin homogenizer). The cells are ruptured by shearing forces and simultaneous decompression. Depending on the type of machine, its capacity ranges from 50 to 5000 L/h. The rigid cell walls of small bacteria are only partially ruptured at the pressures up to 55 MPa (550 bar) achieved by this method. Higher pressures, however, would result in further heat exposure (2.2 °C per 10 MPa). Hence, the increased enzyme yield resulting from improved cell disruption could be counteracted by partial inactivation caused by heating and higher shearing forces. Therefore, efficient cooling must be provided.

The *wet grinding* of cells in a high-speed bead mill is another effective method of cell disruption [3.57]–[3.60]. Glass balls with a diameter of 0.2–1 mm are used to break the cells. The efficiency of this method depends on the geometry of the stirrer system. A symmetrical arrangement of circular disks gives better results than the normal asymmetrical arrangement [3.61]. Given optimal parameters such as stirring rate, number and size of glass beads, flow rate, cell concentration, and temperature, a protein release of up to 90 % can be achieved in a single passage [3.57].

### 3.2.1.2. Cell Disruption by Nonmechanical Methods

Cells may frequently be disrupted by *chemical, thermal,* or *enzymatic lysis.* The drying of microorganisms and the preparation of acetone powders are standard procedures in which the structure of the cell wall is altered to permit subsequent extraction of the cell contents. Methods based on enzymes or autolysis have been described in the literature [3.62]–[3.65]. Ultrasound is generally used in the laboratory. In this procedure, cells are disrupted by shearing forces and cavitation. An optimal temperature must be maintained by cooling the cell suspension because heat is generated in the process. Additional problems may arise from generation of free radicals.

## 3.2.2. Separation of Solid Matter

After cell disruption, the next step is separation of extracellular or intracellular enzymes from cells or cellular fragments, respectively. This operation is rather difficult because of the small size of bacterial cells and the slight difference between the density of the cells and that of the fermentation medium. *Continuous filtration* is used in industry. Large cells, e.g., yeast cells, can be collected by *decantation.* Today, efficient *centrifuges* have been developed to separate cells and cellular fragments in a continuous process. Residual plant and organ matter can be separated with simpler centrifuges or filters.

### 3.2.2.1. Filtration

The filtration rate is a function of filter area, pressure, viscosity, and resistance offered by the filter cake and medium. For a clean liquid, all these terms are constant which results in a constant flow rate for a constant pressure drop. The cumulative filtrate volume increases linearly with time. During the filtration of suspensions, the increasing thickness of the formed filter cake and the concomitant resistance gradually decrease the flow rate. Additional difficulties may arise because of the compressibility of biological material. In this case, the resistance offered by the filter cake and, hence, the rate of filtration depend on the pressure applied. If the pressure applied exceeds a certain limit, the cake may collaps and total blockage of the filter can result.

**Pressure Filters.** A *filter press* (plate filter, chamber filter) is used to filtrate small volumes or to remove precipitates formed during purification. The capacity to retain solid matter is limited, and the method is rather work-intensive. However, these filters are highly suitable for the fine filtration of enzyme solutions.

**Vacuum Filters.** Vacuum filtration is generally the method of choice because biological materials are easily compressible. A *rotary vacuum filter* (Figure 15) is used in the continuous filtration of large volumes. The suspension is usually mixed with a filter aid, e.g., kieselguhr, before being applied to the filter. The filter drum is coated with a thin layer of filter aid (precoat). The drum is divided into different sections so that the filter cake can also be washed and dried on the filter. The filter cake is subsequently removed by using a series of endless strings or by scraper discharge (knife). The removal of a thin layer of precoat each time exposes a fresh filtering area. This system is useful for preventing an increase in resistance with the accumulation of filter cake during the course of filtration.

**Cross-flow Filtration.** In recent years, a new method of filtration, cross-flow filtration, has been devised [3.65], [3.68]. In conventional methods, the suspension flows perpendicular to the filter material (Fig. 16 A). In cross-flow filtration, the input stream flows parallel to the filter area (Fig. 16 B), thus preventing the accumulation of filter cake and an increased resistance to filtration. To maintain a sufficiently high filtration rate, this method must consume a relatively large amount of energy,

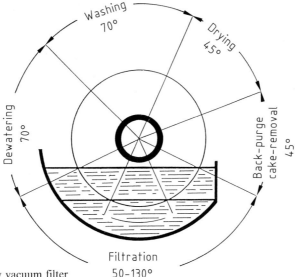

**Figure 15.** Rotary vacuum filter

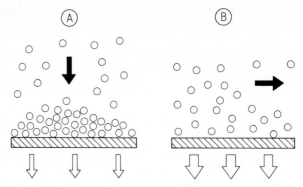

**Figure 16.** Principles of conventional, dead end filtration (A) and cross-flow filtration (B)

in the form of high flux rates over the membranes. With the membranes now available, permeate rates of $30-50 \text{ Lm}^{-2} \text{ h}^{-1}$ can be attained. Indeed, in many cases the use of a separator (see Section 3.2.2.2) is more economical [3.67]. The future of this method depends on the development of suitable membranes, but cross-flow filtration can be conveniently used in recombinant DNA techniques to separate organisms in a closed system.

### 3.2.2.2. Centrifugation

The sedimentation rate of a bacterial cell with a diameter of 0.5 μm is less than 1 mm/h. An economical separation can be achieved only by sedimentation in a centrifugal field. The range of applications of centrifuges depends on the particle size and the solids content (Table 3).

The $\Sigma$ value of a centrifuge is a good criterion for the comparison of centrifuges:

$$\Sigma = F \cdot Z$$

where $F = V/S$, with $V$ the total volume of liquid in the centrifuge and $S$ the thickness of the liquid layer in the centrifuge; i.e., $F$ has the units of an area. $Z = r \cdot \omega^2/g$, where $r$ is the radius of the centrifuge drum, $\omega$ the angular rotation speed, and $g$ the gravitational constant.

**Table 3.** Utilization of different centrifuges

| Type of centrifuge | Solids content, % | Particle size, μm |
|---|---|---|
| Multichamber separator | 0–5 | 0.5–500 |
| Desludging disk separator | 3–10 | 0.5–500 |
| Nozzle separator | 5–25 | 0.5–500 |
| Decanter | 5–40 | 5–50 000 |
| Sieve centrifuge | 5–60 | 5–10 000 |
| Pusher centrifuge | 20–75 | 100–50 000 |

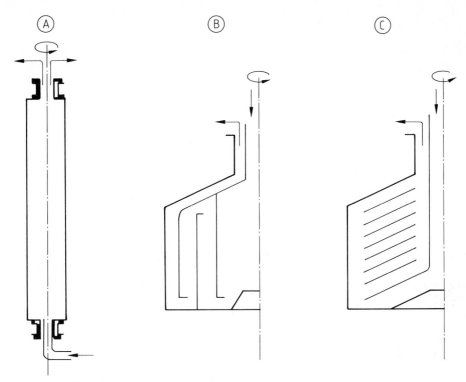

**Figure 17.** Solid-wall centrifuges
A: Tubular bowl
B: Multichamber solid bowl
C: Disk stack

Both sieve centrifuges and solid-wall centrifuges are available. Typical solid-wall centrifuges are shown in Figure 17.

*Decanters* (scroll-type centrifuges) work with low centrifugal forces and are used in the separation of large cells or protein precipitates. Solid matter is discharged continuously by a screw conveyer moving at a differential rotational speed.

*Tubular bowl centrifuges* are built for very high centrifugal forces and can be used to sediment very small particles. However, these centrifuges cannot be operated in a continuous process. Moreover, solid matter must be removed by hand after the centrifuge has come to a stop. A further disadvantage is the appearance of aerosols.

*Separators* (disk stack centrifuges) can be used in the continuous removal of solid matter from suspensions. Solids are discharged by a hydraulically operated discharge port (intermittent discharge) or by an arrangement of nozzles (continuous discharge). Bacteria and cellular fragments can be separated by a combination of high centrifugal forces, up to 15 000 × gravity, presently attainable, and short sedimentation distances. Disk stack centrifuges that can be sterilized with steam are used for recombinant DNA techniques in a closed system [3.69], [3.70].

### 3.2.2.3. Extraction

An elegant method used to isolate intracellular enzymes is liquid–liquid extraction in an aqueous two-phase system [3.71]–[3.75]. This method is based on the incomplete mixing of different polymers, e.g., dextran [*9004-54-0*] and poly(ethylene glycol) [*25322-68-3*], or a polymer and a salt in an aqueous solution [3.76]. The first extraction step separates cellular fragments. Subsequent purification can be accomplished by extraction or, if high purity is required, by other methods. The extractability can be improved by using affinity ligands [3.77], [3.78] or modified chromatography gels, e.g., phenyl-Sepharose [3.79].

### 3.2.2.4. Flocculation and Flotation

According to Stoke's law, the sedimentation rate is directly proportional to the diameter of the particle. The flocculation of bacterial cells to form larger particles can be achieved by the addition of mineral colloids, salts, or organic polymers. The neutralization of charges on the cell surface and the formation of bridges between individual cells lead to agglomeration [3.80]–[3.82]. Agglomerates can then be removed by filtration or centrifugation.

If no stable agglomerates are formed, cells can be separated by *flotation*. Here, cells are adsorbed onto gas bubbles, rise to the top, and accumulate in a froth. This process can be more efficient than flocculation followed by centrifugation, as demonstrated in the separation of single cell protein [3.83].

## 3.2.3. Concentration

The enzyme concentration in starting material is often very low. The volume of material to be processed is generally very large, and substantial amounts of waste material must be removed. Thus, if economic purification is to be achieved, the volume of starting material must be decreased by concentration. Only mild concentration procedures that do not inactivate enzymes can be employed. These include *thermal methods, precipitation,* and to an increasing extent, *membrane filtration.*

### 3.2.3.1. Thermal Methods

Only brief heat treatment can be used for concentration because enzymes are thermolabile. Evaporators with rotating components that achieve a thin liquid film (thin-layer evaporator, centrifugal thin-layer evaporator) or circulation evaporators (long-tube evaporator) can be employed.

### 3.2.3.2. Precipitation

Enzymes are very complex protein molecules possessing both ionizable and hydrophobic groups which interact with the solvent. Indeed, proteins can be made to agglomerate and, finally, precipitate by changing their environment. Precipitation is actually a simple procedure for concentrating enzymes [3.84].

**Precipitation with Salts.** High salt concentrations act on the water molecules surrounding the protein and change the electrostatic forces responsible for solubility. Ammonium sulfate [*7783-20-2*] is commonly used for precipitation; hence, it is an effective agent for concentrating enzymes. Enzymes can also be fractionated, to a limited extent, by using different concentrations of ammonium sulfate. The corrosion of stainless steel and cement by ammonium sulfate is a disadvantage, which causes additional problems in wastewater treatment. Sodium sulfate [*7757-82-6*] is more efficient from this point of view, but it is less soluble and must be used at temperatures of 35–40 °C. The optimal concentration of salt required for precipitation must be determined experimentally, and generally ranges from 20 to 80% saturation.

**Precipitation with Organic Solvents.** Organic solvents influence the solubility of enzymes by reducing the dielectric constant of the medium. The solvation effect of water molecules surrounding the enzyme is changed; the interaction of protein molecules is increased; and therefore, agglomeration and precipitation occur. Commonly used solvents are ethanol [*64-17-5*] and acetone [*67-64-1*]. Satisfactory results are obtained only if the concentration of solvent and the temperature are carefully controlled because enzymes can be inactivated easily by organic solvents.

**Precipitation with Polymers.** The polymers generally used are polyethylenimines and poly(ethylene glycols) of different molecular masses. The mechanism of this precipitation is similar to that of organic solvents and results from a change in the solvation effect of the water molecules surrounding the enzyme. Most enzymes precipitate at polymer concentrations ranging from 15 to 20%.

**Precipitation at the Isoelectric Point.** Proteins are ampholytes and carry both acidic and basic groups. The solubility of proteins is markedly influenced by pH and is minimal at the isoelectric point at which the net charge is zero. Because most proteins have isoelectric points in the acidic range, this process is also called *acid precipitation.*

Precipitation is usually carried out on a small scale. Problems can arise in scaling-up this process [3.85]. The mixing time, the residence time in the reactor (which affects agglomerate formation and enzyme activity), and the shearing forces generated by stirring (which affect the aggregates formed) are critical parameters. When the volume being processed is large, the mixing time is appropriately long and, especially with organic solvents, protein denaturation can occur. Experiments have been conducted to overcome difficulties of this kind by using a continuous process [3.86].

### 3.2.3.3. Ultrafiltration

A semipermeable membrane permits the separation of solvent molecules from larger enzyme molecules because only the smaller molecules can penetrate the membrane when the osmotic pressure is exceeded. This is the principle of all membrane separation processes (Table 4), including ultrafiltration. In reverse osmosis, used to separate materials with low molecular mass, solubility and diffusion phenomena influence the process, whereas ultrafiltration and cross-flow filtration are based solely on the sieve effect. In processing enzymes, cross-flow filtration is used to harvest cells, whereas ultrafiltration is employed for concentrating and desalting.

**Table 4.** Membrane separation processes

| Process | Application | Separation range, $M_r$ |
|---------|-------------|------------------------|
| Cross-flow microfiltration | concentration of bacteria, removal of cell debris | >1 000 000 (or particles) |
| Ultrafiltration | concentration of enzymes, dialysis, fractionation | >10 000 (macromolecules) |
| Reverse osmosis | concentration of small molecules, desalting | >200 |

Difficulties arise from *concentration polarization*. The semipermeable membrane excludes larger molecules, which tend to accumulate near the surface of the membrane because back-diffusion into the solution is limited. As a result of the different rates of diffusion of molecules of different sizes, the separating ability of the membrane changes. Thus, the membrane holds back small molecules more strongly than would be expected from its pore size. This effect limits the applicability of membrane separation. The formation of *gel layers* on the membrane is reduced by maintaining a turbulent flow or a laminar flow with high flow rate. Loss of permeability is also caused by *membrane fouling* or *deposition* on the membrane. In particular, antifoaming agents from fermentation solutions are deposited on the membrane and make concentration of enzymes more difficult.

Membranes are available for ultrafiltration which exclude molecules ranging from 1000 to 300 000 dalton [3.87], [3.88]. Anisotropic membranes consisting of a very thin membrane layer (0.1 – 0.5 µm) and a thicker, porous support layer [3.89] are generally used. The different membranes employed are flat membranes (plate and frame, cassette type, or spiral winding module), hollow fibers, and tubular modules (Fig. 18). Cellulose acetate [9004-35-7] and organic polymers such as polysulfone [25135-51-7], poly(vinylidene fluoride) [24937-79-9], and polypropylene [9003-07-0] have proved to be useful as membrane materials. With the exception of cellulose acetate, these membranes can be cleaned easily with alkali or acid and steam sterilized.

**Desalting.** The desalting of enzyme solutions can be carried out conveniently by diafiltration. The small salt molecules are driven through a membrane with the water molecules. The permeate is continuously replaced by fresh water. In fact, the concentration of salt decreases according to the following formula:

$$\ln \frac{c_0}{c} = \frac{V}{V_0}$$

where $c_0$ = starting salt concentration
$c$ = final salt concentration
$V_0$ = starting volume
$V$ = volume exchanged

Hence, a 99% salt exchange can be achieved when the permeated volume is $5 \cdot V_0$, independent of the starting concentration.

## 3.2.4. Purification

For many industrial applications, partially purified enzyme preparations will suffice; however, enzymes for analytical purposes and for medical use must be highly purified. Special procedures employed for enzyme purification are crystallization, electrophoresis, and chromatography.

### 3.2.4.1. Crystallization

Enzymes can be crystallized from ammonium sulfate solution. In general, however, very pure solutions are required, which explains why crystallization is not commonly used to purify enzymes.

### 3.2.4.2. Electrophoresis

Electrophoresis is used to isolate pure enzymes on a laboratory scale. Depending on the conditions, the following procedures can be used: *zone electrophoresis, isotachophoresis,* or *porosity gradients.* The heat generated in electrophoresis and the interference caused by convection are problems associated with a scaleup of this method. An interesting contribution to the industrial application of electrophoresis is a continuous process in which the electrical field is stabilized by rotation [3.90], [3.91].

### 3.2.4.3. Chromatography

Chromatography is of fundamental importance to enzyme purification (Table 5). Molecules are separated according to their physical properties (size, shape, charge, hydrophobic interactions), chemical properties (covalent binding), or biological properties (biospecific affinity).

A

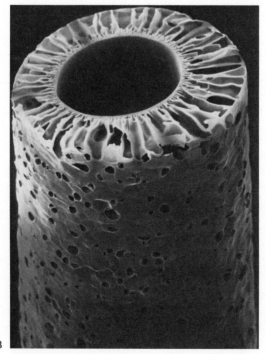

B

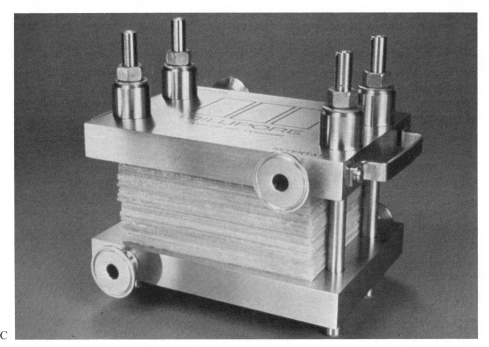

C

**Figure 18.** Different ultrafiltration modules
A: Plate-and-frame module (by courtesy of DDS, Ro-Division)
B: Hollow-fiber module (by courtesy of Amicon)
C: Cassette module (by courtesy of Millipore)

**Table 5.** Chromatographic methods

| Type of chromatography | Principle | Separation according to |
| --- | --- | --- |
| Adsorption | surface binding | surface affinity |
| Distribution | distribution equilibrium | polarity |
| Ion exchange | ion binding | charge |
| Gel filtration | pore diffusion | molecular size, molecular shape |
| Affinity | specific adsorption | molecular structure |
| Hydrophobic | hydrophobic chelation | molecular structure |
| Covalent | covalent binding | polarity |
| Metal chelate | complex formation | molecular structure |

In *gel chromatography,* (also called *gel filtration*), hydrophilic, cross-linked gels with pores of finite size are used in columns to separate biomolecules. Concentrated solutions are necessary for separation because the sample volume that can be applied to a column is limited to ca. 10 % of the column volume. In gel filtration, molecules are separated according to size and shape. Molecules larger than the largest pores in the gel beads, i.e., above the exclusion limit, cannot enter the gel and are eluted first. Smaller molecules, which enter the gel beads to varying extent depending on their size and shape, are retarded in their passage through the column and eluted in order of decreasing molecular mass. The elution volume of a globular protein is proportional to the logarithm of its molecular mass [3.92]. By varying the degree of cross-linking, gels of different porosities and with different fractionation ranges are obtained. Media that collectively cover all molecular sizes are available (Table 6). These include different types of Sephadex, which can be obtained by cross-linking dextran [*9004-54-0*] with epichlorohydrin [*106-89-8*], and Sephacryl, which is prepared by cross-linking allyldextran with *N,N'*-methylenebisacrylamide [*110-26-9*]. Gel filtration is used commercially for both separation and desalting of enzyme solutions.

*Ion-exchange chromatography* is a separation technique based on the charge of protein molecules. Enzyme molecules possess positive and negative charges. The net charge is influenced by pH, and this property is used to separate proteins by chromatography on anion exchangers (positively charged) or cation exchangers (negatively charged) (Table 7). The sample is applied in aqueous solution at low ionic strength, and elution is best carried out with a salt gradient of increasing concentration. Because of the concentrating effect, samples can be applied in dilute form.

The ability to process large volumes and the elution of dilute sample components in concentrated form make ion exchange very useful. The matrix used to produce ion-exchange resins should (1) be sufficiently hydrophilic to prevent enzyme denaturation and (2) have a high capacity for large molecules at fast equilibration. In addition, industrial applications require ion exchangers that give good resolution, allow high flow rates to be used, suffer small changes in volume with salt gradients or pH changes, and can be completely regenerated.

For *hydrophobic chromatography,* media derived from the reaction of CNBr-activated Sepharose with aminoalkanes of varying chain length are suitable [3.93]. This method is based on the interaction of hydrophobic areas of protein molecules with hydrophobic groups on the matrix. Adsorption occurs at high salt concentrations, and fractionation of bound substances is achieved by eluting with a negative salt gradient. This method is ideally suited for further purification of enzymes after concentration by precipitation with such salts as ammonium sulfate.

In *affinity chromatography,* the enzyme to be purified is specifically and reversibly adsorbed on an effector attached to an insoluble support matrix. Suitable effectors are substrate analogues, enzyme inhibitors, dyes, metal chelates, or antibodies. The principle of affinity chromatography is shown in Figure 19. The insoluble matrix (C) is contained in a column. The biospecific effector, e.g., an enzyme inhibitor (I), is attached to the matrix. A mixture of different enzymes ($E_1$, $E_2$, $E_3$, $E_4$) is applied to the column. The immobilized effector specifically binds the complemen-

**Table 6.** Gel filtration media

| Trade name (manufacturer) | Matrix | Fractionation range, $M_r$ |
|---|---|---|
| Biogel (Bio-Rad) | polyacrylamide (P-type) | $100-400\,000$ |
| | agarose (A-type) | $1000-150 \times 10^6$ |
| Ultrogele (IBF, Serva) | agarose/polyacrylamide | $60\,000-1.3 \times 10^6$ |
| | agarose | $25\,000-20 \times 10^6$ |
| Fractogel (Merck) | vinyl polymer | $100-5 \times 10^6$ |
| | (various types) | |
| Sephadex | dextran | $50-600\,000$ |
| (Pharmacia) | (various types) | |
| Sepharose | agarose | $10\,000-40 \times 10^6$ |
| (Pharmacia) | cross-linked agarose | $10\,000-40 \times 10^6$ |
| Sephacryl | sephacryl/bisacrylamide | $5000-1 \times 10^6$ |
| (Pharmacia) | | |
| Glycophase | surface-modified glass | $1000-350\,000$ |
| (Pierce) | (1,2-dihydroxypropyl-substituted) | |

**Table 7.** Ion-exchange resins

| Trade name (manufacturer) | Matrix | Anion- exchange groups** | Cation- |
|---|---|---|---|
| Cellex (Bio-Rad) | cellulose | DEAE | CM |
| | | ECTEOLA | phosphoryl |
| | | QAE | |
| | | TEAE | |
| Bio-Gel A (Bio-Rad) | agarose | DEAE | CM |
| Trisacryl (IBF, Serva) | synthetic polymer | DEAE | CM |
| Fractogel (Merck) | vinyl polymer | DEAE | CM |
| | | | SP |
| Sephadex (Pharmacia) | dextran | DEAE | CM |
| | | QAE | SP |
| Sepharose (Pharmacia) | agarose | DEAE | CM |
|   Sepharose FF* | | | |
|   Q-Sepharose | | | |
|   S-Sepharose | | | |
| Sephacel (Pharmacia) | cellulose | DEAE | |

* FF = fast flow. ** CM = carboxymethyl: $-O-CH_2-COO^- \; Na^+$;
DEAE = diethylaminoethyl: $-O-CH_2-CH_2-N^+(C_2H_5)_2H \; OH^-$;
ECTEOLA = mixture of different amines;
QAE = 2-hydroxypropylamine: $-O-CH_2-CH_2-N^+(C_2H_5)_2-CH(OH)-CH_3 \; Br^-$;
SP = sulfopropyl: $-O-CH_2-CH_2-CH_2-SO_3^- \; Na^+$;
TEAE = triethylaminoethyl: $-O-CH_2-CH_2-N^+(C_2H_5)_3 \; Br^-$.

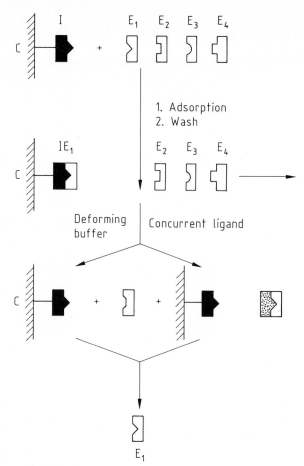

**Figure 19.** Affinity chromatography
For explanation of abbreviations, see text

tary enzyme. Unbound substances are washed out, and the enzyme of interest ($E_1$) is recovered by changing the experimental conditions, for example by altering pH or ionic strength.

Metal ions can also serve as effectors. They are attached to the matrix with the help of complexing agents, e.g., iminodiacetic acid [*142-73-4*]. Here, enzymes are separated on the basis of differing strengths of interaction with metal ions [3.94].

*Immunoaffinity chromatography* occupies a unique place in purification technology. In this procedure, monoclonal antibodies are used as effectors. Hence, the isolation of a specific substance from a complex biological mixture in one step is possible. In this procedure, enzymes can be purified by immobilizing antibodies specific for the desired enzyme. A more general method offers the synthesis of a

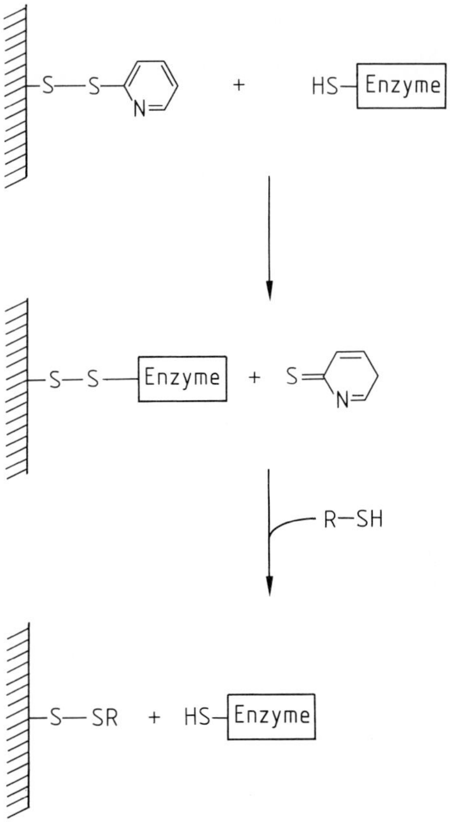

**Figure 20.** Covalent chromatography

fusion protein with protein A by "protein engineering." Protein A is a *Staphylococcus* protein with a high affinity for many immunoglobulins, especially of the IgG class of antibodies. In this way, enzymes that usually do not bind to an antibody can be purified by immunoaffinity chromatography.

*Covalent chromatography* differs from other types of chromatography in that a covalent bond is formed between the required protein and the stationary phases. The principle is illustrated in Figure 20 for an enzyme containing a reactive mercapto group, such as urease (E.C. 3.5.1.5) [*9002-13-5*] [3.95].

**Industrial-Scale Chromatography.** The main applications of industrial-scale chromatography in the 1960s were the desalination of enzyme solutions by use of highly cross-linked gels such as Sephadex G-25 and batch separations by means of ion exchangers such as DEAE-Sephadex A-50. Progress has been made in recent years in improving the stability and hydraulic properties of chromatographic media so that these techniques are now used on a production scale. Important parameters for the scaleup of chromatographic systems are the height of the column, the linear flow rate, and the ratio of sample volume to bed volume [3.96]. Zone spreading interferes with the performance of the column. Factors that contribute to zone spreading are longitudinal diffusion in the column, insufficient equilibration, and inadequate column packing. Longitudinal diffusion can be minimized by using a high flow rate. On the other hand, equilibration between the stationary and the mobile phases is optimal at low flow rates. Because good process economy depends to a large extent on the flow rate, a compromise must be made. In addition, the flow rate is also dependent on particle size; the decisive factor is usually the pressure drop along large columns. Although optimal resolution is obtained only with the smallest particles, the gel must have a particle size that favors a good throughput and reduces processing times. The use of *segmented columns* prevents a large pressure drop in the column (Fig. 21) [3.97]. Above all, the column must be uniformly packed so that the particle-size distribution, porosity, and resistance to flow are the same throughout the column. If this is not done, viscous protein solutions can give an uneven elution profile, which would lead to zone bleeding. The design of the column head is important for uniform distribution of the applied sample. This is generally achieved by symmetrical arrangement of several inlets and perforated inserts for good liquid distribution. The outlet of the column must have minimal volume to prevent back-mixing of the separated components.

**Figure 21.** Chromatographic KS 370 stack columns (by courtesy of Pharmacia)

**Figure 22.** Production-scale chromatography column
(Pharmacia)

Columns of glass, plastic, or in the case of large columns, metal (stainless steel) can be used. Metal columns are usually highly polished because a smooth surface is more resistant to corrosion. Viton or poly(tetrafluoroethylene) (PTFE) is applied as sealing material. Polyethylene tubing can be used to connect small columns. Larger tubing, with a diameter of more than 4 mm, must be made of metal, with special screw connections. No dead angles should be present in which liquid can collect and cause contamination by microbial growth.

Figure 22 shows a typical industrial chromatographic column. To avoid contamination of the column, sterile filters are placed at the entrance and exit. Optimal and efficient production systems not only must meet official standards such as those of the Food and Drug Administration (FDA) or Bundesgesundheitsamt (BGA), but must also comply with rules of good manufacturing practice (GMP), which makes cleaning-in-place (CIP) necessary. The system must be operated so that impurities from the process or from microbial growth can be removed completely. Chromatographic columns can be cleaned only by washing with chemicals. The reagents used for this purpose are described in Table 8. Figures 23 and 24 show the steps involved in processing extracellular and intracellular enzymes, respectively.

## 3.2.5. Product Formulation

In the application of enzymes, stability during transport and storage is a particular concern. Small amounts of highly purified enzyme are conveniently stored as suspensions in ammonium sulfate or poly(ethylene glycol) at $+4\,°C$. Freeze-drying of enzymes is a gentle procedure and can also be employed; however, it is too expensive for large-scale use. Raw enzyme preparations and heat-stable enzymes can be spray-dried. In many cases, stabilizers must be added to the enzymes before drying so that loss of enzyme activity is minimized. To reduce the formation of dust in dry preparations, embedding (prills) or coating (marumerizer) e.g., with poly-(ethylene glycol) or ethoxylated fatty alcohols has proved useful. Another method of enzyme stabilization is immobilization on a solid carrier.

## 3.2.6. Waste Disposal

Because of the generally low concentration of enzyme in the starting material, the volume of material that must be processed is large, and substantial amounts of waste

**Table 8.** Reagents for cleaning-in-place (CIP) of chromatographic media

| Substance | Purification power (protein and fat) | Sterilization and removal of pyrogenes |
|---|---|---|
| Sodium hydroxide | + + | + |
| Acids | − | + |
| Sodium phosphate | + | − |
| Surfactants | + + | − |

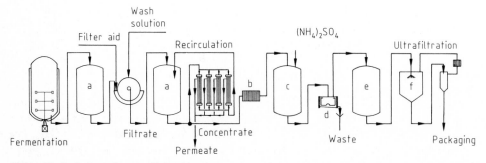

**Figure 23.** Isolation of extracellular enzymes
a) Hold tank; b) Filtration; c) Precipitation vessel; d) Centrifuge; e) Suspension vessel; f) Spray drier

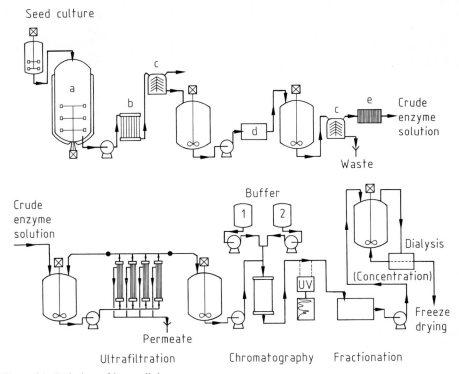

**Figure 24.** Isolation of intracellular enzymes
a) Fermentor; b) Heat exchanger; c) Centrifuge; d) Cell disruption; e) Filter

accumulate. The spent fermentation medium can still contain large amounts of unused nutrients. However, recycling is generally not possible because of the presence of metabolites in the medium. Solid organ remains and mycelium, which are used as animal feed, can be separated. The latter must be carefully checked for undesired metabolites, e.g., antibiotics, before being fed to animals.

In recombinant DNA techniques, the need to maintain absolute containment is of great concern. Waste must be chemically or thermally inactivated before disposal, to ensure that no live organisms escape into the environment.

## 3.3. Immobilization

Synthesis, decomposition, and partial conversion of different compounds in biological systems are catalyzed by enzymes, water-soluble globular proteins. The catalytic activity of enzymes is brought about by their tertiary and quaternary (oligomeric enzymes) structures. All enzymes have a catalytic center, one or several substrate binding sites, and one or several regulatory sites.

The characteristics of enzymes as catalysts are (1) enzymes are able to catalyze reactions at ambient temperature and pressure and in a pH range around neutral, and (2) enzymes have strict substrate specificity, stereospecificity, regiospecificity, and reaction specificity. These facts suggest that energy-saving, resources-saving, and low-pollution processes could be designed by using enzymes, because such processes would operate at relatively low temperature and atmospheric pressure with little byproduct formation. However, the molecular structure of enzymes that is essential for their catalytic activity is liable to be destroyed under conditions such as high temperature, high or low pH, presence of organic solvents, or even conditions suitable for catalysis. The recovery of active enzymes from spent reaction mixtures is another problem when free (nonimmobilized) enzymes are used. Immobilization is one way of eliminating some of the disadvantages inherent to enzymes.

Therefore, the immobilization of biocatalysts – not only enzymes but also cellular organelles, microbial cells, plant cells, and animal cells – is attracting worldwide attention in the practical application of bioprocesses. In general, immobilized biocatalysts are stable and easy to handle compared to their free counterparts. One of their most important features is that they can be used repeatedly in a long-term series of batchwise reactions or continuously in flow systems. At present, applications of immobilized biocatalysts include (1) the production of useful compounds by stereospecific or regiospecific bioconversion, (2) the production of energy by biological processes, (3) the selective treatment of specific pollutants to solve environmental problems, (4) continuous analyses of various compounds with a high sensitivity and a high specificity, and (5) medical uses such as new types of drugs for enzyme therapy or artificial organs.

These processes require the immobilization not only of single enzymes but also of several different enzymes, organelles, or cells that catalyze more complex reactions. Various methods have been developed for the immobilization of biocatalysts, and they are being used extensively today.

### 3.3.1. Definitions

Immobilized enzymes are defined as "enzymes physically confined or localized in a certain defined region of space with retention of their catalytic activities, which can be used repeatedly and continuously" [3.98]. This definition is applicable to enzymes, cellular organelles, microbial cells, plant cells, and animal cells, that is, to all types of biocatalysts. In some cases, these biocatalysts are bound to or within insoluble supporting materials (*carriers*) by chemical or physical binding. In other cases, biocatalysts are free, but confined to limited domains or spaces of supporting materials (*entrapment*).

The European Federation of Biotechnology, Working Party on Applied Biocatalysis (formerly, the Working Party on Immobilized Biocatalysts) has proposed that immobilized cells should be classified as *viable* and *nonviable,* where viable includes *growing* and *nongrowing* or *respiring*. According to the same definition, the terms *dead, resting*, and *living* should not be used [3.99]. However, in this section, the terms *treated, resting,* and *growing* – and in some cases, *living* – are used for classifying the condition of immobilized cells. Treated cells are those subjected to chemical or physical treatment before or after immobilization, and resting cells do not show growth during utilization.

## 3.3.2. History

In 1916, NELSON and GRIFFIN observed that yeast invertase (E.C. 3.2.1.26) [9001-57-4] adsorbed on charcoal was able to catalyze the hydrolysis of sucrose [3.100]. After that, several reports were published on the immobilization of physiologically active proteins by covalent binding on several supports. However, immobilized enzymes were not used in practice until 1953, when GRUBHOFER and SCHLEITH immobilized several enzymes, such as carboxypeptidase, diastase, pepsin, and ribonuclease, on diazotized polyaminostyrene resin by covalent binding [3.101]. Thereafter, MITZ reported the ionic binding of catalase (E.C. 1.11.1.6) [9001-05-2] on DEAE- cullulose in 1956 [3.102]. BERNFELD and WAN described the entrapment of trypsin (E.C. 3.4.21.4) [9002-07-7], papain (E.C. 3.4.22.2) [9001-73-4], amylase, and ribonuclease in polyacrylamide gel in 1963 [3.103], and QUIOCHO and RICHARDS demonstrated the cross-linking of carboxypeptidase A (E.C. 3.4.17.1) [11075-17-5] with glutaraldehyde in 1964 [3.104]. Microencapsulation of carbonic anhydrase (E.C. 4.2.1.1) [9001-03-0] was reported by CHANG in 1964 [3.105] and the preparation of liposomes containing amyloglucosidase (E.C. 3.2.1.3) [9032-08-0] by GREGORIADIS in 1971 [3.106]; both were used in enzyme therapy. During this pioneering period, KATCHALSKI-KATZIR and co-workers made extensive contributions to the theoretical understanding of immobilized enzymes [3.107].

In 1969, CHIBATA and co-workers of Tanabe Seiyaku Co., Japan, were successful for the first time in the industrial application of immobilized enzymes. Fungal aminoacylase (E.C. 3.5.1.14) [9012-37-7] was immobilized on DEAE-Sephadex through ionic binding and used for the stereoselective hydrolysis of *N*-acyl-D,L-amino acids to yield L-amino acids and *N*-acyl-D-amino acids [3.108]. The first industrial application of immobilized microbial cells was also performed by CHIBATA and co-workers in 1973 to produce L-aspartate from ammonium fumarate by polyacrylamide gel-entrapped *Escherichia coli* cells containing a high activity of aspartase (E.C. 4.3.1.1) [9027-30-9]. A fair number of processes using immobilized biocatalysts have been industrialized so far. They are discussed briefly in Section 3.3.5.

## 3.3.3. Methods

For the application of immobilized biocatalysts, their screening to the desired activity and characteristics is most important. In addition, selection of the appropriate combination of supporting material and immobilization method, both of which should be suitable for each biocatalyst, is necessary. No systematic concept is available at present for design of the most appropriate method of immobilization for various biocatalysts. Optimization is carried out in general by trial and error.

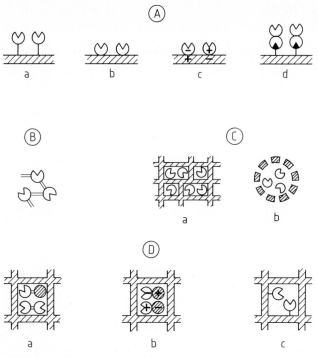

**Figure 25.** Immobilization of biocatalysts

A: Carrier binding methods

   a) Covalent binding; b) Physical adsorption; c) Electrostatic forces; d) Biospecific binding
   (coenzyme, antibody, effector)

B: Cross-linking with bi- or multifunctional reagents

C: Entrapment in gel matrices, microcapsules, liposomes, hollow fibers, or ultrafiltration
   membranes

   a) Lattice type; b) Microcapsule type

D: Combined methods

   a) Entrapment and cross-linking; b) Electrostatic binding and entrapment; c) Entrapment and
   covalent binding

At present, various immobilization techniques are available [3.109], and
Figure 25 illustrates the principles of these methods. Because each method has its
own merits and demerits, selection should be based on the intended purpose, including type of biocatalyst, type of reaction, and type of reactor.

Supporting materials should have adequate functional groups to immobilize the
biocatalysts, as well as sufficient mechanical strength; physical, chemical, and biological stability; and nontoxicity. Easy shaping is also important for applying immobilized biocatalysts to different types of reactors. Furthermore, the economic feasibility should be examined. Some supporting materials are described in the following
sections.

## 3.3.3.1. Carrier Binding

Enzyme proteins have amino acid residues containing chemically reactive groups, ionic groups, and/or hydrophobic groups, as well as hydrophobic domains. These amino acid residues and the hydrophobic domains can participate in the immobilization of enzymes through covalent linkage, ionic binding, or physical adsorption. Various types of insoluble supports (carriers) are utilized as is or after proper modification or activation.

**Covalent Binding.** Amino acid residues that are not involved in the active site or substrate-binding site of the enzymes to be immobilized can be used for covalent binding with supports. These are the ε-amino group of lysine, the mercapto group of cysteine, the β-carboxyl group of aspartic acid, the γ-carboxyl group of glutamic acid, the phenolic hydroxyl group of tyrosine, or the hydroxyl groups of serine and threonine. Hydroxyl, carboxyl, and amino groups are especially excellent targets because of their relative abundance in enzyme molecules.

Enzymes immobilized by covalent binding have the following advantages: (1) because of the tight binding, they do not leak or detach from supports during utilization; (2) immobilized enzymes can easily come into contact with substrates because the enzymes are localized on the surface of supports; and (3) an increase in heat stability is often observed because of the strong interaction between enzyme molecules and supports.

On the other hand, disadvantages of covalent binding are: (1) active structures of enzyme molecules are liable to be destroyed by partial modification; (2) strong interaction between enzyme molecules and supports often hinders the free movement of enzyme molecules, resulting in decreased enzyme activity; (3) optimal conditions of immobilization are difficult to find; (4) this method is not suitable for immobilization of cells; and (5) supports, in general, are not renewable. Hence, this principle is well-suited to expensive enzymes whose stability is significantly improved by covalent binding.

Despite these disadvantages, covalent binding is often applied to the preparation of immobilized enzymes for analytical purposes. Some typical examples of covalent binding methods are described below, but many other techniques have also been reported.

*Cyanogen Bromide.* The cyanogen bromide method was first demonstrated by AXÉN and his co-workers [3.110]. It involves the activation of supports having vicinal hydroxyl groups (**1**), such as polysaccharides, glass beads, or ceramics, with cyanogen bromide [*506-68-3*] (**2**), to yield reactive imidocarbonate derivatives **3**. The subsequent reaction between the activated supports and enzyme molecules gives N-substituted isourea (**4**), N-substituted imidocarbonate (**5**), and N-substituted carbamate derivatives (**6**). This method has been widely used for the immobilization of various enzymes, and CNBr-activated supports, such as CNBr-activated Sepharose, are available. Aminated glass beads and aminated polyacrylamide gel are also used as supports in this method. The insertion of spacers such as hexamethylenediamine

[*124-09-4*] is also possible, to avoid strong interaction between enzyme molecules and supports.

*Acid Azide Derivatives.* The acid azide method, which is used for peptide synthesis, is also applicable to the immobilization of various enzymes [3.111]. For example, carboxymethyl cellulose [*9004-42-6*] (**7**) is converted to the methyl ester **8** and then to the hydrazide **9** with hydrazine. The hydrazide reacts with nitrous acid to form the azide derivative **10**, which can then react with enzyme molecules at low temperature to yield immobilized enzymes **11**. Several supports containing carboxyl groups can be used for this method.

*Condensing Reagents.* Carboxyl groups or amino groups of supports and amino groups or carboxyl groups of enzyme molecules can be condensed directly through the formation of peptide linkages by the action of carbodiimide reagents or Woodward's reagent K (*N*-ethyl-5-phenylisoxazolium-3′-sulfonate) [*4156-16-5*]. Carboxymethyl cellulose is one of the supports used for this method [3.112].

*Diazo Coupling.* Supports having aromatic amino groups (**12**) are diazotized with nitrous acid to form the diazonium derivatives **13**, which react with enzyme molecules to yield immobilized enzymes **14**. 4-Aminobenzyl cellulose [*9032-51-3*] [3.113] and polyaminostyrene [*9060-90-6*] [3.114] are typical supports used for this method.

*Alkylation.* Halogenated groups on supports can easily react with amino groups, phenolic hydroxyl groups, and sulfhydryl groups of the enzyme molecules. For example, halogenated acetyl cellulose (15) [3.115] and triazinyl derivatives of ion-exchange resins (16) or cellulose [3.116] may be used for this method.

$$\text{[support]}-OH \xrightarrow[\text{BrCH}_2\text{COOH/dioxane}]{\text{BrCH}_2\text{COBr}}$$

$$\underset{\textbf{15}}{\text{[support]}-O-COCH_2Br} \xrightarrow{E}$$

$$\downarrow \text{NaI/ethanol}$$

$$\underset{\textbf{15}}{\text{[support]}-O-COCH_2I} \xrightarrow{E} \text{[support]}-O-COCH_2-E$$

$$\text{[support]}-NH_2 \longrightarrow \text{[support]}-NH-\underset{\textbf{16}}{\text{(triazinyl-Cl)}} \xrightarrow{H_2N-E} \text{[support]}-NH-\text{(triazinyl)}(NH-E)_2$$

*Carrier Cross-linking.* Supports and enzyme molecules can be cross-linked with bi- or multifunctional reagents. For example, aminoethyl cellulose [9032-36-4] (17) and amino groups of enzyme molecules are combined through Schiff's base linkage (19) with glutaraldehyde [111-30-8] (18) [3.117]. Diisocyanates represent another group of cross-linking reagents. Several supports containing amino groups are also employed.

$$\underset{\textbf{17}}{\text{[support]}-NH_2} + \underset{\textbf{18}}{OHC(CH_2)_3CHO} + H_2N-E \longrightarrow \underset{\textbf{19}}{\text{[support]}-N=CH(CH_2)_3CH=N-E}$$

**Ionic Binding.** Since catalase was found to be bound to ion-exchange cellulose [3.102], this method has been applied for the immobilization of many enzymes, because the procedure is very simple, the supports are renewable, and the enzymes are not modified. The most notable example is the production of L-amino acids by aminoacylase (E.C. 3.5.1.14) [9012-37-7] immobilized on DEAE-Sephadex [3.108]. Binding of enzymes on supports is affected by the kind of buffers used, pH, ionic strength, and temperature. Several derivatives of cellulose and Sephadex, as well as various ion-exchange resins, can be utilized for immobilization.

**Physical Adsorption.** Biocatalysts often bind to carriers by physical interaction such as hydrogen bonding, hydrophobic interaction, van der Waal's forces, or their combined action. Although biocatalysts are immobilized without any modification, interaction between biocatalyst and support is generally weak and affected by such environmental conditions as temperature or concentration of reactants. Various inorganic supports are often used. Currently, several synthetic resin beads and natural materials (e.g., chitosan beads with micropores of controlled size) having strong adsorption capacities are available. Adsorption followed by cross-linking with glutaraldehyde sometimes stabilizes the activity of immobilized enzymes. Phenoxyacetylated cellulose and glass beads are used as specific supports of a hydrophobic nature. Tannins, which interact strongly with proteins, are also applied as ligands after appropriate immobilization. Cellular organelles and various types of cells can be immobilized by physical adsorption. Supports are renewable under appropriate conditions.

### 3.3.3.2. Cross-linking

Bi- and multifunctional compounds serve as reagents for intermolecular cross-linking of enzymes. Cross-linked enzymes are then insoluble macromolecules. In addition to glutaraldehyde [111-30-8] (20) [3.104] which is the most popular cross-linking reagent, several compounds such as toluene diisocyanate [1321-38-6] or hexamethylene diisocyanate [822-06-0] (21) are used. The yield of enzyme activity is usually low. Microbial cells are also cross-linked with glutaraldehyde to yield cell pellets.

$$OHC(CH_2)_3CHO \;+\; \boxed{E}-(NH_2)_n \;\longrightarrow$$

20

$$-CH{=}N-\boxed{E}-N{=}CH(CH_2)_3CH{=}N-\boxed{E}-$$

$$OCN-R-NCO \;+\; \boxed{E}-(NH_2)_n \;\longrightarrow$$

21

$$\underset{O}{-HNCNH}-\boxed{E}-\underset{O}{NHCNH}-R-\underset{O}{NHCNH}-\boxed{E}-$$

### 3.3.3.3. Entrapment

Entrapped biocatalysts are classified according to the following different types:

*Lattice type:* biocatalysts entrapped in gel matrices prepared from polysaccharides, proteins, or synthetic polymers

*Microcapsule type:* biocatalysts entrapped in microcapsules of semipermeable synthetic polymers

*Liposome type:* biocatalysts entrapped within liquid membranes prepared from phospholipids

*Hollow-fiber type:* biocatalysts separated from the environment by hollow fibers

*Membrane type:* biocatalysts separated from the spent reaction solution by ultrafiltration membranes.

The advantages of entrapping methods are that not only single enzymes but also several different enzymes, cellular organelles, and cells can be immobilized with essentially the same procedures. Biocatalysts are not subjected to serious modification, and immobilization eliminates the effect of proteases and enzyme inhibitors of high molecular mass. However, disadvantages are (1) substrates of high molecular mass can hardly gain access to the entrapped biocatalysts and (2) supports are not renewable. Entrapment within ultrafiltration membranes can avoid the disadvantages inherent in entrapping methods, although inactivated enzyme molecules often precipitate on the membrane surface, which results in decreased permeability to reaction solutions. The lattice-type method is most widely applied for preparing immobilized biocatalysts. Several examples of this technique are mentioned in the following paragraphs.

**Polyacrylamide Gel.** Since BERNFELD and WAN [3.103] reported the entrapment of several enzymes in polyacrylamide gels (**24**), different types of biocatalysts including cellular organelles and microbial cells have been immobilized by this method. This method was also applied to the industrial production of L-aspartate, L-malate, and acrylamide by immobilized microbial cells. In a typical procedure, acrylamide [*79-06-1*] (**22**) and *N,N'*-methylenebisacrylamide [*110-26-9*] (**23**) (cross-linking reagent) are mixed with biocatalysts and polymerized in the presence of an initiator (potassium persulfate [*7727-21-1*]) and a stimulator (3-dimethylaminopropionitrile [*1738-25-6*] (DMAPN) or *N,N,N'N'*-tetramethylenediamine [*110-18-9*]). Although this technique is used for various purposes, acrylamide monomer sometimes inactivates enzymes. Several analogues or derivatives of acrylamide can also be used in this method.

**Alginate Gel.** Several natural polysaccharides, such as alginate, agar, and κ-carrageenan, are excellent gel materials and used widely for the entrapment of various biocatalysts.

Sodium alginate [*9005-38-3*], which is soluble in water, is mixed with a solution or suspension of the biocatalysts and then dropped into a calcium chloride solution to form water-insoluble calcium alginate gels that immobilize enzymes, cellular organelles, or microbial cells [3.118]. However, gels are gradually solubilized in the presence of calcium ion-trapping reagents such as phosphate. Aluminum ions or several divalent metal ions can be substituted for calcium ions.

This method is used widely for immobilization of various biocatalysts because of its simplicity and the availability of sodium alginate.

**κ-Carrageenan Gel.** CHIBATA and co-workers [3.119] have extensively screened gel materials for the immobilization of enzymes and microbial cells in industrial applications and have found κ-carrageenan (**25**) to be the best.

25        ($n = 250-2000$)

κ-Carrageenan in saline is mixed with a solution or suspension of biocatalysts and dropped into a solution of gelling reagent, such as potassium chloride [*7447-40-7*] [3.120]. Various cations, such as ammonium, calcium, and aluminum, also serve as good gelling reagents. Hardening of gels with glutaraldehyde and hexamethylenediamine often stabilizes the biocatalysts entrapped in κ-carrageenan gels.

This method has replaced polyacrylamide gels in the industrial production of L-aspartate [3.121] and L-malate [3.122] by means of immobilized microbial cells.

κ-Carrageenan gels can be solubilized in saline or in water, which enables the number of cells in the gel to be counted.

**Synthetic Resin Prepolymers.** With the application of a variety of bioreactions, including synthesis, transformation, degradation, or assay of various compounds having different chemical properties, entrapment of biocatalysts in gels of controlled physicochemical properties has become desirable. Selection of suitable gels and modification of natural polymers to meet each purpose are usually difficult. For these reasons, a new entrapping method using synthetic resin prepolymers has been developed by FUKUI and co-workers [3.123].

Specific features and advantages of the prepolymer method are (1) entrapment procedures are very simple and proceed under very mild conditions; (2) prepolymers do not contain monomers that may have unfavorable effects on enzyme molecules; (3) the network structure of gels can easily be controlled by using prepolymers of optional chain length; and (4) the physicochemical properties of gels, such as the hydrophilicity–hydrophobicity balance and the ionic nature, can be changed by selecting suitable prepolymers synthesized in advance in the absence of biocatalysts.

Photo-cross-linkable resin prepolymers having hydrophilic (**26**) or hydrophobic (**27**) properties, cationic or anionic nature, and different chain lengths have been developed [3.123], [3.124].

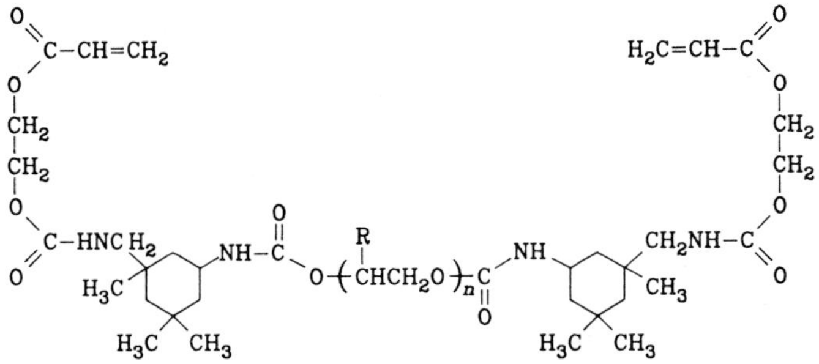

**26**: R = H

**27**: R = CH$_3$

Mixtures of prepolymers and biocatalysts are gelled by irradiation with long-wavelength UV light for several minutes in the presence of a proper sensitizer such as benzoin ethyl ether [574-09-4]. This method has been applied for the pilot-scale production of ethanol by immobilized growing yeast cells.

Entrapment by urethane prepolymers (**28**) with different hydrophilicity or hydrophobicity and chain length is much simpler [3.123], [3.125]. When the liquid prepolymers are mixed with an aqueous solution or suspension of the biocatalysts, the prepolymers react with each other to form urea bonds and liberate carbon dioxide.

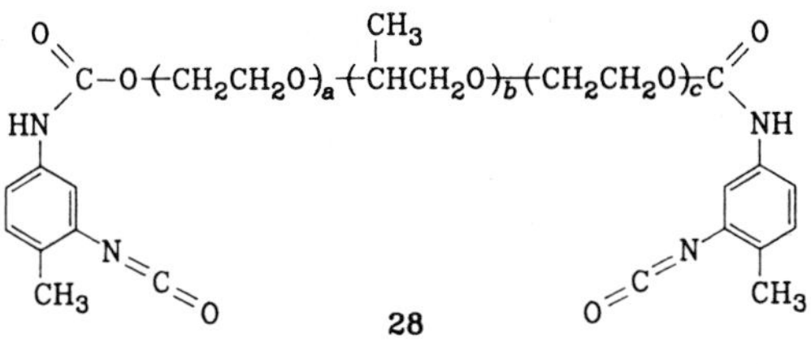

**28**

These prepolymers can immobilize not only enzymes but cellular organelles and microbial cells as well. The hydrophilicity–hydrophobicity balance of these gels has been demonstrated to affect especially the bioconversion of lipophilic compounds in organic solvent systems [3.126].

## 3.3.4. Characterization

The characteristics of immobilized biocatalysts should be described in the literature because different methods and preparations must be compared for their evaluation. However, determination of what kinds of description and what types of parameters should be given is difficult. Methods for estimating these parameters

vary from group to group. Therefore, at present, the results described in the literature cannot be compared. The Working Party on Immobilized Biocatalysts within
the European Federation of Biotechnology has made a proposal for the description
of several properties of immobilized biocatalysts [3.127]. Although the proposal will
be difficult to follow completely, it will be useful for the characterization of immobilized biocatalysts.

The minimum requirements recommended for characterization of an immobilized biocatalyst are as follows:

1) *General description*
    reaction scheme
    enzyme and microorganism
    carrier type
    method of immobilization

2) *Preparation of the immobilized biocatalyst*
    method of immobilization, reaction conditions,
    dry mass yield, activity left in supernatant

3) *Physicochemical characterization*
    biocatalyst shape, mean wet particle size, swelling behavior
    compression behavior in column systems, abrasion in stirred vessels, or minimum fluid-
        ization velocity and abrasion in fluidized beds

4) *Immobilized biocatalyst kinetics*
    initial rates vs. substrate concentration for free and immobilized biocatalyst, effect of pH and
        buffer
    diffusional limitations in the immobilized biocatalyst system (effect of particle size or enzyme
        load on activity)
    degree of conversion vs. residence time (points on a curve)
    storage stability (residual initial rate after storage for different periods)
    operational stability (residual initial rate or transforming capacity of reaction system) after
        operation for different periods

## 3.3.5. Application

Immobilized biocatalysts — enzymes, cellular organelles, microbial cells, plant
cells, and animal cells — have been applied to the production or conversion of
various compounds such as amino acids, peptides and enzymes, sugars, organic
acids, antibiotics, steroids, nucleosides and nucleotides, lipids, terpenoids, fuels, or
commodity chemicals. Some typical examples are described below.

A representative industrial application of immobilized enzymes is the production
of various L-amino acids, such as L-alanine, L-isoleucine, L-methionine, L-phenylalanine, L-tryptophan, and L-valine, by fungal aminoacylase (E.C. 3.5.1.14)
[*9012-37-7*] immobilized on DEAE-Sephadex by ionic binding [3.108]. This enzyme
hydrolyzes stereoselectively the L-isomer of *N*-acyl-D,L-amino acids (**29**) to yield
L-amino acids (**30**) and unreacted *N*-acyl-D-amino acids (**31**); the latter are subsequently converted into the racemic form by heating. In this way, both the L- and
D-isomers are converted completely to L-amino acids.

DL $-$R$-$CH$-$COOH    $+$ H$_2$O $\longrightarrow$     L$-$R$-$CH$-$COOH  $+$  D$-$R$-$CH$-$COOH  $+$  R'COOH
  |                                                      |                           |
  NHCOR'                                                NH$_2$                      NHCOR'
**29**                                                      **30**                           **31**

CHIBATA and co-workers also succeeded in producing L-aspartic acid [*56-84-8*] (**33**) from ammonium fumarate [*14548-85-7*] (**32**) by polyacrylamide gel-entrapped *Escherichia coli* cells containing a high activity of aspartase (E.C. 4.3.1.1) [*9027-30-9*].

HOOC$-$CH$=$CH$-$COOH  $+$  NH$_3$ $\longrightarrow$ HOOC$-$CH$_2$$-$CH$-$COOH
                                                                    |
                                                                   NH$_2$
            **32**                                           **33**

An active and stable preparation was obtained (half-life, 120 d) when the immobilized cells were incubated with a substrate solution for 48 h at 30 °C [3.128]. This process is the first example of an industrial application of immobilized cells and the enzyme partially purified from *E. coli* cells is not as stable even after immobilization. At present, the process can be improved by entrapping the cells in κ-carrageenan gels, followed by hardening with glutaraldehyde and hexamethylenediamine; these immobilized cells have a half-life of 680 d [3.121].

L-Alanine [*56-41-7*] (**34**) is produced from L-aspartic acid [*56-84-8*] (**33**) by κ-carrageenan gel-entrapped cells of *Pseudomonas dacunhae* having a high activity of L-aspartate 4-decarboxylase (E.C. 4.1.1.12) [*9024-57-1*] [3.129]. This process was commercialized in 1982 by the Tanabe Seiyaku Co. in Japan.

HOOC$-$CH$_2$$-$CH$-$COOH $\longrightarrow$ CH$_3$$-$CH$-$COOH $+$ CO$_2$
              |                                    |
             NH$_2$                               NH$_2$
     **33**                                   **34**

Immobilized penicillin acylase (amidase) (E.C. 3.5.1.11) [*9014-06-6*] is now widely used in the production of 6-aminopenicillanic acid [*551-16-6*] (**36**) from penicillin G [*61-33-6*] (**35**). Glucose isomerase (E.C. 5.3.1.18) [*9055-00-9*] is used in the production of high-fructose syrup (**38**) from glucose [*50-99-7*] (**37**).

**35**

**36**

|  | CHO |  | CH$_2$OH |
|--|-----|--|----------|
|  | H$-$C$-$OH | | C$=$O |
|  | HO$-$C$-$H | $\longrightarrow$ | HO$-$C$-$H |
|  | H$-$C$-$OH | | H$-$C$-$OH |
|  | H$-$C$-$OH | | H$-$C$-$OH |
|  | CH$_2$OH | | CH$_2$OH |

            **37**                          **38**

A process for production of L-malate (**40**) from fumarate (**39**) has been developed by CHIBATA and co-workers by using *Brevibacterium ammoniagenes* cells containing fumarase (E.C. 4.2.1.2) [*9032-88-6*] entrapped in polyacrylamide gels [3.130].

$$\text{HOOC-CH=CH-COOH} + \text{H}_2\text{O} \longrightarrow \text{HOOC-CH}_2\text{-CH-COOH}$$
$$\qquad \textbf{39} \qquad\qquad\qquad\qquad \textbf{40} \quad \overset{|}{\text{OH}}$$

The entrapped cells were treated with bile extract to suppress the formation of byproduct succinate, which leads to the industrial application of this system in 1974. Later *B. flavum* was substituted for *B. ammoniagenes* and κ-carrageenan gels for polyacrylamide gels. κ-Carrageenan gel-entrapped cells showed a half-life of 160 d [3.122].

The latest use for immobilized biocatalysts is the production of cheap commodity chemicals such as acrylamide (**42**) from acrylonitrile (**41**).

$$\text{CH}_2\text{=CHCN} + \text{H}_2\text{O} \longrightarrow \text{CH}_2\text{=CHC} \overset{\nearrow \text{O}}{\underset{\searrow \text{NH}_2}{}}$$
$$\qquad \textbf{41} \qquad\qquad\qquad\qquad \textbf{42}$$

Polyacrylamide gel-entrapped cells of *Corynebacterium* species are used for the production of acrylamide at low temperature by Nitto Chemical Industry in Japan. YAMADA and his co-workers have found that in *Pseudomonas chlororaphis*, nitrile hydratase [*82391-37-5*] participates in this reaction [3.131].

In addition to the immobilized enzymes and treated cells mentioned above, other immobilized living or growing cells are now being investigated extensively for the production of more complex compounds by using the metabolic activities of these cells. Plant and animal cells as well as microbial cells are targets of immobilization because these cells have self-regenerating and self-proliferating catalytic systems [3.132]. Genetically improved cells are also useful biocatalysts.

Bioconversion of lipophilic compounds in organic solvent systems is another subject of interest industrially. Immobilization in or on proper supports stabilizes biocatalysts even in the presence of organic solvents [3.126].

Thus, the applications of immobilized biocatalysts are now expanding along with the development of immobilization techniques and the improvement of biocatalysts.

# 4. Industrial Uses of Enzymes

## 4.1. Survey of Industrial Enzymes [4.1]–[4.3]

Chapter 4 describes properties and actions of the most import industrial enzymes. About 12 categories of enzymes are used in industry. Most of them are hydrolytic enzymes used for the depolymerization of natural substrates with high molecular mass.

Most industrial enzymes are produced by microorganisms, the largest group being proteolytic enzymes from bacteria (59%) followed by carbohydrases (20%).

Animal enzymes include rennets (for cheese making) and pancreatic proteases and lipases; the proteases are used primarily in leather processing. Among the plant enzymes, papain is the largest single industrial product.

Industrial enzymes are usually mixtures of different enzymes. Commercial products are standardized with diluents.

The next sections treat the following technical enzyme preparations: amylases, including amyloglucosidase; cellulases; hemicellulases; pectinases; proteases; lipases; and oxidoreductases.

**Units.** Many different units are used to measure the activity of hydrolytic enzymes. These units vary from one manufacturer to another. The following list contains a few of these [4.32, pp. 552–557]:

MWU: modified Wohlgemuth unit, measures the amount of enzyme that converts 1 mg of soluble starch to a definite size dextrin in 30 min under the conditions of the assay

SKB: unit according to the method of SANDSTEDT, KNEEN, and BLISH, measures the amount of enzyme required to dextrinize 1 g of $\beta$-limit dextrin to a definite size in 1 h under the conditions of the assay [4.5]

GAU: glucoamylase unit, measures the amount of enzyme needed to liberate 1 g of reducing sugar per hour as glucose under the conditions of the assay

NU: Northrop unit, measures the amount of enzyme required to give 40% hydrolysis of a defined casein substrate under the conditions of the assay

HU: hemoglobin unit, measures the amount of enzyme needed to liberate 67.08 mg of nonprotein nitrogen under the conditions of the assay [4.5a]

DU: diazyme unit (dextrinogene unit), measures the amount of enzyme used to convert one pound of starch to dextrose in 72–96 h at pH 4 and 60 °C

AU: Anson unit is used to measure the activity of detergent enzyme preparations. Crystalline subtilisin has 25–30 Anson units per gram [4.11]; the unit is U/g (measured on hemoglobin)

## 4.1.1. Amylases

Amylases are enzymes that catalyze the hydrolysis of starch. They are extensively distributed in nature and widely employed in industry, particularly the food industry in the liquefaction of starch and starch-containing raw materials. The action of amylases can be represented schematically as follows [4.4]:

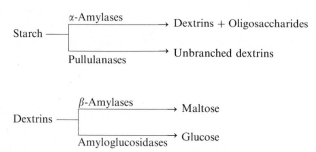

Amylases are classified as endoamylases or exoamylases (see also Section 2.2).

*Endoamylases.* α-Amylases (E.C. 3.2.1.1) [*9000-90-2*], α-1,4-glucano hydrolases, belong to the class of endoamylases. They are also known as dextrinogenic or liquefying amylases. The α-1,6-glycosidic bonds of amylopectin are cleaved by α-1,6-glucano hydrolases, such as isoamylase (E.C. 3.2.1.68) [*9067-73-6*] or pullulanase (E.C. 3.2.1.41) [*9075-68-7*].

*Exoamylases.* The exoamylases are also known as saccharifying or saccharogenic amylases. β-Amylases (E.C. 3.2.1.2) [*9000-91-3*] attack gelatinized starches and dextrins, to release maltose. Glucamylases (E.C. 3.2.1.3) [*9032-08-0*], also known as amyloglucosidases, remove glucose from the nonreducing ends of starch chains or dextrins.

**α-Amylases.** α-Amylases, together with β-amylases, are present in malt. However, the industrial enzymes are produced from pancreas or from cultures of microorganisms. The various types of α-amylases differ from each other not only in specificity, but also in optimal pH range and stability to heat.

The use of gelatinized starch as substrate has an accelerating effect on the enzyme. The course of hydrolysis can be followed by the rapid fall in viscosity of the substrate solution and by the gradual disappearance of the color produced by iodine. The reaction rate depends on the pH, temperature, and degree of polymerization of the substrate (Table 9).

*α-Amylases from Bacteria.* α-Amylases are available in either liquid or powder form. These preparations contain varying amounts of other enzymes such as proteinases and β-glucanases. Enzyme concentrates, standardized with sodium chloride or starch, are marketed as powders with activities of 20 000 to 30 000 SKB/g (pH 5.0) (for the definition of SKB, see p. 77). Liquid preparations containing 3000 to 12 000 SKB/g are stabilized with sodium chloride or glycerol. α-Amylase from *Bacillus amyloliquefaciens* has an especially high activity of 500 000 SKB per gram of protein.

**Table 9.** Endoamylases

| Enzyme | Source | pH range (pH optimum) | Optimal temperature, °C (inactivation temperature, °C) |
|---|---|---|---|
| Pancreatic amylase | porcine pancreas | 5.5–8.5 (6.0–7.0) | 40–45 (75) |
| Bacterial α-amylase | *Bacillus subtilis* <br> *Bacillus amyloliquefaciens* | 4.5–9.0 (6.5–7.5) | 70–85 (95) |
| Thermostable bacterial α-amylase | *Bacillus licheniformis* | 5.8–8.0 (7.0) | 90–105 (120) |
| Fungal α-amylase | *Aspergillus oryzae* <br> *Aspergillus niger* | 4.0–7.0 (5.0–6.0) | 55–60 (80) |
| Malt α-amylase | barley malt <br> wheat malt | 3.5–7.5 (4.5–7.0) | 60–70 (85) |

Bacterial α-amylases are divided into standard amylases and heat-stable amylases, depending on their stability to heat. *Standard* bacterial α-amylase from *Bacillus subtilis* or *B. amyloliquefaciens* requires temperatures of 70–85 °C for optimal activity. *Heat -stable* bacterial α-amylase from, e.g., *B. licheniformis,* has a temperature optimum between 90 and 105 °C. However, this optimum depends on the pH, the concentration of calcium ions, and the substrate concentration. A typical reaction is carried out on a 30% starch suspension at pH 7.0 and 103–105 °C for 7–11 min in the presence of 70 mg/L of calcium. The enzyme is inactivated by lowering the pH to 3.5–4.5 and the temperature to 80–90 °C for a period of 5–30 min.

Bacterial α-amylases are generally stabilized with calcium ions; some types are also activated by sodium chloride. Indeed, amylases freed from calcium by electrodialysis or complex formation are inactive. Heavy metals such as copper, iron, or mercury inhibit enzyme activity. The breakdown of starch yields dextrins. For instance, hydrolysis of potato starch by standard bacterial α-amylases gives the following dextrin spectrum: G1 (glucose) 5%, G2 (maltose) 7%, G3 (maltotriose) 13%, G4 (maltotetraose) 4%, G5 (maltopentaose) 16%, G6 (maltohexaose) 22%, higher dextrins 33%.

In industry, the *DE value* (dextrose equivalent) is used to characterize the degradation of starches (see also Section 2.2). Values of approximately 20 DE have been obtained by using bacterial amylases, and values of approximately 40 have been obtained with fungal α-amylases.

*α-Amylases from Fungi.* Fungal amylases are obtained from *Aspergillus niger* or *A. oryzae*. They are less resistant to heat than bacterial α-amylases and are inactivated before starch gelatinizes. This is important in the making of bread and biscuits.

The following dextrin spectrum is typical of the action of fungal α-amylases: Gl 8%, G3 32%, G4 18%, G5 9%, G6 4%, higher dextrins 17%.

Fungal α-amylase preparations used in the food industry are standardized with edible substances such as sugar or starch and have activities between 500 and 100 000 SKB/g. Proteinases are also present in these preparations.

*Malt Amylases.* Malt is produced by the malting of barley, wheat, or rye; it also contains β-amylase and varying amounts of proteinases and β-glucanases. Its α-amylase activity is 30–90 SKB/g. As a result of the β-amylase present, dextrin samples obtained on digestion of starch contain maltose in almost quantitative yield.

*Pancreatic Amylases.* Pancreatic amylases have mercapto groups (SH) which are rapidly inactivated by heavy-metal ions. In the presence of chloride ions, optimal activity is observed at pH 7.0. Apart from malt, pancreatic amylase preparations were formerly used on a large scale in desizing.

*Isoamylase (Pullulanase).* The α-1,6-glucosidic bonds of starch are cleaved by isoamylases. The enzyme of yeast and mold is known as isoamylase (E.C. 3.2.1.68) [9067-73-6], and the enzyme obtained from, e.g., *Aerobacter aerogenes* or *Klebsiella aerogenes* is called pullulanase (E.C. 3.2.1.41) [9075-68-7]. When amylopectin is treated with pullulanase, linear amylose fragments are obtained. Pullulanase-containing enzyme preparations are used together with amyloglucosidases primarily in the complete saccharification of starches and dextrins to glucose. Optimal activity is attained at 60 °C and ca. pH 6 or, in combination with amyloglucosidases, at 45 °C and pH 4–5.

**β-Amylases.** β-Amylase (E.C. 3.2.1.2) [9000-91-3], present in grain, soybeans, and sweet potatoes, attacks amylose chains, effecting successive removal of maltose units from the nonreducing ends. In the case of amylopectin, this process stops two to three glucose residues from the α-1,6-branching points (the designation β-amylase refers to the cleaved maltose which has an anomeric carbon atom in the β-configuration). The optimal temperature is 55 °C at pH 5.1–5.5. β-Amylase is used in alcohol production and in breweries as a replacement for malt. Obtained in quantity from barley, it is used to saccharify starch in the production of maltose syrup.

**Glucamylases.** Glucamylase (E.C. 3.2.1.3) [9032-08-0], also called amyloglucosidase, belongs to the group of exo-α-1,4-D-glucan glucohydrolases. It attacks starch chains, releasing glucose from the nonreducing ends. The optimal pH and temperature of glucamylases from different microorganisms vary (Table 10). These enzymes can hydrolyze starch to give glucose in theoretically 100% yield. The α-1,6-branching points are cleaved ca. 30 times slower than the α-1,4-bonds. The reaction rate decreases with decreasing chain length of the dextrin substrate, and maltose is attacked most slowly. Amyloglucosidases from various microorganisms differ greatly from each other with respect to the relationship between maltose and amylose cleavage.

Industrial preparations of glucamylases contain, in addition to α-amylases, varying amounts of trans-glucosidases which, at high substrate concentrations, transfer a glucose moiety to another molecule, e.g., maltose. As a result of transglucosidase

**Table 10.** Glucamylases

| Source | pH range (optimum) | Optimal temperature (inactivation), °C |
|---|---|---|
| *Aspergillus niger* (different types) | 3.0–6.0 (4.5–5.0) | 55–60 (90) |
| *Aspergillus awamori* | 2.0–7.0 (3.0–5.0) | 55–65 (85) |
| *Rhizopus niveus* | 3.0–6.0 (3.0–5.5) | 55–60 (70–80) |

impurities, the maximum saccharification achieved is reduced from 94% glucose to less than 90% glucose.

Most commercial glucamylase preparations are liquids stabilized with sodium chloride or glycerol. They have activities of 200–600 U/g (for the definition of U, see p. 20). The specific activity is approximately 4000 U per gram of protein.

*Immobilized Glucamylase.* Glucamylase can be attached to various supporting materials with high yield either by absorption or by covalent bonding. Dextrins can then be hydrolyzed continuously at 55 °C. The cleavage of oligosaccharides at high substrate and enzyme concentrations (matrix-bound) is reversible; thus, in comparison with soluble enzyme preparations, the back-reaction in matrix-bound enzymes reduces saccharification by about 3%.

## 4.1.2. Cellulases [4.6]

Cellulases are enzyme complexes, the components of which participate in the stepwise breakdown of native cellulose or cellulose derivatives to glucose. According to Wood and McCrae (1979) the hydrolysis of cellulose requires the synergistic reaction of three different hydrolytic enzymes, and the first attack is simultaneously effected by at least two of them. The three enzymes are:

exocellulase or exobiohydrolase  (E.C. 3.2.1.91)
endocellulase or endoglucanase  (E.C. 3.2.1.4)
$\beta$-glucosidase or cellobiase      (E.C. 3.2.1.21)

The first step in the degradation of cellulose is the hydrolysis of some amorphous regions on the cellulose microfibers by endoglucanase. New free ends of the cellulose chain are produced that are then the substrate for exobiohydrolase (often called $C_1$ *activity*). This enzyme removes cellobiose from the nonreducing end of the polyglucan chain. In this way, the substrate for the action of endoglucanase (often called $C_x$ *activity*) is formed. The exobiohydrolase has the highest affinity to cellulose. The purified enzyme is able to hydrolyze 80% of the microcristalline cellulose.

During this sequence of reactions released cellobiose is hydrolyzed by the third enzyme, cellobiase. This step is extremely important, because the accumulation of cellobiose inhibits cellulose degradation by the other enzymes. On the other hand, $C_x$ cellulase degrades soluble fragments or derivatives of cellulose (carboxymethyl cellulose or hydroxyethyl cellulose), thus forming cellobiose.

The action of cellulases on insoluble cellulose depends to a large extent on the pretreatment of the substrate, such as heating, alkaline oxidation or mechanical degradation. In addition, xylans, $\beta$-glucans or lignin bound to cellulose must be degraded, before cellulose-containing raw material can be hydrolyzed.

*Exobiohydrolase,* also called 1,4-$\beta$-glucan cellobiohydrolase (E.C. 3.2.1.91) [*37329-65-0*], removes cellobiose from the nonreducing end of the glucan chain and from cellodextrins. It is formed by cultures of *Trichoderma* or *Aspergillus*. Exobiohydrolase from different preparations of *Trichoderma reesei* has a molecular mass between 42 000 and 60 000.

*Endoglucanases,* also called 1,4-$\beta$-D-glucan-4-glucanohydrolase (E.C. 3.2.1.4) [*9012-54-8*], are formed by numerous types of fungi, including *Aspergillus, Trichoderma, Penicillium,* and bacteria. Molecular masses are between 12 000 and 50 000. The $C_x$ activity is determined by measuring either the viscosity drop of solutions of carboxymethyl cellulose or the increase in the concentration of reducing sugar.

*Cellobiase,* also called $\beta$-glucosidase or $\beta$-D-glucoside glucohydrolase (E.C. 3.2.1.21) [*9001-22-3*], hydrolyzes cellobiose and cellodextrins up to cellohexose to glucose; cellulose and higher cellodextrins are not attacked. The enzyme is formed by some types of *Trichoderma,* but more from *Aspergillus* species especially from *Aspergillus niger*. Molecular masses are between 35 000 (for enzyme from *Trichoderma*) and 218 000 (for enzyme from *Aspergillus oryzae*).

*Cellulases from Humicola.* Recently, cellulases from different *Humicola* species have been found. One of these enzymes with a good stability in the alkaline range has a molecular mass of 70 000 and a pH optimum of 5.0 at 45 °C.

*Cellulases from Bacteria.* With the potential use of cellulases in detergents, cellulases from bacteria with higher stability and activity in the alkaline range have become more important. These endoglucanases are active between pH 5 and 10.5 with carboxymethyl cellulose as substrate. They do not degrade cellulose or phosphoric acid-swollen cellulose. Their molecular mass is around 35 000.

The activity of cellobiase in industrial preparations can be determined by measuring the degradation of cellohexose. Cellulases are used commercially in the degradation of plant materials in detergents, silage, feed additives and as additives for digestive preparations.

Cellulase preparations also contain limited amounts of pectinases, xylanases, and other hemicellulases. Cellulases of *Trichoderma reseei* are active between pH 3 and 7; the optimal temperature is approximately 40 °C. Enzyme preparations obtained from *Aspergillus* are active between pH 4 and 6, and those from *Penicillium* between pH 5 and 8. In fact, the stability in aqueous solution is optimal between pH 3.5 and 6.0. The cellulase complex obtained from bacteria, e.g., from *Erwinia carotovorum,* exhibits optimal activity at pH 7–8.5. Cellulase preparations in powder form lose approximately 15% of their activity per year on dry storage at 20 °C.

## 4.1.3. Hemicellulases

Hemicelluloses are polysaccharides which bear no structural or chemical relationship to cellulose or starch. They are made up predominantly of heteropolymeric xylans, arabans, arabinoxylans, mannans, and galactomannans, but also contain homopolymeric $\beta$-glucans. Enzyme mixtures capable of degrading polymers of this type are called hemicellulases. Inulinase and dextranase are also occasionally included in this category. Hemicellulases have been found in mollusks, plants, molds (*Aspergillus, Trichoderma*), yeast, and bacteria.

**Dextranase.** Dextranase (E.C. 3.2.1.11) [*9025-70-1*], found in *Penicillium* species, effects the cleavage of the $\alpha$-1,6-D-glucosidic bonds of dextran. It is used in the sugar industry to reduce the viscosity of dextrans, which accumulate as a result of infection with dextran-forming bacteria. The optimal conditions for activity of this enzyme are pH 5–5.5 and 45–55 °C.

**Inulinase.** Inulinase (E.C. 3.2.1.7) [*9025-67-6*], found in *Aspergillus* and yeast, is active at pH 3.0–6.0 and at 35–50 °C. This enzyme complex can break down inulin to fructose in a yield of 99% in 24 h. Inulinase can be employed in the production of fructose in countries where large amounts of inulin-containing plants, such as Jerusalem artichoke, are grown.

**$\beta$-Glucanases from Fungi.** $\beta$-Glucanases used in industry are made from *Aspergillus niger* and *Penicillium emersonii*. These preparations, primarily employed in breweries, also contain small amounts of cellulases, laminarinase, or $\alpha$-amylases. A typical *Aspergillus* enzyme is active between pH 4 and 6 and has a temperature optimum of 60 °C at pH 5.0. The enzyme from *P. emersonii* is active over a wider pH range and can even be used at temperatures above 70 °C. The activity of these enzymes is determined, with $\beta$-glucan from barley as substrate, by measuring either the increase in the concentration of reducing sugar or the decrease in viscosity.

**$\beta$-Glucanases from Bacteria.** Industrial preparations of $\beta$-glucanases are obtained from cultures of *Bacillus subtilis* and contain small amounts of $\alpha$-amylases and proteinases. Like $\beta$-glucanases from fungi, the bacterial enzymes are also used in breweries. The optimal activity is between pH 6 and 7.5; rapid inactivation takes place above 60 °C. However, bacterial $\beta$-glucanases which are more stable to heat (optimal activity up to 80 °C) have also been described.

**Xylanases and Pentosanases.** Commercial preparations of xylanases and pentosanases are obtained from *Aspergillus* species and, above all, from *Trichoderma*. They are mixtures of different $\beta$-glycosidases which hydrolyze hemicelluloses and pentosans at varying rates. Optimal conditions for activity are pH 3.5–6.0 and 50–55 °C. These preparations are used commercially, in combination with pectinases, in fruit and vegetable processing. Pentosanase preparations are employed in the processing of rye.

## 4.1.4. Pectinases

Pectins are polymers of galacturonic acid and its methyl ester and occur, together with heteropolymeric carbohydrates, in plants. In fruit and vegetables that are used to produce juices, 60–80% of the pectin's carboxyl groups are esterified with methanol. Carbohydrates, such as arabans, arabinoxylans, and galactans are associated with pectins. Water-soluble pectins, having a high capacity to bind water, fill the intercellular spaces of fruit. Insoluble pectins, so-called protopectins, are localized in the matrix and are attached to heteropolymeric carbohydrates associated with the cellulose of the cell wall. For a survey on pectinases, see [4.7].

Enzyme preparations with pectinase activity have been used since 1930 in the production of fruit juices. Three groups of enzymes are involved in the cleavage of pectins: (1) pectin methylesterases; (2) depolymerizing enzymes, which attack the interior bonds of pectin chains (endoenzymes); and (3) exoenzymes, which liberate galacturonic acid units starting from the ends of the pectin molecule.

**Pectin Methylesterases.** Pectin methylesterases (E.C. 3.1.1.11) [9025-98-3] attack pectins to yield pectic acid and methanol. Pectin methylesterase, also called pectin-esterase (PE), is a very specific enzyme, effecting successive removal of methanol units from the reducing ends of pectin. Hydrolysis, catalyzed by various pectineste-rases, proceeds with a yield of 98%. The pectic acid thus formed gelatinizes in the presence of $Ca^{2+}$ ions. This process is used in France in the depectinization of fermenting apple juice to produce cider. The activity of pectinesterase is determined either by titrating the carboxyl groups released or by measuring the methanol liberated. Pectinesterase occurs in fruit and vegetables, especially citrus fruit and tomatoes. The properties of this enzyme, however, differ from those of pectinesterase obtained from molds or bacteria.

Mold pectinesterase is active between pH 3 and 5; the bacterial enzyme, between pH 7 and 8. In fact, the optimal pH values for stability of two mold enzymes are 3 and 4.5. Inactivation is caused by heating to 80°C. Commercial pectinesterase is a stabilized liquid containing 50–100 U/g. It is employed in the production of apple cider and must be free from depolymerizing pectinases.

**Pectin Depolymerases.** Pectin depolymerases occur in plants, but industrially, they are isolated from cultures of molds and, occasionally, bacteria. Depolymerases or endopectinases split the $\alpha$-1,4-bonds of the pectin chain. In fact, two mechanisms for bond cleavage have been postulated: hydrolytic splitting of the glycosidic bond and transeliminative cleavage:

**Table 11.** Optimal pH values for some pectin depolymerases

| Source | Type* | Optimal pH |
|---|---|---|
| *Aspergillus niger* | Endo-PMG | 5.2 |
| *Aspergillus niger* | Endo-PMG | 5.5 |
| *Flavobacterium pectinovorum* | Endo-PMG | 7.4–8.2 |
| *Aspergillus niger* | Endo-PG | 4.7–4.8 |
| *Aspergillus* species | Endo-PG | 3.7–4.2 |
| *Coniothyrum diplodiella* | Endo-PG | 4.4 |
| *Aspergillus sojae* | Endo-PTE | 5.5 |
| *Bacterium polymyxa* | Endo-PTE | 8.9–9.1 |

* For abbrevations, see text.

Depolymerases are classified on the basis of these mechanisms of cleavage and substrate specificity as follows:
1) Endo-pectin transeliminase (E.C. 4.2.2.3) [*9024-15-1*], PTE, pectin lyase
2) Endo-pectic acid transeliminase (E.C. 4.2.2.1) [*37259-53-3*], PATE
3) Endo-polygalacturonase (E.C. 3.2.1.15) [*9032-75-1*], PG
4) Endo-polymethylgalacturonase (PMG)

The activity of these enzymes can be followed by the rapid decrease in the viscosity of pectin solutions. In fact, the viscosity falls below 50% of the starting value after 2–3% of the glycosidic bonds have been cleaved. In addition, the number of reducing groups formed, determined later- on in the reaction, is an index of enzyme activity. The alcohol test can be used to determine the last stage of the reaction, complete depectinization, when no flocculation occurs on addition of 50% ethanol. The rate of transelimination can be followed photometrically.

Liquid industrial preparations of pectinase are stabilized with salts, sugars, or glycerol. Pectinase in powdered form is standardized with lactose, sucrose, or maltodextrin to give a definite activity. The activities of preparations supplied by different producers cannot be compared with each other. Usually only a statement on the amounts of pectinase generally required in practice is made. International units do not exist. Table 11 shows optimal pH values of pectin depolymerases.

The enzyme supplied as a liquid preparation loses approximately 10% of its activity per year at storage temperatures up to 15 °C. Apart from a mixture of the various pectinases, commercial preparations contain small amounts of other important enzymes, such as hemicellulases, cellulases including arabanase, proteinases, and amylases.

## 4.1.5. Proteinases [4.8]

Enzymes used for industrial digestion of proteins contain predominantly endopeptidases, also known as proteinases or proteases. They act on the interior peptide bonds of proteins and peptides. Proteinases may be classified according to the structural features of their active centers.

1) Some proteinases contain *mercapto groups* at their active centers, these groups being essential for enzymatic activity, e.g., papain, bromelain, ficin, and certain enzymes from microorganisms. These enzymes are inhibited by oxidizing agents and heavy metals.
2) *Metalloproteinases* require such metal ions as zinc, magnesium, or cobalt for activity. The "neutral" bacterial proteinases belong to this class of enzymes; they are inhibited by chelating agents, e.g., ethylenediaminetetraacetic acid (EDTA).
3) Proteinases that contain *histidine residues* at their active centers are inactivated by alkylating agents such as DIFP (diisopropyl fluorophosphate which attacks the serine residue). Examples are trypsin, chymotrypsin, and proteinases from bacteria and molds which are active under alkaline conditions.
4) Acidic proteinases of the pepsin group have *asparagyl* or *glutamyl residues* at their active centers. Apart from pepsin, a number of proteinases from molds, which are active under acidic conditions, belong to this category.

**Pancreatic Proteinases.** Commercially important mixtures of *trypsin* [9002-07-7] and *chymotrypsin* [9004-07-3] are obtained from pancreas and also contain peptidases, amylases, and lipase. In addition, *Pancreatin,* obtained on careful dehydration of the pancreas, contains the inactive precursors of the proteinases, trypsinogen and chymotrypsinogen; enterokinase or some acidic fungal enzymes can convert these zymogens to active enzymes.

At pH 8.5–9.0 (30–40 °C), the reaction rate of proteinases on such substrates as casein, hemoglobin, or gelatin is optimal; at pH 7.5, temperatures of 45–50 °C can be used. The proteinase mixture is most stable in aqueous solution, in the pH range 3–5. Heating to 90 °C at pH 6–8 leads immediately to total inactivation. Calcium ions (1–2 g of $CaCl_2$ per liter) and high substrate concentrations (e.g., collagen solutions) have a stabilizing effect.

Determination of the proteinase in commercial products is carried out with casein (Löhlein–Volhard, Kunitz, and Haghihara methods) or a chromogen (e.g., arginine-4-nitroanilide) as substrate. Pancreatins are often measured by using casein, according to the N.F. (National Formulary) method.

**Pepsin.** Pepsin (E.C. 3.4.4.1) [9001-75-6] is obtained from porcine and bovine gastric mucosa; it is used, in combination with rennet (rennin), in the commercial production of cheese and digestive preparations. At pH 2–4, the rate of reaction on proteins is optimal. The maximum allowable temperature is 50 °C.

**Rennin.** Rennin (E.C. 3.4.4.3) [9001-98-3], rennet, obtained from the stomach of the calf, is a very specific enzyme. It liberates a soluble macropeptide ($M_r$ 800) form $\kappa$-casein of cow's milk. The insoluble para-$\kappa$-casein clots and aggregates to give a solid casein jelly, the rennet curd.

Commercial products are stabilized with preservatives and sodium chloride. They contain varying amounts of pepsin (usually 50:50 mixtures).

To determine rennet strength, either the time required or the amount of enzyme necessary to clot a definite quantity of milk is measured. Microbial proteinases (see acidic proteinases) or plant enzymes can be used instead of animal rennet. For a survey, see [4.9].

**Papain.** Industrial preparations of papain (E.C. 3.4.22.2) [*9001-73-4*] are made by drying the juice of the fruit *Carica papaya*. Apart from papain and chymopapain, other proteinases are also present in these preparations. Granules of raw papain, purified products in powdered form, and liquid preparations (stabilized with glycerol) are used to degrade proteins in making cookies, in the beer industry, and in tenderizing meat. Raw papain is not standardized. The activity at pH 7.5, determined on hemoglobin, is 4000–6000 mU/mg, and crystallized papain has an activity of approximately 100 000 mU/mg. Papain cleaves proteins at pH values between 4.5 and 10 (pH optimum 7–8) and low molecular mass substrates or gelatin between pH 4.5 and 5.5. Reaction temperature should not exceed 55–60 °C. Inactivation occurs below pH 3.0 and under the influence of oxygen. Liquid preparations are stabilized with reducing agents such as sulfite or cysteine.

**Bromelain.** Industrial preparations of bromelain (E.C. 3.4.22.5) [*9001-00-7*], found in the fruit and stalk of the pineapple plant, show the same specificity as papain. Products on the market have an activity of 5000–10 000 U/g (measured on hemoglobin). They are, however, less stable than papain.

**Ficin.** Commercial preparations of ficin (E.C. 3.4.22.3) [*9001-33-6*], found in the sap of the fig tree, are similar to papain in specificity, stability, and sensitivity to inhibition. Ficin is used in the clotting of milk in some countries.

**Microbial Proteinases** [4.10]. Industrial preparations of microbial proteinases are usually mixtures of different types of enzymes, which are isolated from cultures of bacteria and mold. These enzymes differ from each other not only in the pH dependence of enzyme activity, but also in the temperature optimum, stability, and sensitivity to inhibition. They generally exhibit low specificity, but collagenase is an exception.

*Proteinases from Bacteria.* Two types of bacterial proteinases are of commercial importance. They differ from each other in pH range (neutral or alkaline), stability, and structure of the active center.

*Alkaline bacterial proteinases* (E.C. 3.4.21.14) [*9014-01-1*], such as subtilisin, are active over the pH range 7–11 and show only slight variations in specificity. They have a catalytically active serine residue and are irreversibly inhibited by alkylating agents. Commercial preparations, the so-called detergent enzyme, are dust-free granules with an activity of 1–6 Anson units per gram [4.11] (crystallized subtilisin has 25–30 Anson units per gram).

*Neutral bacterial proteinases* (E.C. 3.4.24.4) [*9068-59-1*] are metalloenzymes, active over a pH range of 6–9 (Table 12). Examples are enzymes from *Bacillus subtilis*, *B. thermoproteolyticus,* and *Streptomyces griseus*. Commercial preparations are available in powdered form, standardized with starch or salts to 0.5–2 Anson units per gram, or as stabilized solutions. They are employed in the leather industry, in breweries, and in the production of protein hydrolysates.

**Table 12.** Properties of alkaline and neutral bacterial proteases

| Property | Alkaline | Neutral |
|---|---|---|
| pH range | 7.0–11.0 | 6.0–9.0 |
| pH for optimal stability | 7.5–9.5 | 6.0–8.0 |
| Inhibitors | DIFP,[a] PMSF,[b] potato + soy inhibitor | EDTA[c] sodium dodecyl sulfate |
| Specific substrates | different esters, amides, and peptides | few peptides and esters |

[a] DIFP = diisopropyl fluorophosphate.
[b] PMSF = phenylmethanesulfonyl fluoride.
[c] EDTA = ethylenediaminetetraacetic acid.

**Table 13.** Properties of proteases from molds

| Organism | pH range for | | Optimal pH for stability | Inhibitors* |
|---|---|---|---|---|
| | Hemoglobin | Casein | | |
| *Aspergillus saitoi* | 3.0–4.5 | 2.5–3.0 | 2.0–5.0 | NBS |
| *Aspergillus oryzae* (acidic protease) | 3.0–4.0 | 2.5–3.0 | 5.0 | |
| *Aspergillus oryzae* (neutral protease) | 5.5–7.5 | | 7.0 | EDTA |
| *Aspergillus oryzae* (alkaline protease) | 6.0–9.5 | 6.5–10.0 | 7.0–8.0 | DIFP |
| *Paecilomyces varioti* | 3.5–5.5 | 3.0 | 3.0–5.0 | |
| *Rhizopus chinensis* | 5.0 | 2.9–3.3 | 3.8–6.5 | |
| *Mucor pusillus* | 3.5–4.5 | 5.6 | 3.0–6.0 | |

* NBS = *N*-bromosuccinimide;  EDTA = ethylenediaminetetraacetic  acid;  DIFP = diisopropyl fluorophosphate.

*Proteinases from Molds.* Mold proteinases are classified as acidic, neutral, or alkaline on the basis of the pH dependence of enzyme activity. Commercial preparations are usually mixtures containing primarily one of the enzymes mentioned above. The properties of some typical enzymes are shown in Table 13.

Acidic and neutral proteases from *Aspergillus* are used in the production of beverages and in baking. On the other hand, the alkaline enzymes are used predominantly for liming and bating in the leather industry.

## 4.1.6. Lipases [4.12]

Lipases (E.C. 3.1.1.3) [*9001-62-1*] are carboxylesterases that hydrolyze glycerides present as aqueous emulsions. In that respect, they differ from other carboxylesterases, such as proteases and pectinesterase, which act on substrates in aqueous solution. Therefore, lipases can be regarded as enzymes that hydrolyze esters only at the interface between lipid and water in a heterogeneous system.

A distinction is customarily made between enzymes that cleave esters of fatty acids and lower alcohols (aliesterases), and "real" lipases that hydrolyze glycerides. The specificity of lipases varies considerably, and depends on the substrate chain length and on certain positions in the substrate molecule. Lipases can be isolated on a large scale from a variety of sources, but porcine pancreas and certain microorganisms are the sources of choice. Table 14 lists the properties of common lipase preparations.

**Pancreatic Lipases (Pancreatin).** Lipase preparations are obtained from porcine pancreas, but these also contain esterases, proteases or their zymogens, and amylases.

Apart from naturally occurring triglycerides, oils, and fats, simple fatty acid esters and aryl esters are also cleaved by this enzyme. Both tributyrin and triacetin are rapidly hydrolyzed and, therefore, are often employed as analytical substrates.

*Effectors.* Since lipases act only at the interface between fat droplet and aqueous phase, the reaction rate depends on the degree of emulsification (size of droplet and stability). Gum arabic, poly(vinyl alcohol), or sodium deoxycholate promote emulsification. Calcium ions influence both enzyme activity and stability, which is also

**Table 14.** Comparison of properties of various lipase preparations

| Source (species) | pH range (pH optimum) | Temperature, °C | pH for stability |
|---|---|---|---|
| Porcine pancreas | 6.5–9.5 (7.5–8.5) | 40–45 | 5.5–7.5 |
| *Rhizopus species* | 6.0–7.5 (7.0) | 35–40 | 4.0–8.0 |
| *Mucor javanicus* | 5.5–8.0 (7.0) | 40–45 | 4.5–6.5 |
| *Aspergillus niger* | a) 3.0–7.0* b) 7.5–9.0 | 40–50 | 5.0–7.0 |
| *Pseudomonas* | a) 4.0–5.0** b) 7.0–8.5 | 50–60 | 4.5–10.0 |
| *Candida cylindracea* | 5.0–7.5 | 40–45 | 4.5–8.5 |

\* pH 3.0–7.0 with short-chain fatty acids; pH 7.5–9.0 with fats and oils.
\*\* Two pH optima for the two isoenzymes.

dependent on the sodium chloride concentration (optimum at ca. 0.5 g/L). Animal lipases preferentially catalyze the hydrolysis of fatty acids with more than 12 carbon atoms and predominantly at the C-1 position of glycerol. The reaction rate decreases considerably in the substrate order tri-, di-, and monoglycerides.

**Pregastric Lipases.** Lipases from animal sources include pregastric lipases from goats, sheep, and calves. These enzymes preferentially catalyze the hydrolysis of short-chain fatty acids in milk fat and are used in the production of specially flavored cheeses.

**Microbial Lipases.** The conversion of natural fats and oils into those with specific characteristics has been carried out by chemical and physical methods. Because lipases do not only hydrolyze fats but also synthesize fats, new applications have been developed: hydrolysis of fats, transesterification, stereospecific hydrolysis of racemic esters, and the use of lipases in detergents. Commercial preparations of microbial lipases are produced by fermentation of different fungi and bacteria. The industrial products are mixtures of lipases and esterases. The chain length of the fatty acid and its position in the glycerol molecule significantly affect the specificity of these enzymes.

*Lipases from Aspergillus species.* Different lipases from *Aspergillus niger* or *Aspergillus oryzae* with varying specificity towards long-chain and short-chain fatty acids have been described. The molecular masses are between 20 000 and 25 000; the pH optimum is between 4.5 and 6.5.

The lipase acts on coconut oil, linseed oil, and olive oil with yields of 48–93%. Special types of such lipases are also used in cheese ripening.

*Lipase from Candida cylindracea.* These lipases have molecular masses of 120 000; their isoelectric point is at pH 4.2 and their optimum of activity is between pH 5.2 and 7.2. They hydrolyze olive oil to 95–97%.

*Lipases from Rhizopus.* Different strains of *Rhizopus* are used for the enzyme production, such as *R. arrhizus, R. javanicus, R. niveus,* or *R. delemar.* Most data are available for the enzyme from *R. arrhizus.* Its molecular mass is 43 000, the isoelectric point at pH 6.3. The enzyme is a glycopeptide with 13–14% mannose per molecule. Rhizopus lipases show a 1,3-regiospecificity, the optimum of activity is at pH 5.0–7.0; the temperature optimum is at 30–45°C.

*Lipases from Mucor Species.* This lipase is a good catalyst for the transesterification in the 1,3-position of glycerol. Different types are specific for short- or long-chain fatty acids.

*Lipase from Pseudomonas.* This enzyme has a molecular mass of 29 000 and an isoelectric point at pH 5.8. The enzyme is active and stable at alkaline pH and therefore interesting for the use in detergents. For the same purpose, a lipase from *humicola languinosa* has been recommended.

*Phospholipase from Porcine Pancreas.* The only enzyme in industrial use is phospholipase A. It is used for the transformation of lecithins to lysolecithins.

## 4.1.7. Glucose Isomerase [4.13]

Glucose isomerase (E.C. 5.3.1.18) [*9055-00-9*] is produced by different microorganisms, such as *Bacillus actinoplanes, Arthrobacter,* and *Streptomyces*; it catalyzes the isomerization of glucose to fructose, either as the free enzyme or as a matrix-bound enzyme. At pH 6–7, the reaction temperature can be maintained at 60–75 °C. Magnesium or cobalt ions are required as cofactors, depending on the source of the enzyme. The glucose isomerase from *Actinoplanes missouriensis* is active at a pH of about 7 and at temperatures up to 60 °C in the presence of magnesium ions (approximately 0.8 g of $MgSO_4 \cdot 7H_2O$ per liter). Calcium ions inhibit enzyme activity.

The half-life of matrix-bound glucose isomerase is approximately 1200 h; the conversion achieved is 42 %, which is equivalent to the conversion of about 800 kg of glucose per kilogram of immobilized enzyme. Indeed, the total productivity amounts to 2000–4000 kg of isomerized glucose. The latest developments have resulted in a standard productivity of 22 000 kg dry matter per kilogram of enzyme.

## 4.1.8. Lactases [4.14]

Lactase (E.C. 3.2.1.23) [*9031-11-2*], also called β-galactosidase, hydrolyzes lactose to glucose and galactose. Commercial preparations are isolated from cultures of yeast or mold (*Aspergillus oryzae, A. niger*). They can be used on milk, sweet whey, or sour whey, depending on the properties of the enzyme preparation (see Table 15). Free enzymes are supplied as a stabilized, liquid preparation. Matrix-bound lactase from *A. oryzae* has a half-life of ca. 200 d when used in the digestion of sour whey and a total productivity of approximately 22 000 kg of lactose per kilogram of matrix-bound enzyme [4.15].

**Table 15.** Properties of lactases

| Property | Source | |
|---|---|---|
| | Yeast | *Aspergillus oryzae* |
| pH required for activity | 6.5–7.0 | 4.5–6.5 |
| Optimal temperature | 30–35 °C | 40–50 °C |
| Optimal stability | pH 6.5–7.0  (30 °C) | pH 4.5 |

## 4.1.9. Oxidoreductases

**Glucose Oxidase.** Glucose oxidase (E.C. 1.1.3.4) [*9001-37-0*] is available from *Aspergillus niger, A. oryzae,* and *Penicillium notatum* and catalyzes the oxidation of glucose to gluconic acid in the presence of oxygen:

$$Glucose + O_2 + H_2O \longrightarrow \delta\text{-Gluconolactone} + H_2O_2$$
$$\downarrow {\scriptstyle +\, H_2O}$$
$$Gluconic\ acid$$

Hydrogen peroxide generated is cleaved to water and oxygen by catalase, which is usually present in the same microorganisms. Commercial preparations exhibit a pH optimum between 4.8 and 6.0 at 20 °C. Indeed, they are active over a wide pH range, from 2.7 (acid-stable preparations) to 8.5. Heating to 80 °C for 2 min causes inactivation (90%).

Glucose oxidase is used in analytical chemistry and clinical diagnostics in the quantitative determination of glucose (see Section 5.3.1). It is also employed in the food industry as a preservative (to remove oxygen) for liquid proteins, fruit juices (United States), and mayonnaise. For a survey, see [4.16].

**Lipoxidase.** Lipoxidase (E.C. 1.13.1.13) [*9029-60-1*], also called lipoxygenase, is available from soybean flour or bean flour (France) and catalyzes the oxidation of fatty acids with a conjugated cis–trans diene system to the hydroperoxy derivatives. In the case of linoleic acid, two pH optima have been observed at pH 9.0 and 6.5. The enzyme is used as active bean flour (0.2–1.0%, based on flour) for bleaching.

# 4.2. Enzymes in Starch Processing and Baking

Starch and products derived from starch contribute to essential human nutrition and have many industrial uses. The corn-processing industry, which began in the United States in the 19th century, is a major source of these products [4.17]. The conversion of starch to dextrose and oligosaccharides was based primarily on acid hydrolysis. Conversions at very low and very high pH were known to result in undesirable byproducts and carbohydrate modification, which limited the use of hydrolysis in foodstuffs [4.18]. The application of enzymology to this area, pioneered by JOKICHI TAKAMINE the discoverer of malt diastase, introduced the use of selective degradation and conversion under more moderate conditions [4.19]. Further work indicated that the initial steps in the conversion of starch by acid could be followed by the use of an enzyme to saccharify starch to oligosaccharides and dextrose [4.20].

The growth of this industry accelerated considerably with the discoveries of (1) dual enzyme conversion processes for improved conversion of starch to dextrose and (2) isomerizing enzymes for conversion of dextrose to fructose. Commercial development of these processes provided an important, low-cost sugar substitute derived from corn and other grains [4.21]. Therefore, corn producers who, in 1970, provided some $65 \times 10^6$ t of corn and processed $3 \times 10^6$ t of corn to products, are now providing nearly $250 \times 10^6$ t of corn and processing ca. $20 \times 10^6$ t to products. About 50% of these products are directed to the manufacture of high-fructose corn syrup [4.22].

A second major application for starch processing with enzymes is the production of fuel alcohol. This became important in the 1970s and 1980s when alternatives and partial substitutes for petroleum were sought from agricultural resources. This effort has also aided in considering bioprocess alternatives for the production of other chemicals and chemical intermediates [4.23], [4.24].

A third application of starch-processing enzymes is their use in baking. This area is of interest because it may lead to new application of cells and enzymes in the processing of foodstuffs in general [4.25]–[4.28].

## 4.2.1. Syrups and Sweeteners

Figure 26 shows a typical process used to produce high-fructose corn syrup. Enyzmes play an important role at various stages in this process. Corn kernels, treated with sulfur dioxide and lactic acid bacteria, are softened so that fiber, protein, and oil can be separated from the starch by centrifugal force, based on density and size difference. Enzymes are added only when the starch–water slurry has been prepared. However, interest is increasing in the use of microbial and enzymatic augmentation, which involves such enzymes as cellulases, glucanases, proteases, and pectinases, at this early stage of the process. The objective is to improve stream quality and increase yield of protein or starch; in addition, this could lower the cost of corn wet milling, aid environmental protection, and increase product salability.

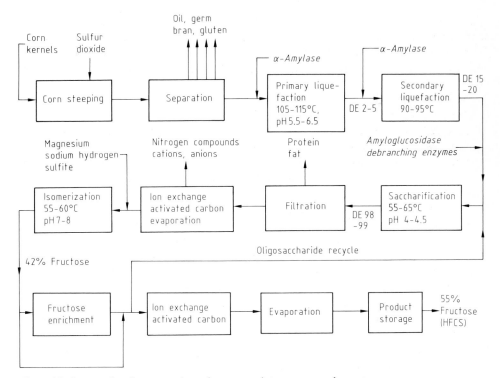

**Figure 26.** Process for the conversion of corn starch to syrups and sweeteners

Many organic and inorganic impurities that remain in the starch milk can affect enzyme performance. Examples include (1) inhibitory cations and anions, (2) factors that affect buffering and pH adjustment, (3) heat-sensitive materials which would result in byproducts, and (4) microbial contamination.

The starch milk is adjusted to a pH of ca. 6.0 and usually subjected to high-pressure steam in a continuous cooker. A small amount of *α-amylase* is added at this point to augment swelling, water absorption, and gelatinization and subsequent thinning of the starch. More amylase is added, and liquefaction is continued until the starch is degraded to polymer units of 15–20 DE (dextrose equivalents, 1 DE has the same reducing power as a 1% aqueous solution of pure dextrose). Total time to reach this level of conversion is up to 2 h in continuous processes. Starches from different sources (e.g., corn, wheat, potato, cassava, and barley) respond differently to this treatment; some can be gelatinized and degraded more easily, based on gelatinization temperature and the presence of fat, protein, and varying glycosidic linkages [4.29]–[4.31]. The starch must be broken down completely to limit the amount of dextrins, small oligosaccharide polymers of glucose, and to prevent molecules from recombining into forms not as susceptible to degradation.

The thermostable α-amylase currently in use commercially can withstand temperatures near the boiling point of water for a sufficient period of time to complete the conversion. Typical conditions and results for conversion are shown in Tables 16 and 17. Present technology is based on α-amylase derived from bacteria (*Bacillus licheniformis* is most common). Other thermophilic varieties such as *B. stearother-mophilus* are currently being tested [4.32].

An important development is the use of amylases that function at a pH slightly below 6.0. Depending on process circumstances, lowering the pH results in improved

**Table 16.** Representative conditions in the conversion of starch to high-fructose corn syrup

| Process step | Enzyme concentration, % | Conversion attained, DE * | Temperature, °C | pH | $Ca^{2+}$ concentration, mg/L | $Mg^{2+}$ concentration, mg/L |
|---|---|---|---|---|---|---|
| Starch conversion | | | | | | |
| Primary liquefaction | α-amylase (0.01–0.02) | 2–5 | 105–115 | 5.5–6.5 | 50–100 | |
| Secondary liquefaction | α-amylase (0.1–0.2) | 15–20 | 90–95 | 5.5–6.5 | 50–100 | |
| Saccharification | amyloglucosidase (0.1) | 98–99 | 55–65 | 4–4.5 | 50–100 | |
| Isomerization | glucose isomerase | 42% (fructose) | 55–60 | 7–8 | 2 | 25–100 |
| Enrichment | | 55, 95% Fructose | | | | |

* DE = dextrose equivalent; for explanation, see text.

**Table 17.** Enzyme properties and sources

| Enzyme | Source | pH range | Temperature range, °C | Typical pH; temperature, °C | Activators | Function |
|--------|--------|----------|------------------------|------------------------------|------------|----------|
| Fungal α-amylase E.C. 3.2.1.1 [9000-90-2] | *Aspergillus niger* *Aspergillus oryzae* *Aspergillus awamori* | 4–6 | 50–70 | 4.5; 65 | | limited hydrolysis of starch |
| Bacterial α-amylase Mesophilic | *Bacillus subtilis* *Bacillus amylo-liquefaciens* | 6–7 | 40–80 | 6; 75 | $Ca^{2+}$ | limited hydrolysis of starch |
| Thermophilic | *Bacillus licheniformis* *Bacillus stearo-thermophilus* | 5.5–6.5 4.5–6.5 | 90–120 | 5.8; 110 6; 115 | $Ca^{2+}$ | high-temperature liquefaction of starch |
| β-amylase E.C. 3.2.1.2 [9000-91-3] | *Bacillus polymyxa* wheat soybean | 4–7 | 55–75 | 5; 60 | | formation of low-DE syrups and in baking |
| Pullulanase E.C. 3.2.1.41 [9075-68-7] | *Bacillus* species *Aerobacter aerogenes* | 3.5–4.5 | 55–65 | 4; 60 | $Ca^{2+}$ | debranching, 1,6-bond degradation |
| Amyloglucosidase E.C. 3.2.1.3 [9032-08-0] | *Aspergillus niger* | 3.5–4.5 | 55–65 | 4; 60 | | saccharification, primarily 1,4-bond degradation |

quality, less byproduct formation, and easier processing. As indicated in Tables 16 and 17, calcium is necessary to stabilize and activate α-amylase. Compounds from corn such as phytin (phosphate-bearing ring structures) can competitively bind the calcium and thus decrease enzyme performance. Calcium concentrations used commercially must ensure performance and avoid the necessity of further refining to remove the cation.

The partially converted stream is then adjusted to ca. pH 4.0 and cooled to ca. 60 °C to permit saccharification to occur. *Amyloglucosidase,* sometimes augmented with *pullulanase,* is added for saccharification. The pullulanase (debranching enzyme) allows more rapid degradation of 1,6-glycosidic bonds and higher conversion to glucose when the stream contains more solids. The use of high solids lowers costs since less water must be removed in later stages [4.33], [4.34]. The presence of an α-amylase component in the saccharification stage can also benefit and augment amyloglucosidase in the disruption of 1,4- and 1,6-bonds. Conversions that require

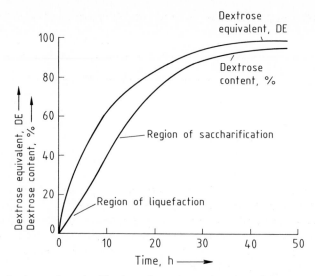

**Figure 27.** Liquefaction and saccharification of starch as a function of time

too much time favor reversion (i.e., re-formation of 1,6-bonds). These reversion reactions yield byproducts such as isomaltose, which are considered impurities, and reduce the yield of glucose [4.35]. When transglucosidase is present as an impurity in amyloglucosidase, it will also promote such undesirable reversion reactions; however, this impurity is removed in the manufacture of amyloglucosidase. Figure 27 presents a typical performance curve for conversion of starch to dextrose [4.36].

The possible use of immobilized enzymes (amyloglucosidase and pullulanase) in saccharification has been explored. To date, yields from such systems are lower than those obtained in batch saccharification at comparative solids contents (30–35%) and temperature levels (60 °C). Equivalent yields can only be obtained at lower temperatures (favoring microbial contamination) and higher dilution (greater cost). The corn wet-milling industry would benefit from such technology only if new methods were introduced to improve process economics and aid in quality maintenance [4.37]–[4.40]. Currently, the immobilized enzyme-assisted saccharification of the oligosaccharide recycle stream from fructose enrichment (see Fig. 26) is being investigated.

In a typical process, the effluent from saccharification is filtered to remove protein and fat, and then purified by treatment with activated carbon and ion exchange. The result is a solution of at least 95% dextrose. This solution can be either crystallized to yield pure dextrose or sent on for isomerization–refining to fructose corn syrup.

Processes to prepare intermediate-level conversion syrups can be based on a combination of acid and enzyme conversion. Alternatively, a series of enzymes, including $\beta$-amylase (plant or microbially derived), amyloglucosidase, and debranching enzymes can be used in immobilized form to yield these syrups. Typical product profiles for this type of syrup are indicated in Table 18 [4.41], [4.42].

**Table 18.** Profiles of high-maltose syrup (contents, wt %) prepared by carbon-immobilized enzymes [4.39]

|  | Glucose | Maltose | Maltotriose | Maltotetraose and higher |
|---|---|---|---|---|
| Immobilized enzyme system |  |  |  |  |
| $\beta$-Amylase | 1 | 51 | 14 | 34 |
| $\beta$-Amylase – pullulanase |  | 60 | 8 | 32 |
| $\beta$-Amylase – pullulanase – $\alpha$-amylase | 2 | 69 | 18 | 11 |

The use of such syrups could increase considerably in the future, but enyzme sales must be large enough to create incentive for such development. Current markets include confectionery and fruit processing. Prospects might improve if there is a demand for such an intermediate level, defined feedstock in the preparation of important chemicals from carbohydrates. However, the demand for the latter depends on oil pricing and availability.

Lower conversion syrups, traditionally prepared by acid or acid – enzyme processes, are used where properties such as retained viscosity (pastes, glues, thickeners) are important. Here, continuous processes are employed to limit color formation, reversion products, and yield loss [4.43]. The exclusive use of enzymes for this purpose may be a trend to minimize such difficulties, control conversion to a definite final DE, and control the distribution of carbohydrate polymers [4.44].

## 4.2.2. Fuel Alcohol

Starch conversion to produce fuel alcohol (ethanol) is analogous to the conversion to syrups and sweeteners. A variety of grains and tubers are used worldwide for alcohol production. For example, in the United States, ground corn is the most common source, while cassava along with sugar is used in Brazil. When ground corn is used as raw material, as opposed to starch, a large amount of undissolved solids must be handled.

As indicated in Figure 28, bacterial thermostable *amylase* and *amyloglucosidase* are used to convert feedstock to a fermentable substrate best suited for assimilation by yeast in ethanol production [4.45]. The use of genetically modified yeast, which synthesizes the amyloglucosidase as part of its growth and metabolism, has been reported [4.46]. This can be advantageous in a new facility to save capital cost; however, the way to obtain the best combination of saccharification and fermentation time must be evaluated on a case-by-case basis. Such recombinant DNA technology may be used in beverage brewing [4.47].

Feedstock for this process can include recycled yeast and distillers' grains (backset), mixed with ground corn and enzyme. The mixture is hydrated, slurried, and subjected to intensive liquefaction to ensure conversion of all the starch in the presence of undissolved solids. This can involve several stages and temperatures.

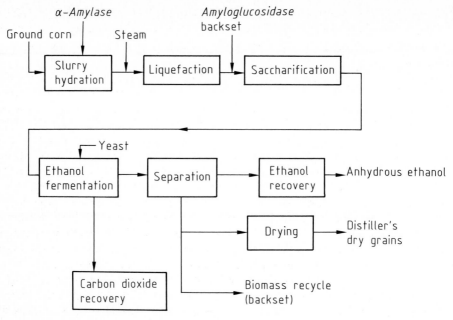

**Figure 28.** Conversion of ground corn to ethanol

After secondary liquefaction, the material is saccharified by treatment with amyloglucosidase. The saccharification time can vary and is shorter than the time required for the analogous step in syrup manufacture since complete conversion to glucose prior to fermentation may be neither necessary nor desirable, because excess glucose could cause excessive biomass formation at the expense of ethanol yield from the feedstock.

If raw materials other than corn are used, other enyzmes and process steps may be required. For example, barley necessitates the use of $\beta$-glucanases to break down the viscous glucans present, and this can be augmented with proteases, much as mashing is accomplished in the preparation of syrup for alcoholic beverage fermentation [4.47].

## 4.2.3. Baking

Cereal grains such as wheat or rye are used as sources of flour for baking. Enzymes may be either endogenous to the grain or added as supplements for several purposes. For example, proteases can be used to degrade gluten in a controlled manner, and amylases can be used to obtain the desired profile of minimal dextrins, fermentable sugars for yeast metabolism, and carbon dioxide formation. Free peptides and amino acids resulting from the action of protease can combine with dextrins to yield desired Maillard-type browning reactions; the beneficial result of this, if not excessive, is brown crust, good color, and added flavor. Proteases can reduce the dough viscosity caused by gluten and, thereby, improve processing. The desired softening effect of proteases can vary; greater softness is desirable for biscuits and waffles [4.48].

Both types of amylase, $\alpha$-amylase and $\beta$-amylase, are used in baking. $\beta$-Amylase is generally available in sufficient quantity in the grain, but $\alpha$-amylase is often deficient and must be supplemented. The amount of additional enzyme must be sufficient for desired gas production, volume control, and color. It must not be added in excess because this can result in excessive dextrin formation, leading to loaf stickiness, dark color, and possibly insufficient product strength. The $\alpha$-amylase supplement can be derived from malted flour (enzyme formation through fermentation in the grain); however, fungal $\alpha$-amylase is commonly used because this $\alpha$-amylase and cereal $\beta$-amylase are mesophilic and cease activity as the temperature approaches and passes through the gelatinization point (70–80 °C). This provides a means to control dextrin formation. The malt amylases are thermally more stable and can lead to excessive breakdown of the starch. The enzymes derived from *Aspergillus oryzae* act in the pH range 4.5–5.5 [4.49]. The pH of the dough can vary with the product, so a pH near 7.0 may result and alter the enzyme dose. In the formation of bread, the temperature may increase from 30 °C to 90 °C over a period of 20 min. Dextrin formation begins to decrease at ca. 70 °C as enzyme activity declines. Expansion is also limited by the inactivation of yeast as these temperatures are reached. Levels of fungal amylase range from 5–250 SKB units per 100 g of flour (for explanation of SKB, see p. 77), depending on the application.

**Other Enyzmes in Baking.** *Proteases* are usually derived from bacterial sources (neutral pH variety), but fungal proteases can also be used in certain cases. The levels employed to aid in browning, machining, and dough extension range from 2 g per 100 kg of flour in crackers, through 15 g per 100 kg of flour in biscuits, and up to 50–100 g per kg of flour in waffles. The dose depends on the pH of the dough [4.49]. *Lipoxygenases* from soybean have been explored to improve whitening, and *phospholipase* (animal source) has been studied as an antistaling agent. *Pentosanases* (hemicellulases) have been used to cleave residual pentosans in flour in certain cases to aid in antistaling and water absorption.

## 4.2.4. Characterization

Enzymes commonly used for starch processing are generically classified as amylases. As shown in Figure 29, a distinction exists between *endoamylases* that hydrolyze random or specific internal glycosidic bonds and exoamylases that start cleavage at the nonreducing end of the molecule. *Debranching enzymes*, which are now coming into greater use, are classified as direct or indirect, and cleave 1,6-glycosidic bonds. The indirect variety (amylo-1,6-glucosidase) first requires the removal of an oligosaccharide by transglucosylase, leaving a glucose residue bound to a tetrasaccharide through a 1,6-bond (Fig. 30) [4.50]. Pullulanases or isoamylases are direct-

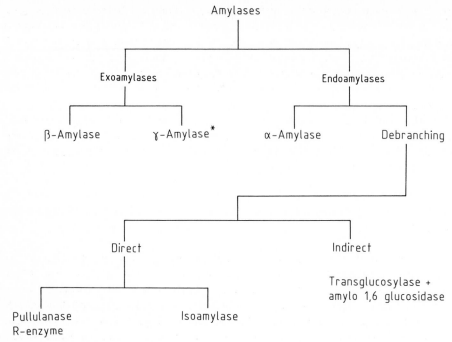

**Figure 29.** Classification of amylases
    * Amyloglycosidase

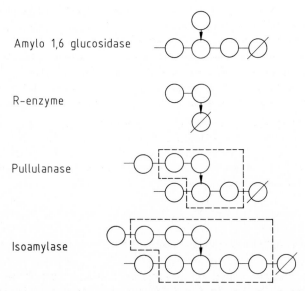

**Figure 30.** Carbohydrate structure requirements for hydrolysis of 1,6-bonds by debranching enzymes
↓ Cleavage by enzyme mentioned          Ø Reducing end

acting debranching enzymes; the differences in oligomer structure necessary for cleavage of 1,6-bonds are indicated by the dotted lines in Figure 30; i.e., pullulanases cleave preferentially at α-maltosyl-maltose residues whereas isoamylases cleave preferentially at α-maltotriosyl-maltotetraose residues. The R-enzyme, derived from broad bean, cleaves the 1,6-bond of a trisaccharide branched structure.

The varieties of amylase offer many alternative ways to improve the conversion of starch; as indicated in Table 17 (p. 95), only some of these methods have been explored commercially. Opportunities exist to combine such enzymes or to fractionate existing enzymes and characterize individual components, thereby perhaps discovering new activity characteristics. Only some of these have been accepted for use in food processing [4.51].

## 4.2.5. Analysis

Product quality and consistency require sufficient analytical monitoring. In starch processing, the solids content is measured by refractometry with proper temperature control. The presence of protein as a contaminant can be detected by the Kjeldahl method. If separation and identification of proteins are desired, isoelectric focusing, electrophoresis, and ion-exchange chromatography are used; however, these methods are not common as quality tests in industry. The dextrose equivalent (DE), defined as the availability of free carbohydrate ends able to reduce $CuSO_4$ solutions is measured to determine the degree of conversion of starch. More detailed knowledge of carbohydrate composition requires the use of high performance liquid chromatography (HPLC) and gas chromatography. The presence of undesirable color can be measured spectrophotometrically at several wavelengths [4.52]. The relative amount of fructose present can be determined by optical rotation with a polarimeter and by HPLC. Atomic absorption methods are used to determine cations. The ionic purity of streams can be assessed by measuring electrical conductivity. For enzyme assays, see Chapter 2 and Section 5.5.

*Retrogradation* is a common difficulty in baking [4.53]–[4.55]. Research continues into several assay methods of potential use in starch processing and baking. These include enzyme sensors [4.56], determination of structures [4.57], chromatographic analysis of starch components subjected to α-amylase isozymes [4.58], ultrasonic detection of enzyme hydrolysis of starch [4.59], and a method that avoids iodine staining of starch [4.60].

## 4.2.6. Trade Names and Manufacturers

Representative trade names and manufacturers of starch-processing and baking enzymes are listed in Table 19. More information is given in [4.32]. Manufacturers have their own methods of defining activity; laboratory evaluation is required to compare products.

**Table 19.** Typical trade names and manufacturers

| Enzyme | Trademark | Manufacturer |
|---|---|---|
| Bacterial α-amylase | Tenase | Miles Laboratories |
| | Optiamyl | Miles Kali-Chemie |
| | BAN | Novo-Nordisk |
| | Rapidase | Gist-brocades |
| Bacterial thermostable α-amylase | Termamyl | Novo-Nordisk |
| | Taka-therm II | Miles Laboratories |
| | Taka-lite | Miles Laboratories |
| | Opti-therm | Miles Kali-Chemie |
| | Maxamyl | Gist-brocades |
| Amyloglucosidase | Diazyme | Miles Laboratories |
| | Optidex | Miles Kali-Chemie |
| | AMG | Novo-Nordisk |
| | Spezyme | Finn Biochemical |
| | Amigase | Gist-brocades |
| Debranching Enzyme | Dextrozyme (contains AMG) | Novo-Nordisk |
| | Promozyme | Novo-Nordisk |
| β-Amylase | β-Amylase | Amano |
| | Spezyme BBA | Finn Biochemicals |
| Fungal amylase | Biozyme | Amano |
| | MKC fungal amylase | Miles Kali-Chemie |
| | Clarase | Miles Laboratories |
| | Fungamyl | Novo-Nordisk |
| Fungal protease | Amano "A" | Amano |
| | Fungal protease | Miles Laboratories |
| Neutral protease | HT Proteolytic | Miles Laboratories |
| | Neutrase | Novo-Nordisk |

# 4.3. Glucose Isomerization

Enzymatic isomerization of glucose [50-99-7] to fructose [57-48-7] in starch processing is carried out on an industrial scale worldwide, but predominantly in the United States. The commercial product obtained, high-fructose corn syrup (HFCS), typically contains 42 or 55% fructose, based on dry substance. It is used as an alternative sweetener to sucrose or invert sugar in the food and beverage industry.

Enzymatic glucose isomerization was first established on an industrial scale in 1967 by Clinton Corn Processing Co. in the United States, using in-house enzyme technology. Around 1974, immobilized glucose isomerase became commercially available. With the increasing acceptance of HFCS, especially in the soft drink industry, the glucose isomerization process was rapidly adopted by practically all major starch-processing companies in the Western World between 1975 and 1980. A substantial increase in HFCS consumption occurred around 1978 with the introduction of *fructose enrichment*, a chromatographic separation of fructose and glucose which makes possible the production of HFCS having increased fructose content and sweetness. In 1988, the total amount of HFCS produced worldwide exceeded $7 \times 10^6$ t, based on dry matter.

**Reaction.** Glucose can be reversibly isomerized to fructose. The kinetics of the reaction is well-described in the literature [4.61], [4.62].

Glucose          Fructose

Industrial isomerization is catalyzed by the enzyme glucose isomerase. The reaction may also be base-catalyzed, but this type of catalysis is nonspecific, producing D-mannose as a major byproduct [4.63], and has never been industrially important.

The equilibrium conversion of glucose to fructose under industrial process conditions is around 50 % (Table 20), and the enthalpy of the slightly endothermic reaction is 5 kJ/mol [4.64].

**Enzyme.** Glucose isomerase is produced by several microorganisms as an intracellular enzyme [4.65]. The commercially important varieties show superior affinity to xylose and are classified as xylose isomerase (E.C. 5.3.1.5) [*9023-82-9*]. The first patent on enzymatic isomerization of glucose was issued in 1960 [4.66]. Important research was carried out in Japan [4.67], resulting in the first patented reusable glucose isomerase [4.68]. The research and development leading to the current glucose isomerization process has been reviewed by several authors [4.65], [4.69]–[4.71]. The glucose isomerases used commercially today are all immobilized and granulated to a particle size between 0.1 and 1.5 mm. Details of immobilization are given in Table 21; for general treatment of immobilization, see Section 3.3. The enzymes are rather similar to each other with respect to dependence on temperature (Fig. 31), pH (Fig. 32), and metal ion activation. Typically, addition of $Mg^{2+}$ to the feed syrup is

**Table 20.** Chemical equilibrium for glucose isomerization

| Temperature °C | % Fructose at equilibrium |
|---|---|
| 55 | 50.0 |
| 60 | 50.7 |
| 65 | 51.5 |
| 70 | 52.4 |
| 80 | 53.9 |
| 90 | 55.6 |

**Table 21.** Examples of commercial glucose isomerases

| Trade name/manufacturer | Enzyme source | Immobilization technique | Reference |
|---|---|---|---|
| Maxazyme/Gist-brocades (The Netherlands) | *Actinoplanes missouriensis* | occlusion in gelatin followed by cross-linking with glutaraldehyde | [4.73] |
| Optisweet/Miles Kali-Chemie (Federal Republic of Germany) | *Streptomyces rubiginosus* | absorption of purified enzyme on silica followed by cross-linking with glutaraldehyde | [4.74] |
| Spezyme/Finnsugar (Finland) | *Streptomyces rubiginosus* | electrostatic binding of purified enzyme on DEAE-cellulose agglomerated with polystyrene and $TiO_2$ | [4.75] |
| Sweetzyme/Novo-Nordisk (Danmark) | *Bacillus coagulans* | cross-linking of homogenized cell material with glutaraldehyde | [4.76] |
| Swetase/Nagase (Japan) | *Streptomyces phaechromogenes* | binding of heat-treated mycelia to anion-exchange resin | [4.77] |

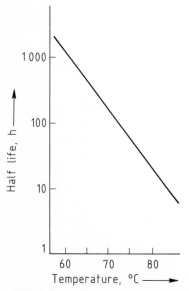

**Figure 31.** Temperature–stability profile of immobilized glucose isomerase
Reaction parameters: 45 wt% glucose; average pH, 7.5; $Mg^{2+}$ concentration, $4 \times 10^{-4}$ mol/L

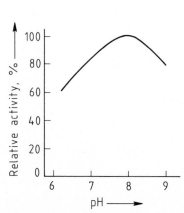

**Figure 32.** Activity–pH profile of glucose isomerase from *Bacillus coagulans*
Reaction parameters: 45 wt% glucose; 60 °C; $Mg^{2+}$ concentration, $16 \times 10^{-4}$ mol/L

recommended, and $Co^{2+}$ and $Fe^{3+}$ are potential activators; $Ca^{2+}$ acts as an inhibitor by displacing $Mg^{2+}$ in the enzyme molecule [4.72]. Table 21 lists examples of commercial glucose isomerases.

**Process.** Glucose isomerization can only be made economically feasible by immobilizing the enzyme. A relatively high reaction temperature is necessary to obtain a reasonable fructose yield. The pH must be ca. 7 or higher to secure satisfactory enzyme activity and stability. Under these conditions, glucose and fructose are rather unstable and easily decompose to organic acids and colored byproducts (carbonyl compounds). To limit byproduct formation, reaction time must be minimized; this can be done economically only by using high concentrations of immobilized isomerase. Figure 33 shows a flow sheet of a typical process.

Isomerization of glucose to HFCS on an industrial scale is carried out almost exclusively in continuous fixed-bed reactors [4.69], in which purified glucose (dextrose) syrup from the saccharification stage of a starch-processing plant is passed through a bed of granular, immobilized glucose isomerase. The enzyme granules must be rigid enough to prevent bed compaction during operation.

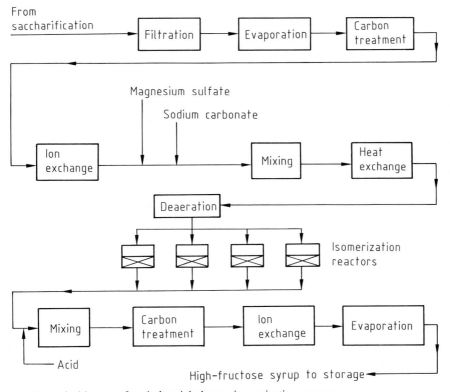

**Figure 33.** Typical layout of an industrial glucose isomerization process

At the temperature commonly used in industrial isomerization, the equilibrium ratio of fructose to glucose is ca. 0.50, but to avoid excessive reaction time, the conversion is normally limited to ca. 0.45. Space velocities between 0.2 and 2 bed volumes per hour are typical.

*Feed Syrup.* The main criteria for selecting the feed syrup specifications are optimization of enzyme productivity and limitation of byproduct formation (Table 22).

**Table 22.** Typical feed syrup specifications

| | |
|---|---|
| Temperature | 55–60 °C |
| pH | 7.5–8.0 |
| Dry-substance content | 40–50 wt% |
| Glucose content | $\geq 95\%$ * |
| $SO_2$ | 0–100 ppm |
| Calcium ion | $\leq 1$ ppm |
| $MgSO_4 \cdot 7\ H_2O$ (activator) | 0.15–0.75 g/L** |
| Conductivity | $\leq 100\ \mu S/cm$ |
| UV absorbance (280 nm) | $\leq 0.5$ |

* Based on dry matter.   ** Dependent on type of glucose isomerase.

The use of immobilized enzymes requires highly purified substrates to prevent rapid deactivation and clogging of the enzyme bed. Insoluble impurities in the glucose feed syrup (fat, protein) are removed by filtration, and soluble impurities (peptides, amino acids, and salts) by treatment with activated carbon, followed by ion exchange [4.77], [4.78].

The dry-substance content of the feed syrup is adjusted to 40–50%. Higher syrup concentration will result in a reduced isomerization rate due to diffusion resistance in the pores of the immobilized enzyme. A deaeration step removes dissolved oxygen that would increase byproduct formation. The pH is adjusted to the productivity optimum of the enzyme. Low pH is preferable for the sake of monosaccharide stability, but exposure to pH below ca. 7.0 for longer periods of time inactivates commercial glucose isomerases. Small amounts of organic acids are formed during the isomerization; therefore, adjustment of the feed pH with sodium carbonate buffer is advantageous.

The isomerization temperature is normally 55–60 °C. Lower temperatures lead to increased risk of microbial infection. Higher temperatures increase the isomerization rate but reduce enzyme and monosaccharide stability. Except for cases of limited production capacity, high enzyme stability and low byproduct formation are given highest priority.

*Enzyme Decay.* During operation, the immobilized enzyme loses activity. When the feed syrup is carefully purified, the most likely explanation for this activity decay is heat denaturation of the enzyme [4.79]. Most commercial enzymes show an exponential decay as a function of time (Fig. 34).

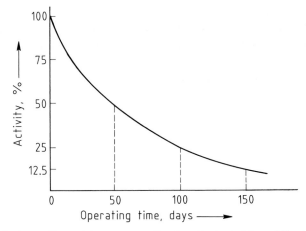

**Figure 34.** Activity (syrup flow rate) versus operating time for typical immobilized glucose isomerase

Typically, a reactor load of glucose isomerase is replaced after three half-lives, i.e., when the activity has dropped to around 12.5% of the initial value. The most stable commercial glucose isomerases have half-lives of more than 100 days in industrial practice.

To maintain a constant fructose concentration in the product syrup, the feed flow rate is adjusted according to the actual activity of the enzyme. With only one isomerization reactor in operation, excessive variations in syrup production rate would result. To avoid this, several reactors containing enzyme of different age are operated in combination. With a system of eight reactors, the variation in total syrup flow can thus be limited to ±13% of the average. The isomerization reactors may be connected to operate in parallel or in series.

*Reactor design* for glucose isomerization in the United States has been described recently [4.77]. Reactor diameter is normally between 0.6 and 1.5 m. Typical bed height is 2–5 m. Minimum bed height–diameter ratio for one reactor is 3:1 to ensure good flow distribution. For several reactors in series, this restriction becomes less severe. Plants producing more than 1000 t of HFCS (based on dry matter) per day typically use at least 20 individual reactors.

**Product.** Around two-thirds of the HFCS produced worldwide is used in the chromatographically enriched form [4.77] containing 55% fructose (based on dry matter) for sweetening nonalcoholic beverages. The 42% HFCS obtained directly by enzymatic isomerization is used mainly in the baking, canning, and dairy industries.

Because of the high hygroscopicity of the fructose component, HFCS cannot replace sucrose in the manufacture of hard candy. Table 23 lists typical properties of HFCS, and Table 24 demonstrates its economic importance [4.83 a]. During 1987, commercial products of pure fructose derived from HFCS were introduced (in syrup as well as crystalline form).

**Table 23.** Typical properties of HFCS

|  | Fructose content * | |
|---|---|---|
|  | 42 wt% | 55 wt% |
| Dry substance, wt% | 71 | 77 |
| pH | 3.5–4.5 | 3.5–4.5 |
| Ash, sulfated, wt% | 0.05 | 0.05 |
| Color, C.I.R.F. ** | 0.003 | 0.003 |
| Glucose, wt% | 53 | 41 |
| Fructose, wt% | 42 | 55 |
| Viscosity at 25 °C, Pa · s | 15 | 85 |
| Appearance | clear, colorless | clear, colorless |
| Relative sweetness (sucrose = 100) | 90 | 100–110 |

\* Based on dry matter.  \*\* Corn Industries Research Foundation (United States).

**Table 24.** Commercial importance of HFCS in 1985

|  | Replacement of industrial sucrose, % | Annual production, based on dry matter, $10^6$ t |
|---|---|---|
| United States | 60 | 5.7 |
| Japan | 30 | 0.7 |
| Europe* | 2 | 0.3 |

\* In Europe, production of HFCS is limited by EEC regulations.

Reviews of HFCS application are given in [4.80]–[4.83]. Both 42 and 55% HFCS are used almost exclusively as a reduced-cost replacement for liquid sucrose and invert sugar; the price of HFCS is typically 10–20% lower than that of sucrose, based on sweetening power.

# 4.4. Proteolytic Enzymes

Proteolytic enzymes, proteases, or proteinases, are enzymes which under appropriate conditions, specifically hydrolyze the peptide bonds of proteins. All proteases have characteristic properties with regard to pH and temperature, ion requirements, specificity, activity, and stability. The specificity depends on the amino acids involved in the peptide bond to be hydrolyzed. These biochemical parameters determine the application of a protease. However, its commercial feasibility also depends on the development and production costs of the enzyme, its market, and the economics of application.

Table 25 surveys the uses of industrial proteases. This section deals in detail with the use of proteases in detergents, leather treatment, the synthesis of human insulin,

**Table 25.** A survey of industrial applications of proteolytic enzymes

| Market | Enzyme | Application |
|---|---|---|
| Detergents | alkaline proteases | extensive use in laundry detergents for protein stain removal |
| Dairy (see Section 4.6) | calf rennet | coagulation of milk protein (cheese production) |
| | fungal proteases | replacement of calf rennet |
| | chymosin | active component of calf rennet; production by genetically engineered microorganisms is being developed |
| | proteases | production of enzyme-modified cheese; processing of whey protein |
| Leather | trypsin | bating of leather |
| | other proteases | dehairing and dewooling of skins |
| Beverages (see Section 4.7) | papain | removal of turbidity in beverages |
| Baking (see Section 4.2) | neutral protease | dough conditioner |
| Meat and fish (see Section 4.5) | papain | meat tenderizing |
| | several proteases | recovery of protein from bones and waste fish |
| Food processing | several proteases | modification of protein-rich material, i.e., soy protein or wheat gluten |
| Sweeteners | thermolysin | reverse hydrolysis in aspartame synthesis |
| Medicine | trypsin | removal of dead tissue and dissolution of blood clots |
| | chymopapain | treatment of certain types of hernia |
| Insulin | carboxypeptidase trypsin | conversion of hog insulin into human insulin |

and the production of aspartame. Other applications mentioned in Table 25 are described in different sections of this article.

The worldwide requirement of enzymes for individual applications varies considerably. Some enzymes are produced in large quantities, up to thousands of tons, but are of relatively little unit value, ie, $ 10–20 per kilogram. These proteases are referred to as industrial bulk enzymes; Table 26 estimates worldwide sales [4.84]–[4.86]. Other proteases are used in organic synthesis (Section 4.10) in amounts up to tens of tons. The market for medical proteases (Section 5.4) is small in terms of tonnage, but their price per unit activity is very high.

**Table 26.** Worldwide sales of industrial bulk proteases in 1985 [4.84]–[4.86]

| Application | Enzyme | Origin | Sales, $10^6$ $ |
|---|---|---|---|
| Detergents | alkaline proteases | microorganisms, e.g., *Bacillus alkalophilus* | 140 |
| Cheese | rennet | stomach of suckling calves | 50 |
| manufacturing | microbial rennet | microorganisms, e.g., *Mucor miehei* | 30 |
| Leather | trypsin | animal pancreas | 10 |

## 4.4.1. Detergent Enzymes

As early as 1913, enzymes were used in laundry detergents by the German chemist OTTO RÖHM, who patented the application of pancreatic enzymes in a presoaking detergent composition [4.87]. These enzymes, e.g., trypsin and chymotrypsin, are active around pH 7–9, which was achieved by presoaking – in these days with sodium carbonate. The washing effect of trypsin and chymotrypsin would be negligible in modern detergents at higher pH.

The first detergent with a bacterial protease, Bio 40, did not appear on the market until 1959 and was produced by Gebrüder Schnyder in Switzerland [4.88]. Proteases were used routinely in laundry detergents after the enzymes had been screened systematically for optimal conditions: pH and temperature characteristics, activity, and stability.

The Danish company Novo introduced Alcalase, an alkaline protease produced by *Bacillus licheniformis*. This was followed a few years later by Maxatase, made by Gist-brocades. Alcalase was formulated in a detergent product, still known as Biotex, which was marketed successfully and stimulated the development of many enzyme-based detergents [4.88].

The rapid growth of enzyme detergents was temporarily set back in the early 1970s, when workers in detergent factories developed allergies to the enzyme preparations. Enzyme manufacturers solved this problem by developing dust-free protease formulations [4.89].

In 1985, approximately 70 % of the heavy-duty laundry detergents (for domestic use) in Europe contained enzymes, compared to 15 % in the United States (Fig. 35).

Presently, the major enzyme suppliers, produce a variety of differently formulated proteases for use in all types of liquid and powder detergents, with or without bleach and phosphates. In addition to proteases, other enzymes are also used in detergents (e.g., α-amylases), and the use of enzymes such as cellulases [4.90] and lipases [4.91] is being developed.

**Requirements.** The compositions of the detergent and washing liquor, as well as the washing characteristics, determine the functional enzyme properties of a laundry detergent. In addition, the formulated enzyme preparation must be stable on storage. Table 27 shows the usual composition of typical West European powder and liquid detergents. The shelf life of an enzyme is affected by temperature, pH, water activity, bleach, and the presence of denaturing agents such as nonionic and anionic surfactants. For use in powder detergents, enzymes are formulated with fillers and binding agents into small beads or granules, sometimes surrounded with a layer of inert material such as a wax. To avoid segregation of enzyme particles in the powder

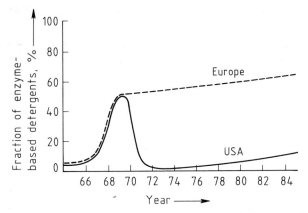

**Figure 35.** Market development of enzyme-based laundry detergents in Europe and the United States

**Table 27.** The average composition of a heavy-duty Western European powder detergent and of a liquid detergent

| Ingredient | Powder, wt% | Liquid, wt% |
| --- | --- | --- |
| Anionic surfactants | 2–10 | 8–10 |
| Nonionic surfactants | 0.5–6 | 18–20 |
| Soap | 1–5 | 7–12 |
| Sequestrants | 30–50 | |
| Ethylenediaminetetraacetic acid | | 1 |
| Bleach + activator | 20–30 | |
| Triethanolamine | | 7–13 |
| Ethanol | | 4–6 |
| Propylene glycol | | 4–6 |
| Enzymes | 0.5 | 1 |
| Perfume | 0.2 | 0.4 |
| Optical brighteners | 0.4–0.8 | 0.2 |
| Sodium sulfate | to make 100% | |
| Water | | to make 100% |

detergent, e.g., during transport, the particle size of the enzyme beads must be more or less identical to the size of the detergent particles. Furthermore, the particles must dissolve rapidly when the wash liquor is prepared.

The average liquid detergent has a high water activity, in combination with an alkaline pH and the presence of denaturing ionic compounds; these conditions favor autodigestion of proteases, unless the pH is lowered to 7–8.

**Characteristics.** Composition of the wash liquor and laundering practices differ from country to country [4.92]. In Europe, detergent concentrations of 4–10 g/L are used. The detergent formulation contains bleach, and the washing process is characterized by a heating-up phase followed by a constant-temperature phase. In the United States and Japan, the detergent concentration is lower. Bleach and builders are often added separately or not at all. Washing is done at constant temperatures, significantly lower than those in Europe.

Enzyme suppliers have developed proteolytic enzymes that fulfill all the requirements of detergent formulation, stability, and activity under washing conditions. The types of stain subjected to proteolytic attack are composed of protein and other soil. The protein acts as a binder and fixes the other soil components to the fabric. The protease acts synergistically with the detergent: limited hydrolysis is sufficient to make the stain susceptible to removal by the detergent components. Detergent proteases are endoproteases: they split the protein into oligopeptides.

Powdered laundry detergents, which constitute the largest part of the detergent market, function at a high pH [4.92] (Fig. 36). The relationship between temperature and activity is another important characteristic (Fig. 37). The activity rises with increasing temperature until it decreases suddenly at higher temperatures due to enzyme inactivation. Because many powder detergents contain such sequestering agents as polyphosphates, nitrilotriacetic acid (NTA), or zeolites, enzymes [4.93] whose activity depends on loosely bound metal ions (e.g., $Mg^{2+}$ or $Ca^{2+}$) cannot be used in detergents. Finally, the enzymes must not be denatured by the action of nonionic and anionic surfactants.

**Activity Measurement.** Formulated commercial enzyme preparations [4.94] contain only a small percentage of enzyme protein (on a weight basis). They are sold on the basis of activity. In general, the supplier will specify in the product sheets the analytical method used to determine activity. In most cases, the enzyme is incubated at specified pH and temperature with large substrate molecules, e.g., casein or hemoglobin. At the end of incubation, the amount of hydrolyzed and, therefore, solubilized protein is measured spectrophotometrically by UV absorption, usually preceded by acid precipitation of the larger fragments. Other methods detect the oligopeptides, produced by coupling with a color reagent such as trinitrobenzene-sulfonic acid [4.95]. Because of large variations among detergent formulations, measurement of activity in the presence of detergent components is meaningless. In addition, the major detergent manufacturers perform their own tests of an enzyme's activity, stability, and performance before it is added to a detergent formulation.

**Washing Performance.** The contribution of an enzyme to the washing performance of a detergent is determined in the laboratory by test washings in specially developed washing machines. The West European washing process is simulated in a so-called Launder-ometer; the American, in a Tergotometer [4.96], [4.97].

For laboratory testing, different types of artificial stains such as milk, blood, cocoa, grass, egg, or grease are used. The stains are applied to different fabrics such as cotton or polyester. Specific stains are developed and produced in the laboratory. Other stains are commercially available. A well-known test fabric used to study the removal of protein is EMPA 116 [4.98], a cotton cloth soiled with milk, blood, and ink, the ink being glued to the fabric by milk and blood proteins.

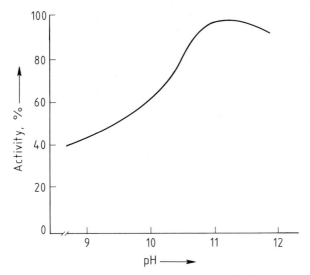

**Figure 36.** Activity–pH profile of Maxacal, a commercial high-alkaline protease
Activity is determined by incubating the enzyme for 48 min at 40 °C with casein as substrate. The reaction is terminated by adding trichloroacetic acid; the amount of acid-soluble material, measured spectrophotometrically at 260 nm, represents the protease activity (DU).

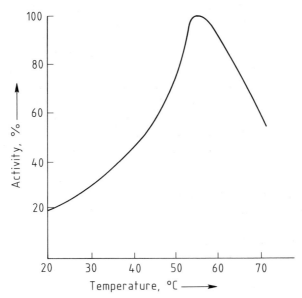

**Figure 37.** Temperature–activity profile of Maxacal
Activity is determined at pH 10 as described in Figure 36.

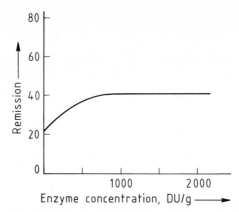

**Figure 38.** Contribution of increasing amounts of Maxacal to the removal of protein-based stains
Increasing amounts of protease, expressed in Delft units (DU) per gram of detergent, were incubated
for 40 min at 40 °C in a Launder-ometer with the test fabric EMPA 116 in the presence of a typical
European laundry detergent without a bleaching system.

Protein hydrolysis is required for the efficient removal of ink. The amount of ink
left on the cloth after washing is determined by measuring the remission of the cloth.
This is an indicator of the washing performance contributed by the enzyme to the
detergent (Fig. 38). Such laboratory tests are used to screen and study the fundamental characteristics of detergent enzymes [4.99]. Since this type of testing on
artificial stains in washing machines is only indicative of certain effects, detergent
enzymes are tested in a later developmental phase on garments with real stains in real
washing machines.

**Future Trends.** The development of new detergent compositions will also have an
important impact on the development of new proteases and other types of detergent
enzymes [4.92], [4.100]. In several countries, environmental considerations led to the
partial or complete substitution of phosphates by other sequestrants [4.92], [4.101].
As a result, detergent compositions were changed. The pH was increased to 10, and
enzymes with altered pH–activity profiles were required. An example is shown in
Figure 36 (p. 113).

The tendency to reduce the amount of energy required for washing, i.e., the
substitution of biochemical energy for thermal and mechanical energy, will affect the
use of enzymes in detergents [4.89], [4.92], [4.93]. Washing at lower temperatures
requires the development of new bleaching systems [4.102]. Traditional bleaching
systems like perborate are effective only above 60 °C. Combinations of perborate
with bleach activators such as tetraacetylethylenediamine or such new bleaching
compounds as magnesium monoperoxyphthalate, are active above 40–50 °C. In
washing without a heating-up phase, traditional enzymes may not be fully compati-
ble with the low-temperature bleaching systems. In the past, this was not a problem
since the detergents used for constant-temperature processes (United States) did not

contain a bleaching agent. At present, a strong tendency exists to incorporate a bleaching system into detergents for these washing processes. In that case, bleach-tolerant enzymes must be developed for specific markets.

The reduced activity of some of the nonbiological ingredients of the detergent is another consequence of low washing temperatures [4.92], [4.100]. In principle, enzymes other than proteases could compensate for the reduced efficacy of the detergent. For example, removal of fatty stains at low temperatures may be improved by addition of lipases, i.e., lipid-degrading enzymes.

A tendency can be noted toward the development of multifunctional detergents, i.e., one product with the characteristics of several – the ideal being a product that contains a presoak function, a wash function, builders, a bleaching system, and a softener. One compatibility problem inherent in this type of formulation is the presence of anionic and cationic surfactants in one detergent. However, the cationic surfactants, which have a softening function, may be replaced by an alkaline cellulase which also acts as a softener, especially on cotton fabrics [4.103]. In addition, the cellulase has soil-loosening and antiredeposition effects.

Another trend is the increasing appearance of liquid detergents on the market, especially in the United States, where liquids comprise 25 % of the laundry detergent market.

Because the relatively low pH of enzyme-based liquid detergents may impair the removal of starch stains, $\alpha$-amylases are already being used in liquid detergents in some markets [4.104]. The currently available bleaching systems are not stable in water–containing liquid detergents. A biological enzyme–containing bleaching system may represent an alternative, e.g., methanol oxidase with methanol as substrate [4.105].

Except for high-alkaline proteases and $\alpha$-amylases, the other enzymes mentioned are still being developed. Their commercial success will depend on both technical feasibility and economics of application. Protein engineering techniques are likely to become a powerful tool for the improvement of existing enzymes.

## 4.4.2. Enzymes in Leather Production

The processing of skins and hides for the production of leather is a traditional art. Several different proteases can be used in the individual stages of the manufacturing process [4.106]. The rationale behind the use of proteases for processing hides and skins lies in the fact that protein is the major building block of hair and skin. Hair is composed of $\alpha$-keratin — fibrous, insoluble protein molecules containing a large fraction of cysteine residues and having an $\alpha$-helix conformation. The $\alpha$-keratin is arranged in bundles of fibrils. Different skin layers are composed of collagen, $\alpha$-keratin (epidermis), and some elastin. In addition, albumine, globuline, glycoproteins, and other globular proteins are present. Collagen contains a large fraction of glycine, alanine, proline, and hydroxyproline. It is arranged in a triple-helix conformation [4.107]. The use of enzymes with different specificities has made possible selective hydrolysis of the noncollagenous constituents of the skin. The stages of leather manufacturing are (1) curing, (2) soaking, (3) dehairing and dewooling, (4) bating, and (5) tanning.

*Curing.* To preserve hides and skins before the actual processing, they are steeped in a brine bath and dried in the sun. Salt is added to the flesh side.

*Soaking.* At the tannery, hides and skins are washed and soaked in surfactants and antimicrobial compounds. Alkaline proteases are used to remove nonfibrillar proteins such as albumins and globulins. This is important for the next process step. Proteases used by the detergent industry are generally suitable for this purpose since they are relatively resistant to the increased pH of about 10 used in the soaking bath. Trypsin, a pancreatic protease, is also used. Small amounts of amylase and lipase activity in trypsin preparations are particularly advantageous for skins with a fatty flesh side [4.108].

*Unhairing and Dewooling.* Unhairing was traditionally done by soaking the skins for several days in a strongly alkaline bath (10% lime). Now the use of alkaline proteases more than halves the amount of lime required. As a result, the final quality of the leather is improved and significantly less wastewater is produced.

*Liming* is normally an 18-h process using 2% $Na_2S$ and 3% lime. Enzymatic soaking leads to a 24-h process in the lime yard with the main target of speeding up soaking. Dewooling is performed by applying a so-called dewooling paint to the flesh side of the skin and keeping the skin at 20–35 °C for 10–20 h before the wool is pulled. Dewooling paint is composed of hydrated lime, sodium chlorite, alkaline proteases, and water. An alternate method used for fine-wooled skins requires the application of a powder to the flesh side and maintaining the skins for 24 h at 25 °C. The powder is composed of sodium sulfate and chloride, ammonium sulfate and chloride, and neutral proteases produced by microorganisms. The soaking and dewooling steps are sometimes combined. Fungal lipases improve the removal of wool grease [4.108]

*Bating.* In the bating step, hides and skins are deswollen and treated with enzymes and chemicals to make them soft and supple, and to prepare them for tanning. Traditionally, animal feces were used (as an enzyme source) for bating. Today, trypsin and small amounts of alkaline and neutral proteases are used primarily [4.109]. Bating conditions such as pH, temperature, time, amount of enzyme, and composition of the enzyme preparation are determined by the type of leather to be produced.

*Tanning.* After treatment with an acid solution for deliming, skins and hides are colored by chrome tanning. Although enzymes are not directly involved in this step, the enzymatic treatments of earlier steps influence the quality of tanning [4.106]. In the near future, the application of enzyme systems will increase. The systems will be optimized for action on different types of skin, and enzyme systems other than the ones based on protease will be developed, e.g., for the removal of gum and fat [4.108]. Furthermore the technology of using enzyme preparations will be improved by speeding up the diffusion of the entire enzyme into the skin by the application of pressure.

## 4.4.3. Enzymatic Synthesis of Aspartame

The synthesis of aspartame, a low-calorie sweetener, has attracted much attention, especially since its approval by the FDA [4.110]. Aspartame is a dipeptide consisting of L-aspartic acid and the methyl ester of L-phenylalanine. Its sweet taste depends on the L-conformation of the two amino acids, the presence of the methyl ester, and the correct coupling of the amino acids [4.111]. Whereas α-aspartame is 150–200 times sweeter than sucrose, β-aspartame has a bitter taste.

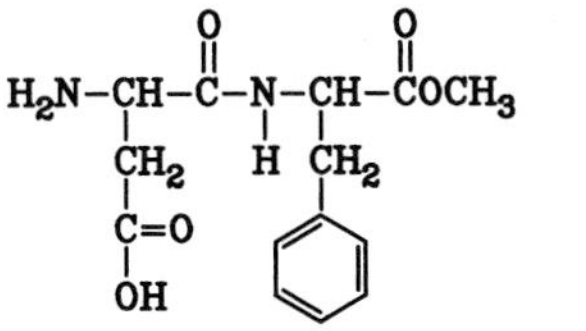

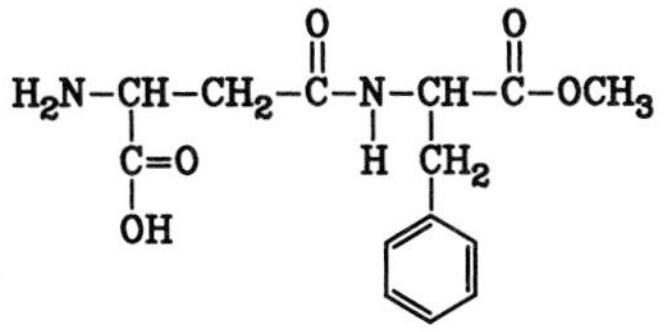

$$H_2N-CH-\overset{\overset{O}{\|}}{C}-N-CH-\overset{\overset{O}{\|}}{C}OCH_3$$

with $CH_2$, $H$, $CH_2$ ; $C=O$ ; $OH$ ; phenyl ring

$$H_2N-CH-CH_2-\overset{\overset{O}{\|}}{C}-N-CH-\overset{\overset{O}{\|}}{C}-OCH_3$$

with $C=O$, $OH$ ; $H$, $CH_2$ ; phenyl ring

α–Aspartame [22839–47–0]    β–Aspartame [22839–61–8]
sweet                                       bitter

Aspartame is currently produced by a chemical process in which the required stereospecificity adds to the production costs. These disadvantages can be overcome by an enzymatic approach [4.112]. The enzyme thermolysin (E.C. 3.4.24.4) [9060-59-1], a neutral metalloprotease from *Bacillus thermoproteolyticus,* is particularly suited for the synthesis of aspartame. Under appropriate conditions, it catalyzes the condensation of N-protected L-aspartic acid and D,L-phenylalanine methylester. Reactions occur exclusively at the α-carboxylic group of aspartic acid and with the L-isomer of phenylalanine methyl ester:

$$Z-NH-CH-\overset{\overset{O}{\|}}{C}-OH \;+\; H_2N-CH-\overset{\overset{O}{\|}}{C}-OCH_3$$

with $CH_2$, $C=O$, $OH$ ; $CH_2$, phenyl ring

L - Z - Asp            D,L - Phe - OCH₃

↓ Thermolysin

$$Z-NH-CH-\overset{\overset{O}{\|}}{C}-N-CH-\overset{\overset{O}{\|}}{C}-OCH_3 \;+\; H_2N-CH-\overset{\overset{O}{\|}}{C}-OCH_3$$

with $CH_2$, $C=O$, $OH$ ; $H$, $CH_2$, phenyl ring ; $CH_2$, phenyl ring

L - Z - Asp -L - Phe - OCH₃            D - Phe - OCH₃

↓ Hydrogenation            ↓ Racemization

L - Asp -L - Phe - OCH₃            D,L - Phe - OCH₃

α - Aspartame

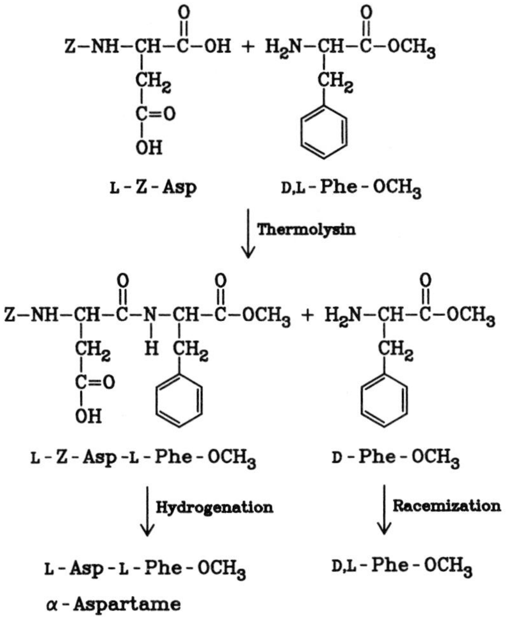

where Z = benzyloxycarbonyl

For economic reasons, the enzyme must be recycled, e.g., by precipitation, immobilization, or use of a membrane reactor. Therefore three different production methods be can distinguished [4.113]:

1) *Synthesis in an Aqueous System.* Accumulation of the reaction product in the resin of an immobilized enzyme or clogging of the reactor membrane would be disadvantageous.
2) *Synthesis in a Two-Phase System.* The reaction product is formed in the water phase and then transferred immediately to an organic phase.
3) *Synthesis in an Organic Solvent.* This process seems particularly suitable for continuous operation with an immobilized enzyme.

An industrial production unit based on the enzymatic synthesis of aspartame, developed by Toyo Soda (Japan) and DSM (The Netherlands), has recently started production.

### 4.4.4. Synthesis of Human Insulin from Hog Insulin

Human and hog insulin differ in only one amino acid, having threonine or alanine, respectively, at the C-terminal end of the B-chain. Diabetes mellitus patients, who have been treated with hog insulin for a long time, may produce antibodies against this type of insulin, caused either by impurities, e.g., proinsulin, or by the presence of the wrong amino acid in the B-chain.

A number of insulin manufacturers have developed enzymatic methods to convert hog insulin into human insulin [4.114]–[4.116]. Two enzymes are required: carboxypeptidase A for the removal of the wrong amino acid, alanine, from the C-terminal end of the B-chain and trypsine for coupling threonine to this chain. Carboxypeptidase A (E.C. 3.4.17.1) [*11075-17-5*] from bovine pancreas, is a protease that exoproteolytically degrades proteins starting at the C-terminus [4.116]. Hydrolysis of the C-terminal end of the A-chain may complicate the first reaction. In the next step, threonine is coupled to the C-terminal end of the B-chain by trypsin (E.C. 3.4.21.4) [*9002-07-7*] (from pancreas), an endoproteolytic enzyme. The carboxyl group of threonine is esterified with a *tert*-butyl group to protect it from unwanted side reactions [4.114].

In 1982, Novo (Denmark) brought human insulin made by enzymatic synthesis to the market. Two months later Eli Lilly (United States) followed with human insulin made by genetic engineering. In the long run, the process based on genetic engineering will turn out to be more profitable.

## 4.5. Meat Processing

Cooked meat is considered tender if it can be masticated easily and, at the same time, retain the desired texture. Tenderness is influenced by a number of factors which are not yet well understood. This applies not only to the biochemical reactions involved in rigor mortis but also to their termination. Indeed, numerous endogenous enzymes take part in this process including endogenous proteinases, particularly the *cathepsins*. These enzymes change muscle protein during maturation or aging of meat.

SCHWIMMER studied the mechanism of the maturation process in detail [4.117]. Natural maturing of carcasses or pieces of meat takes place in ca. 10 d at $1-2\,°C$. Slow maturation has the advantage of producing very tender meat but the disadvantages of moisture loss and shrinkage. Optimal meat maturation in special cold-storage depots ($1-2\,°C$, $83-86\%$ humidity) leads to a water loss of up to $7\%$ in $3-4$ weeks. Since 1940, attempts have been made to use exogenous enzyme preparations as meat tenderizers. Proteinases capable of digesting connective tissue and muscle protein have been chosen for this purpose.

Enzymes used on a commercial scale are papain (E.C. 3.4.22.2) [*9001-73-4*], bromelian (E.C. 3.4.22.5) [*9001-00-7*], and ficin (E.C. 3.4.22.3) [*9001-33-6*]. The main problem associated with enzyme use is their even distribution in the tissue. Factors influencing this distribution are diffusion, time, salt content, and enzyme concentration.

If preparations are sprinkled only on the surface of the meat (as recommended for kitchen use) or if pieces of meat are dipped into an enzyme solution, generally only the surface is tenderized and the interior remains tough. In the kitchen, after enzyme application (e.g., $2\%$ NaCl with $0.002\%$ bromelain), the meat is repeatedly poked to make it easier for the enzyme to penetrate. The main effect of the proteinase is exerted only during cooking.

Commercial methods can be divided into premortem and postmortem procedures. In postmortem treatment, a proteinase solution is spread into the carcass by repeated injections, possibly under pressure. In the premortem method, the Swift technique developed in 1960, a very pure, sterilized papain solution is injected intravenously $2-10$ min before the animal is slaughtered.

Still another method involves the injection of a papain solution that has been reversibly inactivated by oxidation. The enzyme is reactivated in the last stages of cooking by the liberation of mercapto groups.

Pancreatic proteinases are used in the maturation of fish. Bacterial proteinase is employed to dissolve bone meat or segments of meat.

# 4.6. Dairy Products

The use of biocatalysts in food chemistry, especially in making dairy products, is one of the oldest examples of biotechnology and goes back thousands of years. A typical case is the use of either isolated biocatalysts (rennet) or cell cultures (lactobacilli, streptococci, micrococci) in the production of cheese.

Biocatalysts are responsible for numerous reactions, such as the formation of aromatic compounds, changes in matrix and structure and variation of pH. These processes are generally referred to as "fermentation."

A number of fermented products are made available today by use of purely biological procedures; an example is the enormous variety of cheeses. Milk itself contains a series of enzymes, such as peroxidases, proteases, acid phosphatase (E.C. 3.1.3.2) [*9001-77-8*], and xanthine oxidase (E.C. 1.1.3.22), which are partly heat-stable and survive pasteurization [4.118]–[4.124].

Milk may also contain enzymes from contaminating microorganisms. For instance, psychotropic microorganisms release lipases (E.C. 3.1.1.3) and proteinases into the environment [4.125]–[4.129]. When biochemical conversions are carried out by adding cell cultures or isolated enzymes, the presence of these endogenous enzymes must be taken into consideration.

Table 28 lists a number of isolated enzymes used in the production of dairy products. Enzymes marked by an asterisk are also used immobilized on a supporting structure. Indeed, recent advances in this technology have dramatically increased the efficiency of dairy processing. A series of review articles and books deal with the application of biocatalysts in the production of food and dairy products [4.130]–[4.144].

## 4.6.1. Enzymes from Rennet and Rennet Substitutes

The classical example of the use of a biocatalyst is in cheese making, where rennet is added to gel the casein in milk. Today, other enzymes are employed in this step.

**Rennet Enzymes.** Rennet, a mixture of chymosin (E.C. 3.4.4.3) [*9001-98-3*], also called rennin, and pepsin (E.C. 3.4.4.1) [*9001-75-6*], is obtained from the gastric mucosa of young mammals, e.g., calves and lambs [4.145], [4.146]. The pepsin-to-chymosin ratio of different rennet preparations can vary considerably because chymosin is present only in the stomachs of unweaned mammals and is later replaced by pepsin. The content of pure chymosin depends on the age and species of the animal. Only chymosin is able to convert specifically casein from the sol to the gel state. Pepsin and other proteolytic enzymes are much less specific and give rise to a number of degradation products which tend to taste bitter. For this reason, pure chymosin and high-quality rennet are important.

Chymosin is a highly specific endoproteinase. It splits only $\varkappa$-casein into a glyco-macropeptide and para-$\varkappa$-casein by selectively cleaving the 105–106 bond between phenylalanine and methionine [4.130], [4.131], [4.147]–[4.174]. Optimal chymosin activity is measured between pH 5 and 6. The enzyme destroys the protective colloidal function of $\varkappa$-casein. As a result, the surface of the casein micelle becomes more hydrophobic, leading to gel formation. Therefore, the proteolytic activity and specificity of rennet preparations are of great importance, and both the chymosin and the pepsin contents of rennet preparations must be determined accurately; in addition, the proteolytic activity is measured by analyzing the casein cleavage products [4.175].

**Rennet Substitutes.** Because of the limited availability of pure chymosin, other proteolytic substitutes have been employed in cheese making, e.g., enzymes formed

**Table 28.** Enzymes used in the processing of dairy products

| Enzymes | Origin |
|---|---|
| **Rennet Enzymes** | |
| Chymosin* (rennin, calf rennet) (E.C. 3.4.4.3) [*9001-98-3*] | calf, lamb, kid |
| Pepsin* (E.C. 3.4.4.1) [*9001-75-6*] | calf, hog, chicken |
| **Rennet Substitutes** | *Mucor miehei, Mucor pusillus, Endothia parasitica,Bacillus subtilis* |
| **Other Rennet Substitutes** | *Mucor racemosus, Bacillus cereus, Bacillus mesenterius, Bacillus polymyxa, Bacillus licheniformis, Bacillus megaterium, Irpex lacteus, Absidia racemosa, Rhizopus* species, *Aspergillus oryzae, Aspergillus nidulans, Candida* species |
| **β-1,4-Galactosidases*** (E.C. 3.2.1.23) [*9031-11-2*] | *Kluyveromyces lactis, Kluyveromyces fragilis, Saccharomyces lactis, Aspergillus niger, Aspergillus oryzae, Bacillus stearothermophilus, Bacillus subtilis* |
| **Other Enzymes** | |
| Glucose isomerase (E.C. 5.3.1.18) [*9055-00-9*] | *Bacillus* species |
| Lysozymes (E.C. 3.2.1.17) [*9001-63-2*] | chicken egg, microbial |
| Peroxidase (E.C. 1.11.1.7) [*9003-99-0*] | bovine milk |
| Catalase* (E.C. 1.11.1.6) [*9001-05-2*] | bovine liver |
| Superoxide dismutase (E.C. 1.15.1.1) [*9054-89-1*] | bovine milk, erythrocytes |
| Thiol oxidase* (E.C. 1.8.3.2) [*9029-39-4*] | |
| Acid phosphatases* (E.C. 3.1.3.2) [*9001-77-8*] | potato |
| Alkaline phosphatases* (E.C. 3.1.3.1) [*9001-78-9*] | calf intestine |
| Lipases (E.C. 3.1.1.3) [*9001-62-1*] | pancreas, *Streptococcus* species, *Penicillium* species, *Aspergillus* species |
| Glucose oxidase* (E.C. 1.1.3.4) [*9001-37-0*] | |
| **Other Proteinases** | |
| Papain* (E.C. 3.4.22.2) [*9001-73-4*] | *Carica papaya* |
| Ficin* (E.C. 3.4.22.3) [*9001-33-6*] | *Ficus carica* |
| Trypsin* (E.C. 3.4.4.4) [*9002-07-7*] | bovine, hog |
| Chymotrypsin* (E.C. 3.4.4.5) [*9004-07-3*] | bovine, hog |

* Enzyme is also used in immobilized form.

by microorganisms, which are considered to be safe according to food laws. Examples are preparations from *Mucor miehei* and *M. pusillus* [4.176]–[4.184], *Bacillus subtilis* [4.185]–[4.192], and *Endothia parasitica* [4.193]–[4.197]. Apart from microbial enzymes, proteolytic enzymes from other sources have also been tested for their ability to coagulate milk, e.g., plant proteinases such as papain (E.C. 3.4.22.2) [*9001-73-4*], ficin (E. C. 3.4.22.3) [*9001-33-6*], and bromelain (E.C. 3.4.22.5)

[*9001-00-7*] [4.198]–[4.200]. Microbial enzymes are readily available and, therefore, gradually replacing chymosin in cheese making. However, the search for new substitutes continues. Techniques of genetic engineering have also been applied to the production of biologically active milk-coagulating enzymes. *Escherichia coli* and *Lactobacillus lactis* have been manipulated genetically and made to produce large amounts of chymosin in its zymogen form, prochymosin. Activation of the zymogen to give chymosin and the action of this enzyme on milk in cheese making are currently being studied [4.201]–[4.204]. The specificity of other endopeptidases has been examined by analyzing the peptide fragments formed from casein or by measuring the ratio of digested to undigested protein [4.205]. A further parameter is the bitterness of the casein peptides generated. The next step is to check the ability of the enzyme to coagulate milk. Time and temperature during gel formation are monitored, and gel stability is analyzed by use of rheological parameters. In addition, the loss of milk protein by nonspecific proteolytic degradation is determined, a factor of great importance for the yield of cheese.

Substantial changes in pH occur during the *ripening* of cheese. Hence, the pH dependence of the specificity of endoproteinases used in cheese making must also be tested. Large differences between nonspecific proteinases and highly specific chymosin have been observed during ripening — a factor which can give rise to undesirable cheese flavors.

Mixtures of chymosin and rennet or rennet substitutes are often used in order to reduce nonspecific proteolytic degradation and avoid the formation of bitter peptides. In addition, enzyme preparations (e.g., rennet) must be stabilized with preservatives such as propionic acid and sorbic acid; these preservatives are later found in cheese. This can be avoided only if enzyme preparations that have been sterilized by filtration or lyophilized are employed.

Numerous tests are carried out to control the ripening and quality of cheeses produced with rennet substitutes. In unripened cheese, the growth of microbial flora on the cheese is encouraged after the initial processing. The ripening process, i.e., the growth and metabolic activity of these microorganisms, in turn is largely dependent on the peptide spectrum formed during enzyme treatment.

**Analysis.** Determination of the individual enzyme components of rennet preparations is important. Both classical and modern biochemical techniques, such as electrophoresis, isoelectric focusing, and HPLC are used for this purpose [4.201]–[4.218]. Enzyme impurities in rennet can greatly affect the quality of the finished product. Indeed, special attention must be paid to *lipase content* because the formation of free fatty acids and their metabolism negatively influence the aroma of the product.

**Immobilization.** Many difficulties can be overcome by using immobilized chymosin or rennet substitutes [4.219]–[4.221]. Several methods of immobilizing rennet enzymes have been devised (for general immobilization techniques, see Section 3.3). Because hydrophobic interactions play an important part in the formation of casein gel, gel formation does not occur at low temperature (entropy effect). Hence, the cleavage of $\varkappa$-casein (renneting) can be separated from gel formation. Cleavage is first carried out in a reactor containing the immobilized enzyme at 4 °C. After leaving the reactor, the rennet-treated milk is slowly heated so that gel formation can take place and casein can be separated in a continuous process. However, in practice, the enzyme reactor has been found to have a relatively short life span because milk fat and proteins block the activity of the immobilized enzymes. A further problem with the enzyme reactor is the need to maintain aseptic conditions. All precautions must be taken because milk is an excellent substrate for contaminating microorganisms which can destroy the product [4.222]–[4.226].

**Recycling.** Another way to reduce rennet loss is to recover the enzymes from whey and return them to the cycle. The major portion of rennet enzymes is found in whey and can be isolated, e.g., by bioaffinity chromatography with specific, immobilized inhibitors [4.227]–[4.231]. Despite promising scientific results, this method has not yet been accepted in practice [4.232], [4.233].

## 4.6.2. β-1,4-Galactosidases

The use of β-1,4-galactosidases (E.C. 3.2.1.23) [9031-11-2] in the cleavage of lactose is another important application of biocatalysts. β-1,4-Galactosidases are extensively distributed in nature; they hydrolyze lactose to glucose and galactose [4.134], [4.136], [4.234], [4.235]. This improves the solubility and increases the sweetness of the product.

Biological sources of β-galactosidases are summarized in Table 28. Optimal pH values for these enzymes vary considerably. Enzymes from *Bacillus saccharomyces* or *B. kluyveromyces* (neutral pH optima) and enzymes from different strains of *Aspergillus* (acidic pH optima) are used most often. Only microorganisms that are regarded as absolutely safe by food law may be employed. In addition, the enzymes must be used in a highly pure form, free of all impurities, particularly proteinases.

The raw material for lactose is either whey having a pH of 4–6 or milk with a pH of 6.3–6.8. The optimal pH for enzyme activity must correspond to the pH of the substrate. Yeast and certain strains of *Aspergillus* and *Bacillus* produce β-galactosidases with the required properties (Table 29). The β-galactosidase from *Escherichia coli*, which possesses excellent properties, has not been applied because it is not regarded as absolutely safe by the food law [4.236]–[4.238].

**Table 29.** Properties of some β-1,4-galactosidases

| Source | Optimal pH | Optimal temperature, °C |
|---|---|---|
| *Aspergillus niger* | 3.0–4.0 | 55 |
| *Aspergillus oryzae* | 4.8 | 46 |
| *Escherichia coli* | 6.9–7.5 | 45 |
| *Saccharomyces fragilis* | 6.5 | 50 |

*Soluble Enyzmes.* The cleavage of lactose in milk and whey was first accomplished with soluble enzymes. This process has been developed in the United States to such an extent that small amounts of the enyzme are available on the market. The consumer can add the enzyme to milk in the evening, and after 12 h in the refrigerator, at least 80% of the lactose is hydrolyzed. Soluble enzymes have also been used in the preparation of a variety of dairy products such as cheese and yogurt. A disadvantage of soluble products is that the enzyme remains in the product and hence can only be used once [4.239]–[4.242].

*Immobilized Enyzmes.* Loss is reduced when the enzyme can be kept inside the reactor. Hence, efforts have been made to immobilize the enzyme, thus permitting its long-term use [4.243]–[4.247]. Each $\beta$-1,4-galactosidase requires its own specific carrier and binding system because the surface structures of the enzymes differ considerably. Almost all natural and synthetic polymeric carriers have been tested for their ability to hold this enzyme. Synthetic carriers, e.g., polyacrylamide-based carriers with epoxide groups, are suitable. The polymer matrix can be formulated so that optimal technological parameters, e.g., flow rates, are ensured. Other carriers, such as phenol–formaldehyde resins, silicon dioxide, titanium dioxide, or cellulose acetate fibers, have also proved useful [4.248]–[4.260]. Several systems are available today for the hydrolysis of lactose in a continuous process. A potential problem may arise in highly concentrated lactose solutions, where $\beta$-1,4-galactosidases can transfer galactose residues to lactose to form polysaccharides [4.261], [4.262].

The hydrolysis of lactose can be monitored in many ways. First, the amounts of glucose, galactose, and lactose can be determined by enzymatic analysis [4.263]–[4.266] (see Section 5.3.1). Second, property changes in the product (e.g., freezing point) can be determined [4.267], [4.268].

The glucose generated from lactose may be further converted into fructose by using immobilized glucose isomerase (see Section 4.3) [4.269]–[4.271]. This conversion can be carried out in one reactor containing several segments or in a two-step process. The presence of fructose further increases the sweetness of the product.

## 4.6.3. Sulfhydryl Oxidase

The characteristic taste of heated milk is attributed to mercapto (sulfhydryl) compounds. These compounds are released on heating, when the sulfhydryl oxidase contained in milk is denatured simultaneously. According to SWAISGOOD, sulfhydryl oxidase can oxidize mercapto compounds and banish the unpleasant taste associated with heated milk [4.272], [4.273]. Immobilized sulfhydryl oxidase has also been introduced, and the use of this enzyme may well improve the quality of heated milk.

## 4.6.4. Lysozymes

The enzyme lysozyme [E.C. 3.2.1.17] [9001-63-2] is a component of perishable foods such as eggs. It protects the food from infection by favoring bacteriolysis. Attempts have been made to replace the use of nitrate in cheese making by adding lysozyme from egg white; the enzyme suppresses the growth of *Clostridium thyro-butyricum* [4.274]–[4.280].

The general application of lysozyme or other bacteriolytic enzymes offers a valuable, yet still expensive, alternative to the undesirable use of nitrates. Japanese researchers have also tested microbial enzymes for this purpose [4.281]. Economic problems can be overcome if attempts currently being made to produce certain lytic enzymes by genetic engineering are successful [4.282].

## 4.6.5. Production of Aroma and Texture

Enzymes and enzyme complexes are employed to improve the aroma and texture of dairy products. The nature of the microorganisms used for fermentation is one of the factors contributing to the enormous variety of cheeses, which differ from one another in aroma and texture. Ripening can take a very long time, and attempts have been made to accelerate it by adding enzymes or enzyme complexes taken from the product-specific ripening culture [4.283]–[4.289].

For this purpose, commercially available lipases and proteinases have been used. The metabolism of fat, protein, and lactose generates aroma and texture; therefore, enzyme complexes from product-specific cultures were then tested on suspensions of the individual raw materials, and the formation of aroma compounds was found to be greatly accelerated. The use of enzyme-containing liposomes enabled LAW to control the generation of these compounds [4.290].

## 4.6.6. Membrane Cleansing

The advent of membrane separation techniques in the processing of dairy products was closely followed by the use of proteolytic enzymes to break down proteins adhering to the membrane. Either soluble enzymes or enzymes immobilized on the membrane can be employed for this purpose. These enzymes do not find their way into the product, and proteolytic enzymes ranging from plant endoproteinases to microbial biocatalysts can be used [4.291]–[4.294].

## 4.6.7. Phosphatases

Phosphatases can be employed in processing caseins, which are phosphoproteins. The hydrolysis of phosphoserine residues in casein can be accomplished with both acid and alkaline phosphatases, and has the advantage of reducing the phosphate content of food [4.295].

## 4.6.8. Catalase

Milk can be sterilized by treatment with $H_2O_2$ if this is permitted by food law or under special circumstances. In this instance, excess peroxide can be removed with catalase (E.C. 1.11.1.6) [*9001-05-2*] either in soluble or in immobilized form [4.296], [4.297].

# 4.7. Processing of Fruit, Vegetables, and Wine

[4.298], [4.299]

In 1930, Z. J. KERTESZ and A. MEHLIZ observed that pectinases could be used in making fruit juices. Since then, tremendous technological advances have been made in the fruit juice industry. Apart from clear and cloudy fruit and vegetable juices, the main industrial products today are fruit juice concentrates. This applies not only to pomaceous and stone fruit, but also to berries, citrus fruit, and tropical fruit.

Enzymes have been exploited for the following purposes:

1) to improve the yield of juice by enabling better pressing of the fruit;
2) to liquefy the entire fruit for maximal utilization of raw materials;
3) to improve the yield of substances contained in fruit, e.g., acids, and coloring or aroma substances;
4) to clarify juices in order to improve imperishability; and
5) to break down all polymeric carbohydrates such as pectins, hemicelluloses, and starches.

The results obtained from the utilization of individual enzymes are listed in Table 30.

**Clarification of Juice.** The objective of fruit juice clarification is to remove fragments of plant cells and insoluble particles of skin and seeds. Mixtures of pectinases — pectin transeliminase (PTE), polygalacturonase (PG), and pectin methylesterase (PE) — present in industrial enzyme preparations break down pectin contained in these particles. Soluble starch is hydrolyzed by fungal $\alpha$-amylase or by amyloglucosi-

**Table 30.** Use of pectinases in fruit and vegetable processing

| Process | Enzyme* | Objective |
|---|---|---|
| Clarification | pectinases (PTE, PG, PE) | clear juices, improved filtration |
| Depectinization | pectinases, amylases | juice concentrates, improved stability |
| Precipitation of pectin | pectinesterase (PE) | cider production |
| Digestion of mash | pectinases, cellulases, hemicellulases | improved juice yield and pressing ability |
| Maceration | pectin glycosidase (PG) | stable vegetable pulp |
| Liquefaction | pectinases, cellulases, hemicellulases, amylases | entire fruit, pulp concentrates, dry products |

* PTE = pectin transeliminase; PG = polygalacturonase; PE = pectin methylesterase.

dase to oligosaccharides. Acid-stable pectinases (PG and PE) are employed to clarify citrus juice at pH 2.0–2.4. After enzyme treatment, flocculation of the particles is accelerated by adding gelatin, silica sol, or bentonite.

**Precipitation of Pectin.** The removal of pectins can also be achieved by de-esterification followed by flocculation of the resulting pectic acid with calcium ions. Preparations of pectin-esterase which are free from depolymerases are used for this purpose.

**Production of Concentrates.** Today, fruit juice concentrates are important commercially because they have a longer shelf life, as well as lower storage and transportation costs, than juices. After dearomatization, complete removal of pectin with simultaneous break-down of starch is accomplished by treatment with pectinases. The hot clarification method at 55–60 °C is used for this treatment, after which the fruit juice is filtered and concentrated.

**Digestion of Mash.** Soluble and insoluble pectins are major constituents of all berries. After extraction, a considerable part of the juice may remain behind in the marc of the pressed berries because of gelation. Limited pectin degradation (decrease in viscosity and cell breakdown) is caused by enzymatic treatment of the fruit mash. This method is also applied to the processing of storable fruit, particularly apples. In addition to pectins, other polymers can also be dissolved in this process, e.g., araban, which can then flocculate in the concentrate and make it cloudy. Industrial pectinase preparations containing standardized amounts of arabanase can be used for the degradation of these products.

**Maceration and Liquefaction.** In earlier times, cloudy juices were produced by mechanical processing which included thermal treatment. The addition of special "macerating" enzymes such as polygalacturonase (E.C. 3.2.1.15) [*9032-75-1*] serves to dissolve plant tissue, forming a cell suspension. However, enzyme preparations must be free from "clarifying" enzymes (pectinases) in order to obtain a stable, cloudy product. If the viscous, macerated suspension is subsequently treated with cellulases, cell wall lysis gives rise to almost complete degradation of soluble and insoluble carbohydrates. The ensuing products have a low viscosity and can be concentrated easily.

**Production of Wine.** The application of enzymes to wine production serves to reduce costs and improve quality. The use of enyzmes in various steps of wine production is shown in Figure 39.

*Mash.* Addition of pectinases improves the ability of grapes to be crushed. This applies above all to slimy grapes, unripe fruit, or fruit infected by *Botrytis*. At the same time, a higher quality juice can be expressed. In the production of rosé, the desired color is attained shortly after the mash is subjected to pectinases.

*Must.* The removal of slime and coarse particles from must is accelerated and improved by enzymatic treatment. In addition, treatment with β-glucanases breaks

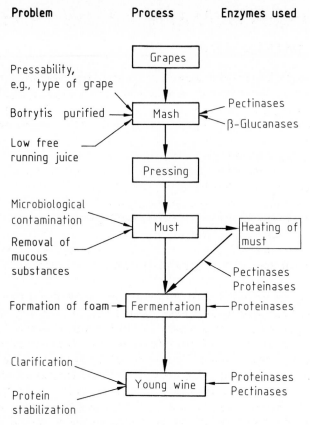

**Figure 39.** Enzymes in the production of wine

down polymeric β-glucans after *Botrytis* infection. If the must is then exposed to proteinases, clouding of the wine at a later stage is prevented.

*Red Wine.* In the event of inadequate color release, thermal wine-making procedures are employed in combination with enzymatic treatment. The advantages are a better extraction of dyes and other components, and both the expression of juice and subsequent clarification are improved. In the production of wine by classical mash fermentation, cloudiness can be clarified by the addition of enzyme combinations – pectinases, hemicellulases, and proteinases – instead of making use of "natural" clarification which often takes several months and requires sulfur dioxide.

# 4.8. Hydrolysis of Protein, Fat, and Cellulose, and Inversion of Sucrose

**Hydrolysis of Protein.** Protein hydrolysis is used to improve the quality of soluble peptide and amino acid mixtures in the production of food and animal feed. Examples of the industrial uses of protein hydrolysis are given in Table 31. In addition, several techniques are available for improving the functional properties of soybean proteins. Both the foaming and the emulsifying power can be increased by treatment with alkaline bacterial proteinases.

In many instances, the bitter-tasting peptides formed during hydrolysis must be subsequently removed, e.g., by charcoal absorption.

Special pancreatic proteinases can be used to digest soybean protein. In fact, the degree of hydrolysis can reach 15%, and the solubility at pH 5.0 ca. 75%, without the occurrence of bitterness. Other applications of protein hydrolysis include the stripping of silver from photographic film (gelatin dispersion) and the use of proteases in the beamhouse method of leather processing.

**Hydrolysis of Fat** [4.300]. Fat hydrolysis, catalyzed by lipases from animals or microorganisms, has several industrial applications: (1) the total hydrolysis of fats and oils, replacing the classical process, (2) the use of alkali-stable microbial lipases in detergents, (3) the use of specific lipases or esterases in cheese ripening and for the production of enzyme-modified cheeses (EMC), (4) in digestive aids. The use of lipases includes the transesterification and the interesterification of fats and oils to produce fats with new properties. Lipases can be used as catalysts for interesterification and transesterification reactions, because the lipase reaction is reversible. In this way, resynthesis of glycerides occurs when lipases are incubated with fats, fatty

**Table 31.** Enzymes in the hydrolysis of proteins

| Substrate | Enzyme | Degree of hydrolysis, % | Application |
|-----------|--------|-------------------------|-------------|
| Fish waste | papain | 10–15 | extraction of fish oil, feeds |
| Fish | pancreatic proteinase | ca. 10 | non-bitter-tasting hydrolysates, dietetics |
| Bone meat | neutral bacterial protease | 2–5 | soups |
| Whey protein | neutral bacterial protease, pancreatic proteinase | 2–8 | improvement of foaming power |
| Casein | pancreatic proteinase | 2–2.5 | soluble, nonbitter casein peptides, improved foam stability |
| Casein | pancreatic proteinase | 37 | medical infusion solutions, microbiological nutrient media |
| Gelatin | pancreatic proteinase, neutral bacterial proteinase | up to 25 | cosmetics |

acids, and alcohols. When the concentration of water in the reaction mixture is decreased, interesterification dominates.

**Hydrolysis of Cellulose.** Efforts to adapt the enzymatic hydrolysis of cellulose and cellulose-containing raw materials to large-scale industrial operation have so far been unsuccessful. A number of attempts have been made, including pilot runs, with hemicellulase preparations, lignases, and cellulases from *Trichoderma*, other fungi, and bacteria. Economical production of monomeric glucose or fermentation to ethanol has not yet been accomplished. Recently, cellulases are used in detergents.

**Inversion of Sucrose.** The inversion products of sucrose — glucose and fructose — compete today with isomerose obtained from starch (see Section 4.2). Yeast invertase (E.C. 3.2.1.26) [*9001-57-4*] has a wide pH optimum (3.5–5.5). If a concentrated sucrose solution is used as substrate, the reaction can be carried out at 65–70 °C. However, at lower sugar concentration, invertase is inactivated at approximately 60 °C.

Matrix-bound invertase has been recommended for the large-scale inversion of sugar. Optimal reaction conditions are, e.g., a 50% sucrose solution, a reactor containing 1 L of matrix-bound invertase, a temperature of 40 °C, and a flow rate of 6 L/h, which results in an inversion of 80%. The productivity is approximately 10 000 kg of 80% invert sugar per kilogram of matrix-bound invertase.

# 4.9. Amino Acids and Hydroxycarboxylic Acids

Biotechnological methods are currently the means of choice for the synthesis of amino acids [4.301], [4.302]. Two grades of proteinogenic amino acids are prepared, namely, animal-feed grade and pharmaceutical grade. For large-scale production of these amino acids, two competing methods are available: (1) fermentation and (2) enzymatic catalysis. Enzymatic processes, which require either cell-free enzymes in a dissolved or immobilized form or cell-bound enzymes, can attain their full utility when inexpensive starting materials are available and when the enzymatic step is carried out with high selectivity in an effective bioreactor. The chief starting materials are racemic amino acid derivatives. However, in recent years, prochiral and chiral precursors have been employed in the commercial production of amino acids.

In comparison, enzymatic processes are of little interest for the production of α-hydroxycarboxylic acids. Amino acid precursors, e.g., α-oxo acids, are sometimes used as substrates.

## 4.9.1. L-Amino Acids from Racemic Precursors

Racemic amino acid derivatives can be converted by hydrolytic enzymes to pure L- or D-amino acids. The D-isomers are usually racemized and returned to the reactor. When racemization occurs as part of the reaction, the racemic mixture is almost completely converted to the desired L-amino acid. L-Methionine and L-valine are made commercially by using the hydrolytic method. In principle, L-alanine, L-phenylalanine, L-tryptophan, and L-cysteine can also be prepared in this way.

**L-Amino Acids from *N*-Acetyl-D,L-amino Acids.** L-Methionine [*63-68-3*] and L-valine [*72-18-4*], required in a highly purified form for diet mixtures and infusion solutions, can be produced efficiently on an industrial scale from *N*-acetyl-D,L-amino acids by enzymatic resolution.

R = (CH$_3$)$_2$CH: valine;  R = CH$_3$S–CH$_2$–CH$_2$: methionine

The enzyme aminoacylase (E.C. 3.5.1.14) [*9012-37-7*] from *Aspergillus oryzae* is employed commercially. The method devised by CHIBATA is based on the use of immobilized acylase in packed columns [4.303]; it has been modified in Germany by replacement of the column with a membrane reactor. The enzymatic reaction is now carried out in a continuous process, the enzyme being kept in the reactor by means of ultrafilter membranes [4.304]. The process is economical because of efficient utilization of the enzyme's activity, efficient racemization, and recycling of undesired *N*-acetyl-D-amino acid. L-Alanine [*56-41-7*], L-phenylalanine [*63-91-2*], and L-tryptophan [*73-22-3*] can also be produced by this method.

**L-Amino Acids from Esters of D,L-Amino Acids.** The optical isomers of amino acids can be produced by treating a racemic mixture of the amino acid ester with either amino acid esterase (E.C. 3.1.1.43) or proteolytic enzymes having esterase activity. The stereospecific hydrolysis of *N*-acetyl-D,L-amino acid esters [4.305] is accomplished by using subtilisin (E.C. 3.4.21.14) [*9014-01-1*].

Despite a series of improvements, e.g., the use of $\alpha$-chymotrypsin (E.C. 3.4.21.1) [*9004-07-3*] in the production of aromatic L-amino acids [4.306], the esterase technique has made little progress in displacing other commercial methods.

**L-Amino Acids from Amides of D,L-Amino Acids.** The resolution of amino acids can also be achieved by catalytic hydrolysis of D,L-amino acid amides with amidases (E.C. 3.5.1.4) [*9012-56-0*] [4.307].

D,L-Amino acid amide → (H₂O, Amidase) → L-Amino acid + NH₃

+ D-Amino acid amide → (Hydrolysis) → D-Amino acid + NH₃

| R | Product |
|---|---------|
| $(CH_3)_2CH$ | valine [*72-18-4*] |
| ⟨benzene⟩–$CH_2$ | phenylalanine [*63-91-2*] |
| ⟨indole⟩$CH_2$ | tryptophan [*73-22-3*] |

This enzymatic reaction proceeds in the presence of either whole cells (e.g., of *Pseudomonas putida*) or raw enzyme extracts in soluble or immobilized form. The method is applicable to the production of both L- and D-amino acids [4.308]. *N*-Acetyl-L-methionine has recently been produced from *N*-acetyl-D,L-methionine amide by immobilized cells of *Erwinia carotovora* in a continuous process [4.309].

**L-Lysine from D,L-$\alpha$-Amino-$\varepsilon$-caprolactam.** An enzymatic method has been developed for industrial synthesis of L-lysine [*56-87-1*], an important animal-feed additive [4.310]. This technique is based on a combination of two enzymatic reactions: (1) selective, hydrolytic ring cleavage of D,L-$\alpha$-amino-$\varepsilon$-caprolactam [*17929-90-7*] (ACL) with $\alpha$-amino-$\varepsilon$-caprolactamase (ACL hydrolase) from *Cryptococcus laurentii* and (2) racemization of D-ACL with a specific ACL racemase found in *Achromobacter obae*.

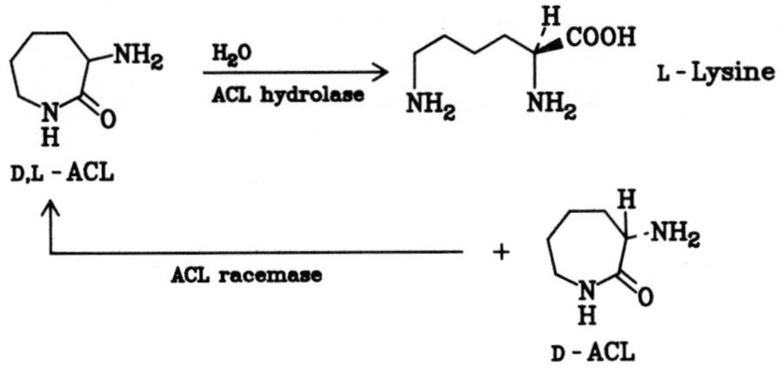

$$\text{D,L - ACL} \xrightarrow[\text{ACL hydrolase}]{\text{H}_2\text{O}} \text{L - Lysine}$$

ACL racemase

D - ACL

A 10% solution of D,L-ACL can be converted to L-lysine in a batchwise process [4.311] by using *C. laurentii* and *A. obae* cells. An almost quantitative conversion takes place in the vessel during 25 h. This method was employed in Japan to produce more than 4000 t of L-lysine hydrochloride [*657-27-2*] annually. The plant has been shut down recently because the process was no longer competitive with the older fermentation procedure.

**L-Cysteine from D,L-2-Amino-2-thiazoline-4-carboxylic Acid.** An enzymatic procedure has been developed for the synthesis of L-cysteine [*52-90-4*], based on a D,L-cysteine derivative D,L-2-amino-2-thiazoline-4-carboxylic acid [*2150-55-2*] (ATC) [4.312]. The enzymes required are found in strains of *Pseudomonas*, e.g., *P. thiazolinophilum*. Ring opening is catalyzed by a specific L-ATC hydrolase, and the intermediate formed is converted to L-cysteine by the action of S-carbamoyl-L-cysteine hydrolase. Unreacted D-ATC is racemized by ATC racemase and returned to the reaction.

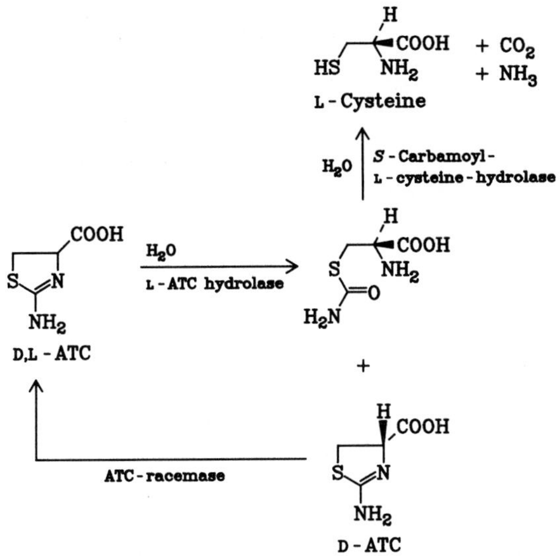

L - Cysteine $+ CO_2 + NH_3$

$H_2O$ | S - Carbamoyl-L - cysteine - hydrolase

$\text{D,L - ATC} \xrightarrow[\text{L - ATC hydrolase}]{\text{H}_2\text{O}}$

ATC - racemase

D - ATC

A molar yield of 95% and a product concentration of 31.4 g/L of L-cysteine has been attained with a mutant of *P. thiazolinophilum*, which contains no cysteine-degrading enzymes [4.313].

**L-Amino Acids and D-Amino Acids from D,L-Hydantoins.** Hydrolysis of 5-substituted hydantoins to *N*-carbamoyl-D-amino acids is catalyzed by microorganisms containing hydantoinase, e.g., *Pseudomonas striata* or alkalophilic *Bacillus* species [4.314]. This reaction is used commercially in the production of D-phenylglycine [*875-74-1*] or D-*p*-hydroxyphenylglycine [*22818-40-2*]. In the last step, the *N*-carbamoyl-D-amino acid is hydrolyzed chemically. Bacterial strains have been found in which both D-hydantoinase, usually identical with dihydropyrimidinase (E.C. 3.5.2.2) [*9030-74-4*], and a D-carbamoylase are present. These strains convert hydantoins directly to the corresponding D-amino acids [4.315]. Examples of such bacteria are *Achromobacter liquifaciens* and a series of *Pseudomonas* species. L-Hydantoinase, on the other hand, is relatively rare. Nevertheless, methods have been devised for the production of aromatic L-amino acids (L-phenylalanine [*63-91-2*], L-tryptophan [*73-22-3*]) by using *Flavobacterium aminogenes* [4.316], various *Bacillus* or *Pseudomonas* species [4.317], and *Arthrobacter* species [4.318] as a source of L-specific hydantoinases and carbamoylases.

A benefit of the hydantoinase-catalyzed reaction is reducing substrate loss; simultaneous racemization ensures that all the hydantoin is converted to either the D- or the L-amino acid.

A nonstereospecific hydantoinase has recently been discovered in *Arthrobacter* species (BH 20) which could be suitable for the synthesis of L-amino acids [4.319].

**L-Amino acids from D,L-α-Hydroxycarboxylic Acids.** The enzymatic conversion of α-hydroxycarboxylic acids to L-amino acids is dependent on a cofactor. This reaction can be carried out in a continuous membrane reactor. It has been applied, e.g., to the conversion of L-lactic acid to pyruvate, which takes place in the presence of L-lactate dehydrogenase (*Lactobacillus* species) and NAD. Pyruvate is then reductively aminated with alanine dehydrogenase (E.C. 1.4.1.1) [*9029-06-5*] from *Bacillus cereus*, to yield L-alanine [4.320]. Simultaneous application of L- and D-specific lactate dehydrogenases results in the conversion of racemic lactic acid to L-alanine [4.321].

$$R-\underset{OH}{\underset{|}{CH}}-COOH \xrightarrow[NAD]{E1/E2} \left[ R-\underset{O}{\overset{\|}{C}}-COOH \right] \xrightarrow[NADH]{E3} R-\underset{NH_2}{\overset{H}{\underset{|}{C}}}-COOH$$

| R | E1/E2/E3* | Product |
|---|---|---|
| CH$_3$ | L-lactate DH (E1) | L-Ala |
| | D-lactate DH (E2) | [56-41-7] |
| | L-alanine DH (E3) | |
| (CH$_3$)$_2$CH–CH$_2$ | L-hydroxy-isocaproate DH (E1) | L-Leu [61-90-5] |
| (CH$_3$)$_2$CH | D-hydroxy-isocaproate DH (E2) | L-Val [72-18-4] |
| CH$_3$–S–CH$_2$–CH$_2$ | L-leucine DH (E3) | L-Met [63-68-3] |
| C$_6$H$_5$CH$_2$ | L-lactate DH (E1) | L-Phe [63-91-2] |
| | D-lactate DH (E2) | |
| | L-phenylalanine DH (E3) | |

*DH = dehydrogenase

**Ala = Alanine; Leu = Leucine; Val = Valine;
Met = Methionine; Phe = Phenylalanine

The isolation of L-hydroxyisocaproate dehydrogenase (L-HIC DH) from *Lactobacillus confusus* and of the D-specific enzyme (D-HIC DH) from *Bacillus casei* or *Leuconostoc oenos* [4.322] has led to new possibilities. L-Leucine and other branched-chain L-amino acids, as well as L-methionine, can now be made from the corresponding synthetic hydroxy analogues by using L-leucine dehydrogenase (E.C. 1.4.1.9) [9082-71-7] in combination with one of the enzymes mentioned above. In the same manner, L-phenylalanine can be made from D,L-phenyllactic acid, with L-phenylalanine dehydrogenase [4.323] serving as the catalyst for the second step. This step requires the presence of poly(ethylene glycol)-bound NADH [4.324], which is regenerated in the first step of the synthesis.

## 4.9.2 L-Amino Acids from Prochiral Precursors

An advantage of the enzymatic asymmetric synthesis of L-amino acids for prochiral precursors is direct formation of the desired enantiomeric form, free of the D-isomer. Apart from ammonia-lyases, the enzymes that catalyze this transformation require cofactors e.g., dehydrogenases (NAD$^+$) or transaminases (pyridoxal 5-phosphate).

**L-Amino Acids from α,β-Unsaturated Compounds.** The enzyme-catalyzed addition of ammonia to fumaric acid is employed commercially in the production of aspartic acid [4.325]. This amino acid is a structural unit of the dipeptide sweetener aspartame (methyl ester of L-aspartyl-L-phenylalanine). In this method, *Escherichia* *coli* cells containing the enzyme L-aspartate ammonia-lyase (E.C. 4.3.1.1) [*9027-30-9*], also known as aspartase, are immobilized inside a fixed-bed reactor in order to limit catalyst loss.

The second structural unit of aspartame, L-phenylalanine, was produced enzymatically by the addition of ammonia to *trans*-cinnamic acid [4.326]. The enzyme L-phenylalanine ammonia-lyase (E.C. 4.3.1.5) [*9024-28-6*] is found primarily in yeast, e.g., *Rhodotorula glutinis* or *R. rubra*.

| R | Enzyme | Product |
|---|---|---|
| HOOC | aspartase | L - aspartic acid [*56 - 84 - 8*] |
| ⬡ | L - phenylalanine ammonialyase (PAL) | L - phenylalanine [*63 - 91 - 2*] |

This method has been used on an industrial scale [4.327], but it is not competetive with the fermentative production of L-phenylalanine.

**L-Amino Acids from α-Oxo Acids.** Specific amino acid dehydrogenases catalyze the stereospecific reductive amination of α-oxo acids to L-amino acids in the presence of the coenzyme NADH.

| R | Enzyme | Product |
|---|---|---|
| $CH_3$ | L - alanine DH | L - alanine [*56 - 41 - 7*] |
| $(CH_3)_2CH-CH_2$ | L - leucine DH | L - leucine [*61 - 90 - 5*] |
| $(CH_3)_2CH$ | L - leucine DH | L - valine [*72 - 18 - 4*] |
| $(CH_3)_3C$ | L - leucine DH | L - *tert*-leucine |
| $C_6H_5-CH_2$ | L - phenylalanine DH | L - phenylalanine [*63 - 91 - 2*] |

This method requires not only coenzyme and enzyme, but also a technology for handling the reaction. The coenzyme is provided by NADH bound to poly(ethylene glycol) ($M_r$ 20 000) [4.324], which can be regenerated efficiently by using formate dehydrogenase (E.C. 1.2.1.43) [*51377-43-6*] from *Candida boidinii*. Appropriate dehydrogenases are generally available, and the technology involved is a continuously operated membrane reactor [4.328]. This method has been applied to the production of L-alanine from pyruvate, catalyzed by L-alanine dehydrogenase from *Bacillus cereus*. The action of L-leucine dehydrogenase from *B. cereus* and *B. sphaericus* is not limited to α-oxoisocaproate; it also catalyzes the reductive amination of the keto analogues of valine, isoleucine, and methionine. Indeed, the synthesis of an amino acid which is not proteinogenic, L-*tert-* leucine [*20859-02-3*] from trimethylpyruvic acid, has also been achieved by this method.

L-Phenylalanine dehydrogenase from *Brevibacterium* species or *Rhodococcus* species [4.323] can be used to synthesize L-phenylalanine from phenylpyruvate and ammonia [4.323]. Transamination, the process of amino group transfer which occurs in the cell, has been applied to the production of amino acids. *Escherichia coli* and coryneform bacterial cells contain transaminases specific for catalyzing the transfer of an amino group from glutamic acid to α-oxo acids which correspond to branched-chain, aliphatic amino acids (L-leucine, L-valine) or to aromatic amino acids (L-phenylalanine) [4.329]. *Corynebacterium glutamicum* ATCC 13032 is rich in both transaminase B and glutamate dehydrogenase, which catalyzes the NADPH-dependent regeneration of glutamate from oxoglutarate and ammonia, and produces L-leucine in a concentration of 25 g per liter of culture medium [4.330].

Cells + CO₂   E1: glutamate dehydrogenase (GDH)
              E2: transaminase B (TRB)

Transamination with L-glutamic or L-aspartic acid to convert phenylpyruvate into L-phenylalanine has been studied intensively. Techniques have been developed that utilize isolated transaminase either in soluble form [4.332] or immobilized [4.331]; whole cells, e.g., a cell suspension of *Pseudomonas pseudoalcaligenes* [4.333] or *E. coli* [4.334]; and immobilized cells [4.335].

Transamination of L-aspartic acid, catalyzed by immobilized *E. coli* cells, has been developed recently [4.336]. Up to 30 g/L of L-phenylalanine have been recovered from this process in a molar yield of 98%. However, this amino acid can be made more efficiently by fermentation than by enzymatic synthesis.

In a variation of this enzymatic method, L-phenylalanine is produced from acetamidocinnamic acid (ACA) [5469-45-4] by strains of *Alcaligenes faecalis* or *Bacillus sphaericus* in two stages: formation of phenylpyruvate by hydrolysis (ACA acylase) of the substrate and subsequent transamination [4.337].

In fact, an ACA acylase has been found in strains of *Rhodococcus* and *Brevibacterium* [4.338] which, together with phenylalanine dehydrogenase, can be used in the commercial production of L-phenylalanine.

## 4.9.3. L-Amino Acids from Chiral Precursors

**L-Alanine from L-Aspartic Acid.** An L-aspartate 4-carboxylyase (aspartate β-decarboxylase; E.C. 4.1.1.12), a pyridoxal 5-phosphate enzyme from microorganisms, catalyzes the β-decarboxylation of L-aspartic acid to form L-alanine [56-41-7] [4.339].

Since 1965, L-alanine has been produced on a large scale by whole cells of *Pseudomonas dacunhae*; product concentrations of 400 g/L of L-alanine are obtained.

**L-Cysteine from β-Chloro-L-alanine.** L-Cysteine [52-90-4] can be obtained from β-chloro-L-alanine by β-replacement. This reaction is catalyzed by L-cysteine hydrogen sulfide lyase (cysteine desulfhydrase) (E.C. 4.4.1.1) [*9012-96-8*] which is present in cells of *Enterobacter cloacae*, for example.

The conversion has an 80% yield and results in L-cysteine concentrations of 50 g/L [4.340].

D-Cysteine synthesis from β-chloro-D-alanine has also been accomplished by using the specific enzyme β-chloro-D-alanine hydrogen chloride lyase. This enzyme, found in *Pseudomonas putida*, catalyzes β-replacement with hydrogen sulfide [4.341].

**L-Citrulline from L-Arginine.** The enzymatic synthesis of L-citrulline [*372-75-8*] from L-arginine relies on hydrolysis by L-arginine iminohydrolase (arginine deimi-nase) (E.C. 3.5.3.6) [*9027-98-9*]. Here, immobilized cells of *Pseudomonas putida* are used in a continuous process [4.342].

## 4.9.4. Synthesis of L-Amino Acids by Enzymatic Carbon–Carbon Bonding

The enzymatic synthesis of L-amino acids by carbon–carbon bonding is cata-lyzed exclusively by enzymes requiring pyridoxal 5-phosphate. This method is becoming increasingly important for efficient synthesis of L-serine, L-threonine, L-tryptophan, and L-tyrosine.

**L-Serine from Glycine and Formaldehyde or Methanol.** The reversible transforma-tion of glycine to serine is catalyzed by serine hydroxymethyltransferase (SHMT) (E.C. 2.1.2.1) [*9029-83-8*] in the presence of pyridoxal 5-phosphate and tetrahydro-folic acid. A method has been devised using SHMT from *Klebsiella aerogenes* as a raw extract [4.342].

An optimally controlled serine bioreactor yields 450 g/L of L-serine [*56-45-1*], with an 88% molar conversion of glycine and a productivity of $8.9 \, \text{g L}^{-1} \text{h}^{-1}$. A glycine-resistant strain of *Hyphomicrobium methylovorum* which produces both

SHMT and methanol dehydrogenase can be used to make up to 34 g/L of L-serine from glycine and methanol [4.343].

**L-Threonine from Glycine and Acetaldehyde or Ethanol.** L-Threonine acetaldehyde-lyase (L-threonine aldolase) (E.C. 4.1.2.5) [*62213-23-4*] catalyzes the reversible cleavage of threonine to glycine and acetaldehyde. This reaction can also represent a synthetic process.

A *Pseudomonas* strain which can be used in the production of L-serine from glycine and methanol is also suitable for the synthesis of L-threonine [*72-19-5*], if methanol is replaced by ethanol [4.344].

**L-Tryptophan from Indole and Pyruvate or Serine.** Two enzymes can catalyze carbon-carbon bonding in the synthesis of L-tryptophan [*73-22-3*]: (1) tryptophan synthase (E.C. 4.2.1.20) [*9014-52-2*], which is responsible for the biosynthesis of L-tryptophan from indole-3-glycerol phosphate; this enzyme catalyzes the formation of L-tryptophan from indole and L-serine; and (2) L-tryptophan indole-lyase (tryptophanase) (E.C. 4.1.99.1) [*9024-00-4*], which converts indole, ammonia, and pyruvic acid to tryptophan [4.345].

Apart from these methods in which the microorganisms employed, such as *Escherichia coli* [4.346], *Pseudomonas,* and *Methylomonas* [4.347], require L-serine, a new technique has been developed in which a specific racemase enables D,L-serine to be used as substrate [4.348].

In general, L-tryptophan can be made more economically, by fermentation. The decision to make it one way or the other is essentially an economic one. A major consideration is the cost of the starting materials indole, serine, or pyruvate.

**L-Tyrosine from Phenol, Pyruvate, and Ammonia.** The synthesis of L-tyrosine [*60-18-4*] from phenol, pyruvate, and ammonia is catalyzed by L-tyrosine phenol-lyase ($\beta$-tyrosinase) (E.C. 4.1.99.2) [*9059-31-8*] [4.349].

Derivatives of tyrosine, e.g., L-DOPA (dihydroxyphenylalanine) [4.350], are obtained by using substituted phenols as substrates.

## 4.9.5. L-Hydroxycarboxylic Acids

**L-Malic Acid from Fumaric Acid.** The reversible hydratation of fumaric acid to yield L-malic acid [97-67-6] is catalyzed by fumarate hydratase (E.C. 4.2.1.2) [9032-88-6], also called fumarase [4.304]. The reaction can be performed with isolated enzyme or with cell cultures, e.g., immobilized cells of *Brevibacterium ammoniagenes* [4.303].

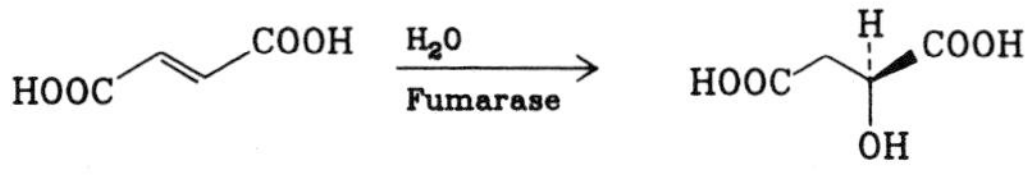

L - Malic acid

Fungal cells of various *Aspergillus* species produce concentrations of up to 400 g/L of L-malic acid [4.351].

**L-α-Hydroxycarboxylic Acids from α-Oxo Acids.** α-Oxo acids can be converted to α-hydroxycarboxylic acids in a pure enantiomeric form by the action of specific dehydrogenases and NADH. In a continuous membrane reactor, regeneration of cofactor in the presence of formate and formate dehydrogenase takes place at the same time [4.352] as the production of hydroxycarboxylic acids.

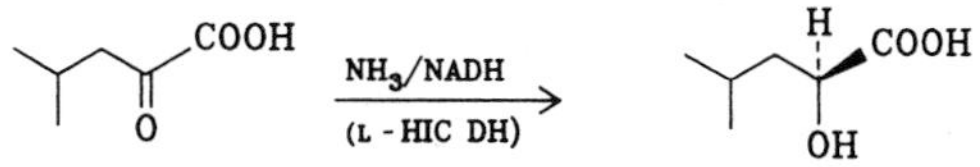

α - Oxoisocaproic
acid

L - α - Hydroxyisocaproic
acid

Examples are the enzymatic production of L-α-hydroxyisocaproic acid [13748-90-8] by using L-α-hydroxyisocaproate dehydrogenase (L-HIC DH) [4.353] from *Lactobacillus confusus*. D-α-Hydroxycarboxylic acids are prepared by the action of D-lactate dehydrogenase [4.354] or D-α-hydroxyisocaproate dehydrogenase (D-HIC DH) from *L. casei* [4.355].

# 4.10. Enzymes in Organic Synthesis

Enzymes are proteins that have catalytic activity. Their potential value as catalysts in organic chemistry has been recognized for many years. Currently more than 2000 enzymes are known [4.356] and several hundred are commercially available. Many others can be obtained through well-developed procedures [4.357]. Enzymes can, in principle, be produced economically in large quantities by recombinant DNA techniques [4.358]. Many of the technical problems that have slowed commercial development of enzyme-catalyzed synthesis — for example, cofactor regeneration, enyzme immobilization, and enzyme stabilization — have been solved. Difficulties in the synthesis of complex biologically important substances represent an increasing problem in many areas of chemistry and pharmacology. For these and other reasons, enzyme-based synthetic chemistry has grown rapidly in recent years [4.359]–[4.364]. This section emphasizes the use of individual enzymes to catalyze reactions useful in synthetic organic chemistry.

## 4.10.1. General Considerations

Enzymes exhibit three characteristic catalytic activities: (1) remarkable acceleration of reaction rates; (2) highly selective mode of action; and (3) susceptibility to regulation by substrates, products, or other species present in solution. *Selectivity* is the most useful of these characteristics. Regulation is most often a nuisance, because it can result in an inhibition of catalytic activity by products.

In considering the application of enzymes in organic synthesis, the availability, specific activity, stability, and lifetime of individual enzymes and their accessory cofactors must be considered. Enzymes need not be particularly pure for most applications as catalysts. Enzymes are normally used in aqueous solutions, but some (lipases) require aqueous–organic interfaces for activity, and many tolerate modest concentrations of organic cosolvents. To enhance their stability and allow their recovery from reaction mixtures, enzymes are often used in immobilized form (see Section 3.3). Of the numerous immobilization methods developed [4.365]–[4.367], glutaraldehyde is the most common in industrial applications [4.365], [4.367]. For laboratory-scale syntheses, covalent attachment of the enzyme to a cross-linked polyacrylamide-co-*N*-acryloxysuccinimide (PAN) polymer is the most general immobilization technique [4.368]. During immobilization, the active site of the enzyme is usually protected by adding a substrate or an inhibitor. Addition of thiols and use of an inert atmosphere during manipulation of enzymes prevent their oxidative inactivation [4.368].

## 4.10.2. Enzymes Not Requiring Coenzymes

Many enzymes do not require cofactors; these enzymes are readily available, inexpensive, stable, and simple to use. They represent the group most widely used industrially in large-scale applications [4.365], [4.367], [4.369] and probably the first group of biocatalysts that will become part of the standard repertoire of organic chemists.

### 4.10.2.1. Esterases, Lipases, and Amidases

Esterases, lipases, and amidases are widely used in kinetic resolutions of racemic mixtures [4.359]–[4.364]. For example, certain epoxyesters can be resolved with hog pancreatic lipase (E.C. 3.1.1.3) [*9001-62-1*] [4.370]; this procedure provides an alternative to the asymmetric epoxidation of allylic alcohols by using transition metals (Fig. 40) [4.371].

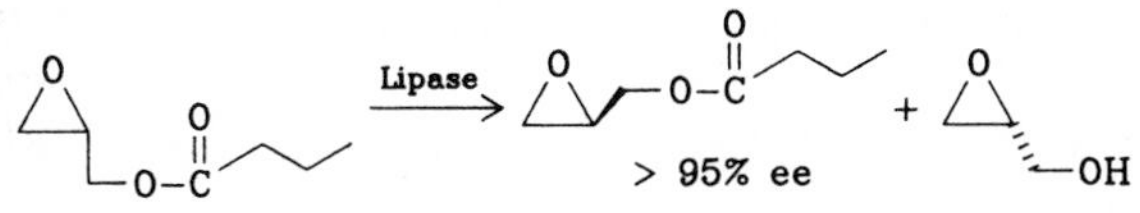

**Figure 40.** Enantioselective hydrolysis of glycidyl butyrate catalyzed by hog pancreatic lipase

Pig liver esterase (PLE) (E.C. 3.1.1.1) [*9016-18-6*] has been applied in an asymmetric synthesis of chrysanthemic [*10453-89-1*], permethrinic [*55701-05-8*], and caronic [*497-42-7*] acids from the corresponding racemic methyl esters [4.372].

SIH et al. have developed a valuable theoretical treatment of such enzymatic transformations, relating the extent of conversion of racemic substrate to enantiomeric excess of the product and enantiomeric selectivity of the enzyme [4.373].

The enzyme-catalyzed asymmetric hydrolysis of meso diesters offers a particularly useful approach to chiral synthons. Synthesis of the antiviral agent showdomycin (**49**), [*16755-07-0*] provides an example (Fig. 41) [4.374]. (+)-Biotin [*58-85-5*] has been prepared from an imidazolone by a similar approach [4.375].

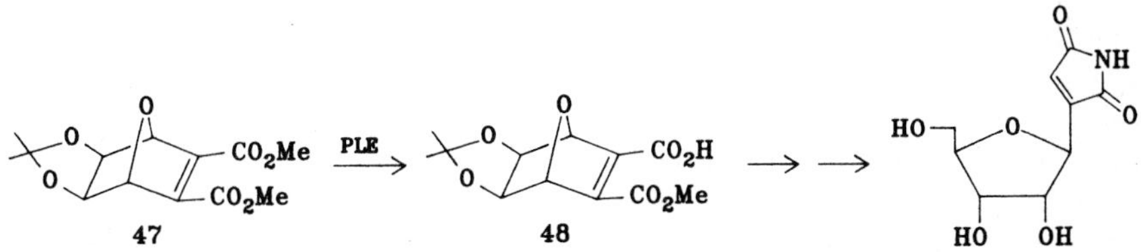

**Figure 41.** Enantioselective hydrolysis of a mesodiester catalyzed by pig liver esterase (PLE)

Lipases are especially useful catalysts, because they operate at the water—organic interface [4.376] and can be applied to water—insoluble substrates. Many are inexpensive and stable, and show broad substrate specificity [4.376].

In addition to hydrolysis, esterase- and lipase-catalyzed transesterification has been investigated [4.377], [4.378], and enantiomeric excesses up to 95% have been obtained.

Amidases have been used primarily to hydrolyze $N$-acylamino acids [4.365], [4.379]. Chymotrypsin (E.C. 3.4.21.1) [*9004-07-3*] and acylase (E.C. 3.5.1.14) [*9012-37-7*] are used in the kinetic resolution of amino acids [4.365], [4.379]. A synthetic route to semisynthetic penicillins is based on the production of 6-aminopenicillanic acid [*551-16-6*] with penicillinase (E.C. 3.5.2.6) [*9001-74-5*] [4.380]. Transacylations catalyzed by proteases have found impressive applications in the synthesis of the penicillin derivatives ampicillin [*69-53-4*], and amoxicillin [*26787-78-0*] [4.381], and of the cephalosporin cephalexin [*15686-71-2*] [4.382].

Amidases also catalyze the formation of bonds in polypeptides and proteins [4.383]–[4.386]: the conversion of porcine to human insulin by a trypsin-catalyzed reaction of porcine insulin with threonine methyl ester [4.384], the thermolysin-

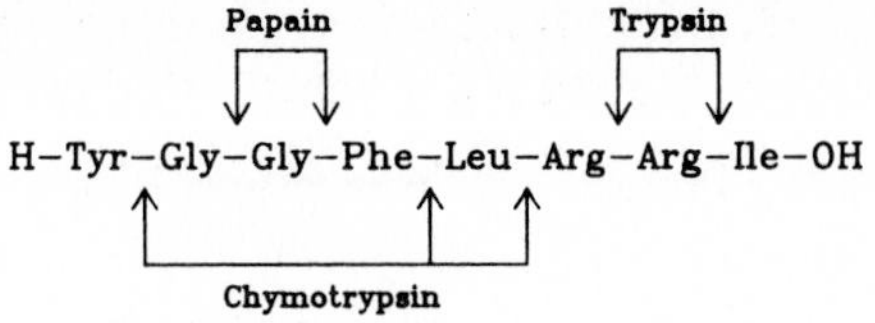

Papain                        Trypsin

H−Tyr−Gly−Gly−Phe−Leu−Arg−Arg−Ile−OH

Chymotrypsin

50: Dynorphin

**Figure 42.** Enzymatic synthesis of dynorphin

catalyzed synthesis of an aspartame precursor [4.385], and the total synthesis of oligopeptides such as dynorphin (**50**) [4.383], [4.386] (Fig. 42) are examples. To effect dehydration, experimental conditions must be carefully optimized. Its sensitivity to conditions limits the general utility of this method to cases in which the demand for the product justifies the effort.

### 4.10.2.2. Aldolases

Aldolases catalyze the cleavage and formation of carbon–carbon bonds in certain carbohydrates [4.387]. Fructose 1,6-diphosphate aldolase (E.C. 4.1.2.13) [9024-52-6] from rabbit muscle condenses dihydroxyacetone phosphate [57-04-5] (DHAP; **52**) with a variety of aliphatic, heterosubstituted, and differentially protected aldehydes (Fig. 43; **51**) [4.388]–[4.390]. This aldolase has been used in the synthesis of rare, nonnatural, and isotopically labeled carbohydrates [4.388], [4.389]. N-Acetylneuraminic acid aldolase (E.C. 4.1.3.3) [9027-60-5] has been used to prepare N-acetylneuraminic acid [131-48-6] and other sialic acids [4.388].

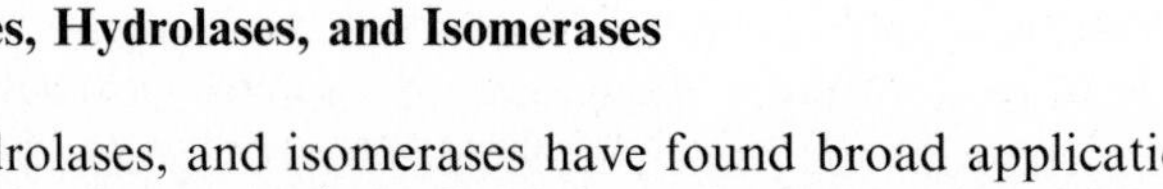

**Figure 43.** Stereospecific aldol addition catalyzed by fructose 1,6-diphosphate aldolase

### 4.10.2.3. Lyases, Hydrolases, and Isomerases

Lyases, hydrolases, and isomerases have found broad application in industrial chemistry, e.g., in the conversion of starch to glucose catalyzed by α-amylase (E.C. 3.2.1.1) [9000-90-2] and glucamylase (E.C. 3.2.1.3) [9032-08-0] (Section 4.2) [4.391], the isomerization of glucose to fructose by glucose isomerase (E.C. 5.3.1.18) [9055-00-9] (Section 4.3) [4.392], and the production of aspartic acid (with aspartase (E.C. 4.3.1.1) [9027-30-9] [4.365], [4.379]) and of malic acid from fumaric acid (with fumarase (E.C. 4.2.1.2) [9032-88-6] [4.365]) (Section 4.9). These enzymes also are used in laboratory synthesis on nonnatural substrates. Galactosidase (α-galactosidase (E.C. 3.2.1.22) [*9025-35-8*]; β-galactosidase (E.C. 3.2.1.23) [*9031-11-2*]), like other glycosidases, has been used in the synthesis of glycosides [4.393]. Epoxide hydrolases can be used to open epoxides regiospecifically [4.394].

## 4.10.3. Enzymes Requiring Coenzymes, but Not Cofactor Regeneration Systems

In many enzymatic systems, required cofactors bind tightly to their respective enzymes and regenerate automatically during the course of the enzyme-mediated reaction: the most important enzyme–cofactor systems exhibiting this type of behavior are those utilizing flavins, pyridoxal phosphate, thiamine pyrophosphate, lipoamide, and certain metal ions as cofactors. The pyridoxal phosphate-containing transaminases have been used for amino acid synthesis by amine transfer from glutamic or aspartic acid to a 2-oxo acid (Section 4.9) [4.395]. The iron-dependent enzyme horseradish peroxidase (E.C. 1.11.1.7) [*9003-99-0*] catalyzes the selective hydroxylation of organic compounds [4.396], and L-DOPA has been prepared this way. ε-Caprolactam [*105-60-2*] has been synthesized from cyclohexanone by using a monooxygenase-catalyzed Baeyer–Villiger oxidation [4.397].

## 4.10.4. Enyzmes Requiring Added Coenzymes

Approximately 70% of all enzymes require nucleoside triphosphates, nicotinamide derivatives, or coenzyme A as cofactors [4.356]. These coenzymes are too expensive to be used stoichiometrically, and methods for their in situ regeneration are now available. Recent reviews summarize methods of cofactor regeneration [4.398].

### 4.10.4.1. Enyzmes Requiring Nucleoside Triphosphates

Regeneration or synthesis of nucleoside triphosphates from the corresponding diphosphates can be achieved by using readily available phosphorylating reagents such as acetyl phosphate [*590-54-5*] [4.399] and phosphoenolpyruvate (**57**) [*73-89-2*] [4.400]: acetate kinase (E.C. 2.7.2.1) [*9001-59-6*] and pyruvate kinase (E.C. 2.7.1.40) [*9001-59-6*] phosphorylate (d)ADP, (d)GDP, (d)CDP and (d)UDP [4.399], [4.401]. The conversion of nucleoside monophosphates to nucleoside diphosphates is not straightforward. Adenylate kinase (E.C. 2.7.4.3) [*9013-02-9*] converts AMP to ADP and has been used to transform CMP to CDP [4.402]. This enzyme, however, is not a practical catalyst for the preparation of other nucleoside diphosphates.

The ATP regeneration schemes have been applied in the synthesis of sugar phosphates [4.398], dihydroxyacetone phosphate [4.389], and *sn*-glycerol 3-phosphate [4.403]. Syntheses of NAD [4.404], ribulose 1,5-diphosphate [4.405], and phosphoribosyl pyrophosphate (PRPP) (Fig. 44; **56**) [4.406], a key intermediate in the biosynthesis of nucleotides, represent more complex examples.

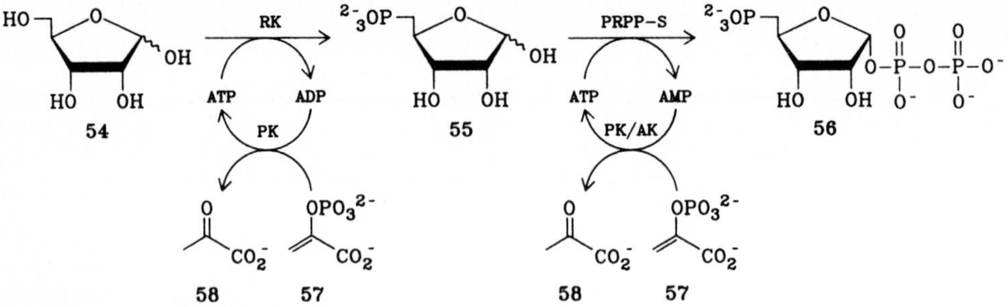

**Figure 44.** Enzyme-catalyzed synthesis of phosphoribosyl pyrophosphate (PRPP)
PK = pyruvate kinase
AK = adenylate kinase
RK = ribokinase
PRPP-S = PRPP synthase

### 4.10.4.2. Enyzmes Requiring Nicotinamide Coenzymes

In contrast to the nucleoside triphosphates, regeneration of nicotinamide cofactors is more difficult, both because these compounds are more expensive and because they are intrinsically unstable in solution [4.364], [4.398]. For reactions in which NAD(P) is used as an oxidant, product inhibition may present a severe problem [4.407]. For regeneration of NADH from NAD, three systems are practical:

1) Formate – formate dehydrogenase (E.C. 1.2.1.2) [9028-85-7] is very efficient, and the only byproduct formed during the reaction is carbon dioxide [4.408]. Degussa used the formate dehydrogenase method to regenerate NADH in the synthesis of amino acids on a large scale [4.409].
2) Glucose – glucose dehydrogenase (E.C. 1.1.1.47) [9028-53-9] is also attractive as a hydride-donating system, although the workup of reaction mixtures in this system is more complicated [4.410].
3) The same problem arises in the use of glucose 6-phosphate dehydrogenase (E.C. 1.1.1.49) [9001-40-5] and glucose 6-phosphate [4.411].

The formate-utilizing enzyme is applicable only to the regeneration of NADH, but carbohydrate-based reductions regenerate both NADH and NADPH. Reductions mediated by NADH have been successfully applied to asymmetric reduction of ketones. Horse liver alcohol dehydrogenase (HLADH) (E.C. 1.1.1.1) [9031-72-5] accepts a variety of substrates [4.412]. For instance, it reduces *cis*-decaline-2,7-dione (**59**) stereospecifically; the product can be converted to (+)-(4*R*)-twistanone [13537-95-6] (Fig. 45; **61**) [4.413]. JONES and co-workers have devised a convenient model that allows assessment of the success of an HLADH-catalyzed reduction [4.414]. Lactate dehydrogenase (L-LDH (E.C. 1.1.1.27) [9001-60-9]; D-LDH (E.C. 1.1.1.28) [9028-36-8]) is another important catalyst. It reduces α-oxo acids enantiospecifically to α-hydroxy acids [4.415]. Both the D- and the L-selective enzymes are available, and both enantiomers of many α-hydroxy acids can be generated

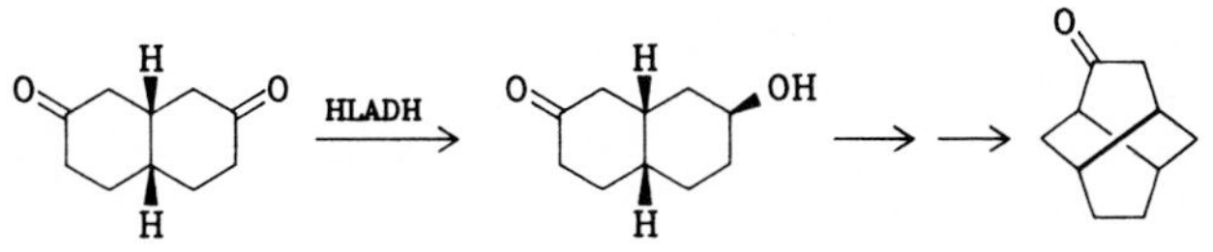

**Figure 45.** Stereoselective reduction of a ketone with horse liver alcohol dehydrogenase (HLADH)

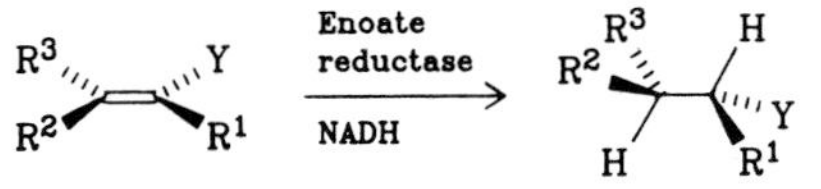

$R^1$ = H, Me, Et, $OCH_3$, F, Cl, Br, NHCHO
$R^2$ = alkyl, aryl, alkyloxycarbonyl
$R^3$ = H, Me, Et
Y = $CO_2$, CHO

**Figure 46.** Stereoselective reduction of olefins catalyzed by enoate reductase

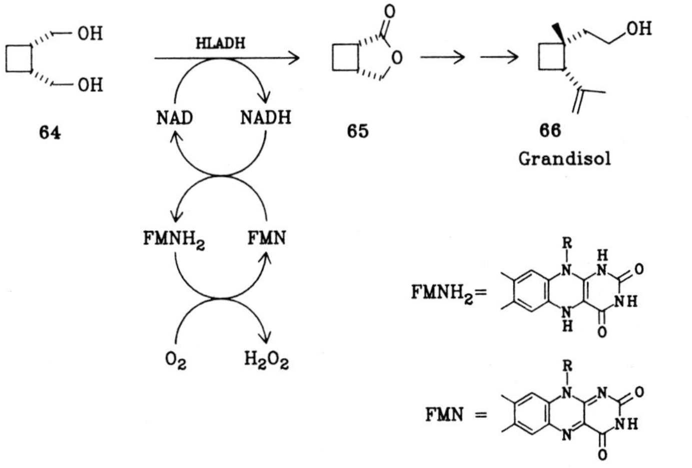

R = $CH_2-(CHOH)_3-CH_2-O-PO_3^{2-}$

**Figure 47.** Horse liver alcohol dehydrogenase (HLADH) catalyzed stereoselective oxidation of a meso diol with in situ NAD regeneration.

(Section 4.9). Enoate reductase (E.C. 1.3.1.31) reduces $\alpha,\beta$-unsaturated carbonyl compounds of type **62** to the saturated derivatives **63** and, thereby, introduces two chiral centers at the same time (Fig. 46) [4.416].

In situ regeneration of NAD has been achieved in most cases with the non-enzyme-catalyzed reoxidation of NADH by oxidized flavin [4.417]. This procedure has been used successfully in preparing chiral lactones from meso diols; a synthesis of grandisol [*26532-22-9*] (**66**) provides an example (Fig. 47) [4.418]. The disadvantages of this system are the large amount of flavin needed, the slow reaction rate, and the requirement for exposure of the enzymes to dioxygen. Probably the best of the proposed alternatives is based on the conversion of ketoglutarate and ammonia to glutamic acid through oxidation of NADH to NAD [4.407].

### 4.10.4.3. Enzymes Requiring Other Cofactors

Currently no straightforward method exists for the synthesis and regeneration of 3′-phosphoadenosine 5′-phosphosulfate [*482-67-7*] (PAPS), an important compound for biochemical sulfations [4.388]. *S*-Adenosylmethionine [*485-80-3*] (SAM), a methyl donor in enzyme-catalyzed reactions, can be synthesized from ATP and methionine [4.419], but the SAM synthetase required is difficult to obtain, and no strategy for SAM regeneration has been proposed. Strategies for regeneration of acetyl-CoA are still in the stage of early development [4.420].

## 4.10.5. Synthesis with Multienzyme Systems

A special feature of enzymatic syntheses is the ability to assemble complex systems of cooperating enzymes and carry out a multistep synthesis in a single reaction vessel. Because of their selectivity, enzymes will accept only the desired substrates. Removal of impurities and byproducts may be less important in enzyme-catalyzed syntheses than in classical multistep transformations.

Some examples of such complex systems include the formation of ribulose 1,5-diphosphate [*2002-28-0*], an important substrate in plant biochemistry, from glucose [4.421], and the construction of lactosamine (**69**), a core dissaccharide common in glycoprotein glycans, from glucose 6-phosphate (**68**) and *N*-acetyl-glucosamine (**67**) (Fig. 48) [4.422]. This latter type of procedure has been extended to oligosaccharides [4.405].

Even more difficult enzyme systems are involved in oxygen-incorporating transformations. These enzymes, especially those of cytochrome P450, can carry out very useful reactions, e.g., selective epoxidation of olefins and functionalization of unactivated hydrocarbons [4.423], [4.424]. These enzymes are in many cases membrane-bound, and they are difficult to obtain and handle. Large-scale applications of these purified enzymes are unlikely to become practical in the near future.

## 4.10.6 Outlook

Enzymatic methods will be used increasingly in research and industry for the preparation of chiral compounds. In addition, they will be usefull in the synthesis of complex molecules needed in immunology, endocrinology, intermediary metabolism, molecular genetics, and plant or insect biology [4.388]. Water-soluble synthetic targets such as carbohydrates and nucleic acids are now commonly manipulated

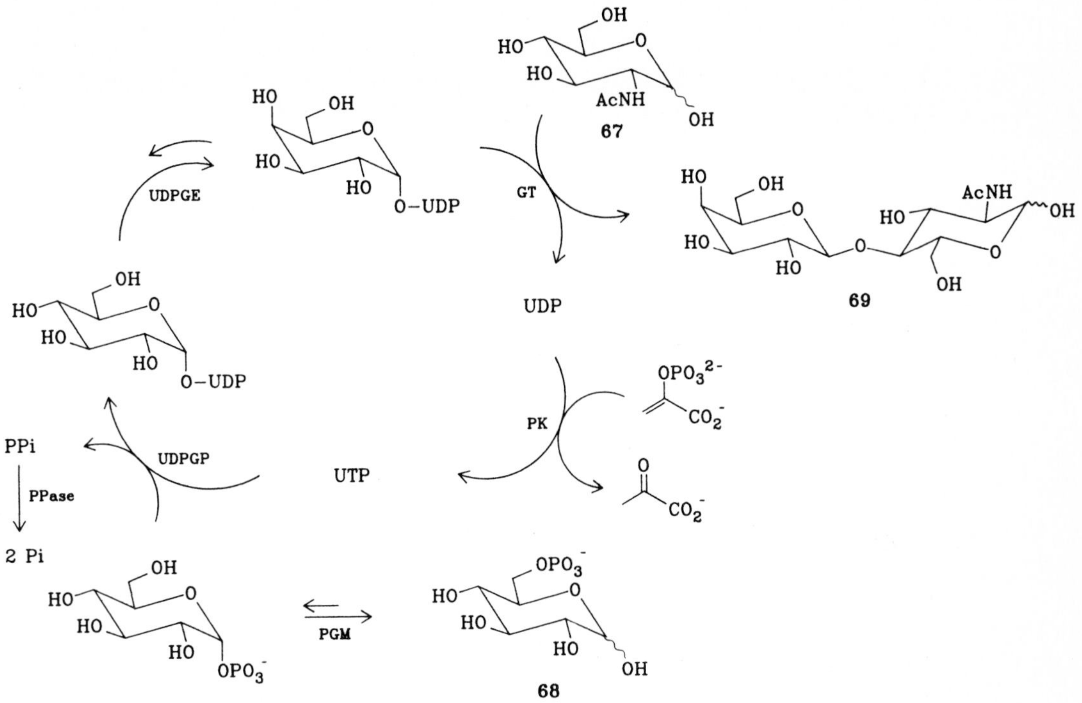

**Figure 48.** Enzymatic synthesis of lactosamine
GT = galactosyltransferase
PK = pyruvate kinase
PGM = phosphoglucomutase
UDPGP = UDP–glucose pyrophosphorylase
UDPGE = UDP–glucose epimerase
PPase = inorganic pyrophosphatase

with enzymes. Enzymology complements both classical synthetic chemistry and biological synthetic techniques. Recombinant DNA and RNA techniques are developing rapidly, and with their development, the design and engineering of synthetic catalysts may become feasible [4.358]. Enyzmes will also be widely used in other synthetic applications [4.388], [4.425] and will continue to find use in isotopic labeling, analysis, and waste treatment.

# 5. Enzymes in Analysis and Medicine

## 5.1. Survey

### 5.1.1. Enzymes in Clinical Diagnosis and Food Analysis

Enzymes can be used as chemical reagents to (1) determine the concentration of substrates, (2) measure the catalytic activity of enzymes present in biological samples, and (3) serve as labels in enzyme immunoassays to determine the concentrations of enzymatically inert substances.

Within the framework of these topics, enzymes serve the following functions:

1) *Analyte Recognition and Transformation.* This is based on the high degree of specificity of most enzymes and is the major advantage of the analytical application of enzymes.
2) *Signal Generation.* In favorable cases of substrate determination, the signal is produced by the primary enzyme reacting with the analyte itself; however, in many systems, the signal must be generated by coupling with an indicator enzyme. In homogeneous as well as heterogeneous immunoassays, the marker enzyme serves to generate a detectable physical signal in response to the antigen–antibody reaction.
3) *Removal of Interferents.* Substances infering with the assay may selectively be converted by specific enzymes. Examples are the conversion of pyruvate by lactate dehydrogenase in the assay of aspartate aminotransferase via coupling to malate dehydrogenase, and of ascorbic acid by ascorbate oxidase in hydrogen peroxide-dependent assays.

The principles and methods of enzymatic analysis are described in detail in [5.1]; a short survey of enzymes as reagents in clinical chemistry is given in [5.2].

The following quality criteria are relevant in the choice of an enzyme to be used for analytical application in clinical diagnosis or food analysis.

**Specificity.** Generally, enzymes are more specific than other chemical reagents [5.3]. However, side activities are often present; unlike contaminating activities, these cannot be removed by purification and may prohibit the use of an enzyme for a particular purpose.

**Purity.** Enzyme purity is characterized by three parameters: (1) contaminating activities, (2) specific activity, and (3) chemical purity. *Contaminating activities* can be separated by further purification. Contaminating activities are usually expressed as percentages of the specific catalytic activity of the reagent enzyme. *Specific activity* is the catalytic activity, expressed in units (U) per milligram of protein (for definition of U, see Chap. 2). High specific activity is not a requirement per se for application of enzymes in analytical systems but is advantageous because the protein load of the assay mixture can be reduced. *Chemical purity,* i.e., the absence of contaminating proteins, is detected with methods such as electrophoresis on polyacrylamide gel or isoelectric focusing. If the contaminants do not interfere with reactions involved in the assay procedure, protein contamination is not important in the analytical application of enzymes.

The requirements for purity depend on the particular assay system in which the enzyme is to be used. Thus, purity is a relative concept.

**Stability.** Enzymes must be stable during long-term storage (e.g., one year at $4\,°C$), in the stock solutions used for the assay (e.g., 8 h at $25\,°C$ or $24–72$ h at $4\,°C$), and in the test mixture during the reaction period at a given incubation temperature (e.g., $10–30$ min at 25 or $37\,°C$). Stability can be improved by addition of stabilizers, by chemical modification of the enzyme (e.g., with poly(ethylene glycol) or dextran, [5.4]), or recently, by optimizing the enzyme's primary structure by protein engineering [5.5]. In addition, enzymes must be free from proteases since these may have a harmful effect on the stability of enzyme preparations.

**Kinetic Properties.** In most assay systems, a low Michaelis constant $K_m$, i.e., a high affinity of the enzyme for its substrate, is advantageous (for definition of $K_m$, see Chap. 2). However, in determining metabolite concentrations by kinetic methods, enzymes with high $K_m$ values are desirable [5.6]; e.g., in the kinetic determination of cholesterol, cholesterol oxidase from *Streptomyces* is used rather than cholesterol oxidase from *Nocardia* (cf. Table 32).

**Inhibition.** To ensure that the reaction proceeds rapidly and completely, the enzyme should not be inhibited by substrates, products, or other components of the sample or the assay mixture.

**pH Optimum.** The proper choice of pH is especially important for coupled assay systems in which the pH optima of the auxiliary and indicating enzymes must be not too different from each other. The pH optimum is also important in determination of an enzyme's catalytic activity in enzyme-coupled systems.

**Solubility and Surface Properties.** To avoid turbidity, good solubility is essential for the use of an enzyme in solution assays. In contrast, enzymes to be used in reagent strips or in solid-phase systems must fulfill special quality requirements concerning their diffusion, matrix-binding, and elution properties.

**Cost.** Many enzymatic analyses compete with chemical or physical methods. Therefore, the price of the reagents — in addition to the cost of instrumentation and working time — is an important consideration, especially for large-scale commercial use of enzymatic analysis. The cost of enzyme production may be substantially lowered by the use of overproducing microorganisms constructed by recombinant DNA technology, as the enzyme source. A detailed account of economic considerations can be found in [5.7].

The quality criteria listed in the previous paragraphs depend on each other, e.g., purity, activity, stability, and price. Therefore, the choice of an enzyme must be optimized in each case with regard to its particular analytical application.

Table 32 (pp. 154–159) lists a selection of enzymes used in diagnosis and food analysis, with some of their molecular properties and applications. Many enzymes are available from various sources; however, only one representative enzyme source is given in Table 32, except for cholesterol oxidase in order to demonstrate the difference between enzymes used for end-point determinations and those used for kinetic substrate determinations. For $K_m$ values and pH optima, which depend strongly on assay conditions, see the appropriate literature. Substrates given include substances reacting directly with the respective enzyme or substances that are measured with the aid of auxiliary enzymes. Only important substrates are given in Table 32, although many enzymes have been used for the determination of additional substances.

References in Table 32 have been taken preferably from standard reference works [5.8]–[5.11]. In some cases, data not published in the scientific literature have been quoted from manufacturer's information. Additional data on the specificity, inhibition, and assay conditions can be found in [5.12]. All enzymes listed can be obtained commercially.

The practical application of enzymes in diagnostics and food analysis is described in detail in Sections 5.2 and 5.3, respectively.

## 5.1.2. Enzymes in Therapy

In principle, enzymes are anything but ideal therapeutic agents because of their inherent adverse properties such as (1) high molecular mass (and, therefore, inability to pass through biological membranes), (2) instability to denaturation and proteolytic attack, (3) inactivation by inhibitors present in biological fluids and tissues, and (4) antigenic properties of heterologous or conformationally disturbed proteins. On the other hand, if these unfavorable factors can be overcome, enzymes might be therapeutically promising because of their expected specificity, biological compatibility, and low level of adverse effects.

At present, enzymes are administered *orally*, e.g., enzymes used as digestive aids and in treatment of pancreas insufficiency; *externally,* e.g., enzymes used for wound cleaning; or *parenterally*, e.g., fibrinolytic agents, coagulation factor substitutes, and asparaginase used for treatment of certain kinds of leukemia. In the latter case, repeated administration of heterologous (nonhuman) enzymes may lead to adverse immunological reactions; this may be prevented by the use of recombinant human

**Table 32.** Enzymes used in diagnostics and food analysis

| Enzyme, E.C. number, and CAS registry number | Source | $M_r$, $10^3$ dalton enzyme (subunit) | $K_m$, mol/L (substrate) | pH optimum | Isoelectric point | Substrate measured | Reference |
|---|---|---|---|---|---|---|---|
| Acetyl-CoA synthetase (acetate–CoA ligase) E.C. 6.2.1.1 [9012-31-1] | yeast | 151 (78) | $2.8 \times 10^{-4}$ (acetate) $3.5 \times 10^{-5}$ (CoA) | 7.6 | – | acetate | [5.14], [5.15] |
| Adenylate kinase (myokinase) E.C. 2.7.4.3 [9013-02-9] | porcine muscle | 21 | $5 \times 10^{-4}$ (AMP) $3 \times 10^{-4}$ (ATP) $1.58 \times 10^{-3}$ (ADP) | ca. 7.5 | 6.1 | AMP | [5.16], [5.17] |
| Alcohol dehydrogenase E.C. 1.1.1.1 [9031-72-5] | yeast | 141 (35) | $1.3 \times 10^{-2}$ (ethanol) $7.8 \times 10^{-4}$ (acetaldehyde) $7.4 \times 10^{-5}$ (NAD$^+$) $1.08 \times 10^{-5}$ (NADH) | 9.0 | 5.4–5.8 | ethanol, other alcohols, aldehydes | [5.18], [5.19] |
| Aldehyde dehydrogenase E.C. 1.2.1.5 [9013-02-9] | yeast | 207 (57) | $9 \times 10^{-6}$ (acetaldehyde) $1.3 \times 10^{-4}$ (NAD$^+$) | 9.3 | – | acetaldehyde, ethanol | [5.20]–[5.22] |
| D-Amino acid oxidase E.C. 1.4.3.3 [9000-88-8] | porcine kidney | 100 (50 or 38–39) | $1.8 \times 10^{-3}$ (d-alanine) $1.8 \times 10^{-4}$ (oxygen) | 9.5 | – | penicillin, hormones | [5.23], [5.24] |
| Amyloglucosidase (glucamylase, glucan 1.4-$\alpha$-glucosidase) E.C. 3.2.1.3 [9032-08-0] | Aspergillus niger | 97 (48) | $2.2 \times 10^{-7}$ (amylopectin) $3.2 \times 10^{-5}$ (amylose) $1.85 \times 10^{-2}$ (maltose) | ca. 5.0 | 4.2 | starch, glycogen | [5.25], [5.26] |
| Ascorbate oxidase E.C. 1.10.3.3 [9029-44-1] | Cucurbita species | 140 (65) | $2.4 \times 10^{-4}$ (l-ascorbate) | 5.6–7.0 | 5.0–5.5 | ascorbate; (removal of ascorbate interference) | [5.27], [5.28] |
| Batroxobin (Bothrops atrox serine proteinase, reptilase) E.C. 3.4.21.29 [9039-61-6] | Bothrops atrox venom (different zoological varieties) | 32–43 | $1.6$–$2.9 \times 10^{-4}$ (B$_z^e$-Phe-Val-Arg-$p$-nitroanilide) | 7.4–8.2 | 6.6 | fibrinogen | [5.29] |
| Catalase E.C. 1.11.1.6 [9001-05-2] | bovine liver | 232 (57.5) | $-^a$ | 6.8–7.0 | 5.4–5.8 | uric acid, cholesterol (indicator enzyme) | [5.30], [5.31] |
| Cholesterol esterase E.C. 3.1.1.13 [9026-00-0] | Pseudomonas fluorescens | ca. 129 (27) | $7 \times 10^{-5}$ (cholesterol oleate) $5.5 \times 10^{-5}$ (cholesterol linoleate) | 7.3–7.6 | 4.5 | cholesterol esters | [5.32] |
| Cholesterol oxidase E.C. 1.1.3.6 [9028-76-6] | Nocardia erythropolis | 59 | $1 \times 10^{-6}$ (cholesterol) | 7.5 | 4.85 | cholesterol | [5.33] |
| Cholesterol oxidase E.C. 1.1.3.6 [9028-76-6] | Streptomyces species | 55 | $6 \times 10^{-4}$ (cholesterol) | 7.0–7.5 | 4.5 | cholesterol | [5.34] |

| Enzyme / E.C. number [registry no.] | Source | Molecular weight (subunit) | $K_m$ values (substrate) | pH | pI | Substance determined | References |
|---|---|---|---|---|---|---|---|
| Choline kinase E.C. 2.7.1.32 [9026-67-9] | yeast | 67–68 | $1.5 \times 10^{-5}$ (choline) $1.4 \times 10^{-4}$ (ATP) | 8.0–9.5 | ca. 6 | phosphatidylcholine | [5.35], [5.36] |
| Choline oxidase E.C. 1.1.3.17 [9028-67-5] | Arthrobacter globiformis | 84 or 71[b] | $1.2 \times 10^{-3}$ (choline) $8.7 \times 10^{-2}$ (betaine aldehyde) | 7.5 | 4.5 | phospholipids | [5.37] |
| Citrate lyase (citrate(pro-3S)-lyase) E.C. 4.1.2.6 [9012-83-3] | Aerobacter aerogenes | 575 (73.8) | $2.1 \times 10^{-4}$ (citrate) | 8.0–9.0 | 8.0 | citrate | [5.38], [5.39] |
| Citrate synthase E.C. 4.1.3.7 [9027-96-7] | porcine heart | 100 (50) | $2.5 \times 10^{-4}$ (citrate) $2.8 \times 10^{-5}$ (CoA) | 8.0 | 5.05 | acetate | [5.40], [5.41] |
| Creatinase (creatine amidinohydrolase) E.C. 3.5.3.3 [37340-58-2] | Pseudomonas species | 94 (47) | $1 \times 10^{-2}$ (creatine) | 8.0 | 4.8 | creatine, creatinine | [5.42], [5.43] |
| Creatininase (creatinine amidohydrolase) E.C. 3.5.2.10 [9025-13-2] | Pseudomonas species | 175 (23) | $3 \times 10^{-2}$ (creatinine) $6 \times 10^{-2}$ (creatine) | 7.8 | 4.7 | creatinine | [5.44] |
| Creatinine deiminase E.C. 3.5.4.21 [37289-15-5] | Corynebacterium lilium | 200 | $6 \times 10^{-3}$ (creatine) | 7.5–9.0 | 4.2 | creatine | [5.44 a] |
| Dihydrolipoamide dehydrogenase (diaphorase) E.C. 1.8.1.4 [9001-18-7] | porcine heart | 114 (57) | $5 \times 10^{-3}$ (lipoamide) $2 \times 10^{-3}$ (lipoate) $2 \times 10^{-4}$ (NAD⁺) $2.7 \times 10^{-4}$ (ferricyanide) | 4.8 (diaphorase) 5.6–6.5 (lipoate reduction) | $5.9$–$7.2^c$ | NADH (indicator enzyme) | [5.45], [5.46] |
| Esterase (carboxylesterase) E.C. 3.1.1.1 [9016-18-6] | porcine liver | 168 (42) | $5 \times 10^{-4}$ (ethyl n-butyrate) $4.4 \times 10^{-4}$ (methyl n-butyrate) $1.5 \times 10^{-4}$ (phenyl n-butyrate) | 8.6–8.8 | 5.0 | triglycerides | [5.47], [5.48] |
| Factor Xa (thrombokinase) E.C. 3.4.21.6 [9002-05-5] | bovine plasma | 47.2 (Xaα); 44.2 (Xaβ) | $5 \times 10^{-4}$ ($B_z^c$-Ile-Glu-Gly-Arg-p-nitroanilide) | 8.3 | – | prothrombin | [5.49], [5.50] |
| Formaldehyde dehydrogenase E.C. 1.2.1.46 [68821-75-0] | Pseudomonas species | 150 (75) | $6.7 \times 10^{-5}$ (formaldehyde) $1.2 \times 10^{-4}$ (NAD⁺) | 7.8 | 5.25 | formaldehyde | [5.51] |
| Formate dehydrogenase E.C. 1.2.1.2 [9028-85-7] | Candida boidinii | 74 (36) | $1.3 \times 10^{-2}$ (formate) $9 \times 10^{-5}$ (NAD⁺) | 7.5–8.5 | – | formate, oxalate | [5.52] |
| β-Fructosidase (invertase, saccharase) E.C. 3.2.1.26 [9001-57-4] | yeast | 270 (135) | $9.1 \times 10^{-3}$ (sucrose) $2.4 \times 10^{-1}$ (raffinose) | 3.4–4.0 | $4.02,\ 4.24^c$ | sucrose | [5.53], [5.54] |
| β-D-Galactose dehydrogenase E.C. 1.1.1.48 [9028-54-0] | Pseudomonas fluorescens | 64 (32) | $7 \times 10^{-4}$ (D-galactose) $2.4 \times 10^{-4}$ (NAD⁺) $2.3 \times 10^{-3}$ (NADP⁺) | 9.1–9.5 | 5.13 | galactose, raffinose | [5.55]–[5.57] |

**Table 32.** (continued)

| Enzyme, E.C. number, and CAS registry number | Source | $M$, $10^3$ dalton enzyme (subunit) | $K_m$, mol/L (substrate) | pH optimum | Isoelectric point | Substrate measured | Reference |
|---|---|---|---|---|---|---|---|
| α-Galactosidase E.C. 3.2.1.22 [9025-35-8] | *Escherichia coli* | 329 (82) | $2 \times 10^{-2}$ (galactose) | 6.0–7.0 | 5.1 | raffinose | [5.58] |
| β-Galactosidase E.C. 3.2.1.23 [9031-11-2] | *Escherichia coli* | 465 (116) | $3.85 \times 10^{-3}$ (lactose) $9.5 \times 10^{-4}$ (2-nitrophenyl-β-D-galactoside) $4.45 \times 10^{-4}$ (4-nitrophenyl-β-D-galactoside) | 8.0 | 4.61 | lactose; (marker enzyme) | [5.59], [5.60] |
| Gluconokinase E.C. 2.7.1.12 [9030-55-1] | *Escherichia coli* | 67 (34) | $7 \times 10^{-5}$ (D-gluconate) $5 \times 10^{-4}$ (ATP) | 8.0 | – | D-gluconate | [5.61] |
| Glucose dehydrogenase E.C. 1.1.1.47 [9028-53-9] | *Bacillus megaterium* | 118 (28) | $4.75 \times 10^{-2}$ (glucose) $4.5 \times 10^{-3}$ (NAD$^+$) | 8.0 (tris-HCl) 9.0 (acetate) | – | glucose | [5.61 a], [5.61 b] |
| Glucose oxidase E.C. 1.1.3.4 [9001-37-0] | *Aspergillus niger* | 160 (79) | $3.3 \times 10^{-2}$ (glucose) $2 \times 10^{-4}$ (oxygen) | 5.5–6.5 | 4.2 | glucose | [5.62], [5.63] |
| Glucose 6-phosphate dehydrogenase E.C. 1.1.1.49 [9001-40-5] | *Leuconostoc mesenteroides* | 103.7 (54.8) | $6.4 \times 10^{-5}$ (glucose-6-phosphate) $1.15 \times 10^{-4}$ (NAD$^+$) $7 \times 10^{-6}$ (NADP$^+$) | 7.8 | 4.6 | glucose-6-phosphate, sucrose, ATP; (marker enzyme) | [5.64], [5.65] |
| Glucose 6-phosphate isomerase E.C. 5.3.1.9 [9001-41-6] | yeast | 120 (60) | $0.7–1.5 \times 10^{-3}$ (glucose 6-phosphate) $2.3 \times 10^{-4}$ (fructose 6-phosphate) | 7.6 | 5.0–5.4$^c$ | fructose | [5.66] |
| α-Glucosidase (maltase) E.C. 3.2.1.20 [9001-42-7] | yeast | 68 | $1.8–2.8 \times 10^{-4}$ (4-nitrophenyl-α-D-glucoside) | 7.5–8.0 | 5.6–5.9$^c$ | α-amylase | [5.25], [5.67], [5.68] |
| β-Glucosidase E.C. 3.2.1.21 [9001-22-3] | *Amygdalae dulces* | 135 (65)$^c$ +90 | $6 \times 10^{-3}$ (2-nitrophenyl-β-D-glucoside) | 4.4–6.0$^c$ | 7.3$^c$ | α-amylase, amygdalin | [5.25], [5.69] |
| Glutamate dehydrogenase E.C. 1.4.1.3 [9029-12-3] | bovine liver | oligomers of 332, e.g., 2200 (55) | $1.8 \times 10^{-3}$ (L-glutamate) $7 \times 10^{-4}$ (2-oxoglutarate) $3.2 \times 10^{-3}$ (ammonia) $7.0 \times 10^{-4}$ (NAD$^+$) $4.7 \times 10^{-5}$ (NADP$^+$) $2.5 \times 10^{-5}$ (NADH) $2.4 \times 10^{-5}$ (NADPH) | 8.5–9.0 (glutamate oxidation) 7.8 (reductive amination) | 4.5 | glutamate, 2-oxoglutarate, ammonia | [5.70], [5.71] |
| Glutamic–oxaloacetic transaminase (aspartate aminotransferase) E.C. 2.6.1.1 [9000-97-9] | porcine heart | 94 (47) | $8.9 \times 10^{-3}$ (L-glutamate) $8.8 \times 10^{-5}$ (oxaloacetate) $3.9 \times 10^{-3}$ (L-aspartate) $4.3 \times 10^{-4}$ (2-oxoglutarate) | 8.0–8.5 | 5.0 | malate | [5.72], [5.73] |

| Enzyme | Source | $M_r$ | $K_m$ values (mol/L) | pH optimum | pI | Determination of | References |
|---|---|---|---|---|---|---|---|
| Glutamic–pyruvic transaminase (alanine aminotransferase) E.C. 2.6.1.2 [9000-86-6] | porcine heart | 115 | $2.5 \times 10^{-2}$ (L-glutamate) $3 \times 10^{-4}$ (pyruvate) $2.8 \times 10^{-2}$ (L-alanine) $4 \times 10^{-4}$ (2-oxoglutarate) | 8.0–8.5 | – | lactate | [5.74] |
| Glycerol dehydrogenase E.C. 1.1.1.6 [9028-14-2] | Enterobacter aerogenes | 340 (56) | $1.7 \times 10^{-2}$ (glycerol) $1.3 \times 10^{-3}$ (dihydroxyacetone) $1.5 \times 10^{-4}$ (NAD$^+$) $1.4 \times 10^{-5}$ (NADH) | 9.0 | – | glycerol, triglycerides | [5.75], [5.76] |
| Glycerol kinase E.C. 2.7.1.30 [9030-66-4] | Bacillus stearothermophilus | 230 (58) | $4.4 \times 10^{-5}$ (glycerol) $6 \times 10^{-5}$ (ATP) | 10.0–10.5 | – | glycerol, triglycerides, lipase | [5.77], [5.78] |
| Glycerol 3-phosphate dehydrogenase E.C. 1.1.1.8 [9075-65-4] | rabbit muscle | 78 (37.5) | $1.1 \times 10^{-4}$ (glycerol 3-phosphate) $3.8 \times 10^{-4}$ (NAD$^+$) | 7.5–8.6 | 6.45 | glycerol, triglycerides | [5.79] |
| Glycerol 3-phosphate oxidase E.C. 1.1.3.21 [9046-28-0] | Aerococcus viridans | 75 | $3.2 \times 10^{-3}$ (L-glycerol 3-phosphate) | 7.5–8.5 | 4.2 | triglycerides | [5.80] |
| Hexokinase E.C. 2.7.1.1 [9001-51-8] | yeast | 104 (52) | $1.0 \times 10^{-4}$ (D-glucose) $7.0 \times 10^{-4}$ (D-fructose) $2.0 \times 10^{-4}$ (ATP) $5.0 \times 10^{-5}$ (D-mannose) | 8.0–9.0 | 4.7 | glucose, fructose, mannose, ATP; creatine kinase | [5.81]–[5.83] |
| 3-Hydroxybutyrate dehydrogenase E.C. 1.1.1.30 [9028-38-0] | Rhodopseudomonas spheroides | 85 | $4.1 \times 10^{-4}$ (D-3-hydroxybutyrate) $8.0 \times 10^{-5}$ (NAD$^+$) | 6.2–6.9 (reduction) 7–9 (oxidation) | – | 3-hydroxybutyrate | [5.84] |
| Isocitrate dehydrogenase E.C. 1.1.1.42 [9028-48-2] | porcine heart | ca. 60 | $2.6 \times 10^{-6}$ (isocitrate) $9.2 \times 10^{-6}$ (NADPH) $1.3 \times 10^{-4}$ (2-oxoglutarate) $1 \times 10^{-7}$ (NADP$^+$) | 7.0–7.5 | 7.4 | isocitrate | [5.85], [5.86] |
| D-Lactate dehydrogenase E.C. 1.1.1.28 [9028-36-8] | Lactobacillus leichmanii | 68 or 80$^d$ | $7 \times 10^{-2}$ (D-lactate) $1.2 \times 10^{-3}$ (pyruvate) $7.1 \times 10^{-5}$ (NADH) | 7.0 | – | D-lactate; glutamic–oxaloacetic and glutamic–pyruvic transaminases | [5.87] |
| L-Lactate dehydrogenase E.C. 1.1.1.27 [9001-60-9] | porcine heart | 140 (35) | $1.5 \times 10^{-4}$ (pyruvate) $3.3 \times 10^{-3}$ (L-lactate) $1.1 \times 10^{-5}$ (NADH) $6.7 \times 10^{-5}$ (NAD$^+$) | 7.0 | –$^c$ | L-lactate, pyruvate, citrate, ADP | [5.88], [5.89] |

**Table 32.** (continued)

| Enzyme, E.C. number, and CAS registry number | Source | $M_r$, $10^3$ dalton enzyme (subunit) | $K_m$, mol/L (substrate) | pH optimum | Isoelectric point | Substrate measured | Reference |
|---|---|---|---|---|---|---|---|
| Lipase (triacylglycerol acylhydrolase) E.C. 3.1.1.3 [9001-62-1] | Pseudomonas species | 32 | $9 \times 10^{-4}$ (tributyrin) | 7.0–8.0 | 4.3 | triglycerides | [5.90] |
| Luciferase, bacterial (alkanal monooxygenase, FMN-linked) E.C. 1.14.14.3 [9014-00-0] | Photobacterium fischeri | 80 (40 + 40) | $4{-}8 \times 10^{-7}$ (FMNH$_2$) | 6.8 | – | NADH | [5.91] |
| Luciferase, firefly (Photinus-luciferin 4-monooxygenase) E.C. 1.13.12.7 [61970-00-1] | Photinus pyralis | 100 (50) | $5 \times 10^{-5}$ (ATP) | 7.5–7.8 | 6.2–6.3 | ATP | [5.92], [5.93] |
| Malate dehydrogenase E.C. 1.1.1.37 [9001-64-3] | porcine heart mitochondria | 67 (34) | $4 \times 10^{-4}$ (L-malate) $3.3 \times 10^{-5}$ (oxaloacetate) | 7.4–7.5 | 6.1–6.4 | malate, oxaloacetate, acetate, citrate; (marker enzyme) | [5.94], [5.95] |
| Mannose 6-phosphate isomerase E.C. 5.3.1.8 [9023-88-5] | yeast | 45 | $0.8{-}1.35 \times 10^{-3}$ (mannose 6-phosphate) | 7.0–7.2 | – | mannose | [5.96] |
| NADH peroxidase E.C. 1.11.1.1 [9032-24-0] | Streptococcus faecalis | 120 (59) | $1.7 \times 10^{-5}$ (NADH) $2.8 \times 10^{-5}$ (H$_2$O$_2$) | 6.0 | – | H$_2$O$_2$ (indicator enzyme) | [5.97], [5.98] |
| Nitrate reductase E.C. 1.6.6.2 [9029-27-0] | Aspergillus species | 200 (97) | $3.2 \times 10^{-4}$ (nitrate) | 7.5 | – | nitrate | [5.99] |
| Peroxidase E.C. 1.11.1.7 [9003-99-0] | horseradish | ca. 40 | –[a] | 6.0–7.0 | –[c] | H$_2$O$_2$ (indicator enzyme and marker enzyme) | [5.98], [5.100] |
| Phosphatase, alkaline E.C. 3.1.3.1 [9001-78-9] | calf intestine | 140 (69) | –[a] | 9.8 | 5.7 | (marker enzyme) | [5.101], [5.102] |
| 6-Phosphogluconate dehydrogenase E.C. 1.1.1.44 [9073-95-4] | yeast | 100 | $1.6 \times 10^{-4}$ (6-phosphogluconate) $2.6 \times 10^{-5}$ (NADP$^+$) | 8.0 | – | gluconate | [5.64], [5.103] |
| Phospholipase C E.C. 3.1.4.3 [9001-86-9] | Bacillus cereus | 23 | $2 \times 10^{-2}$ (phosphatidylcholine) | 8.0 | 7.0 | phosphatidylcholine | [5.104], [5.105] |
| Phospholipase D E.C. 3.1.4.4 [9001-87-0] | Streptomyces chromofuscus | 57 or 50[b] | $1.43 \times 10^{-3}$ (phosphatidylcholine) $5.6 \times 10^{-2}$ (sphingomyelin) | 8.0 | 5.1 | triglycerides, choline phospholipids | [5.106] |

| Enzyme, E.C. number [CAS] | Source | Molecular mass (subunits) | $K_m$ values (substrate) | pH optimum | $^c$ | Assay / related | References |
|---|---|---|---|---|---|---|---|
| Plasmin E.C. 3.4.21.7 [9001-90-5] | human or bovine plasma | 87–91 | $1.7 \times 10^{-4}$ (tosyl-Gly-Pro-Lys-p-nitroanilide) | 7.4 | 6.1–8.4$^c$ | $\alpha_2$-antiplasmin | [5.107], [5.108] |
| Pyruvate kinase E.C. 2.7.1.40 [9001-59-6] | rabbit muscle | 237 (57) | $7.0 \times 10^{-5}$ (phosphoenolpyruvate) $1.0 \times 10^{-2}$ (pyruvate) $3.0 \times 10^{-4}$ (ADP) $8.6 \times 10^{-4}$ (ATP) | 7.0–7.8 | 5.98 | ADP, IDP, creatine kinase, phosphoenol-pyruvate | [5.109], [5.110] |
| Pyruvate oxidase E.C. 1.2.3.3 [9001-96-1] | Pediococcus species | 150 (38) | $1.7 \times 10^{-3}$ (pyruvate) $5 \times 10^{-4}$ (phosphate) | 6.5–7.5 | 4.0 | pyruvate, ADP, glutamic–oxaloacetic and glutamic–pyruvic transaminases | [5.111], [5.112] |
| Sarcosine oxidase E.C. 1.5.3.1 [9029-22-5] | Pseudomonas species | 174 (110, 44, 21, 10) | $4 \times 10^{-3}$ (sarcosine) $1.2 \times 10^{-2}$ (N-methylalanine) $1.3 \times 10^{-4}$ (oxygen) | 8.0 | – | sarcosine, creatinine | [5.113] |
| Sorbitol dehydrogenase (L-iditol dehydrogenase) E.C. 1.1.1.14 [9028-21-1] | sheep liver | 115 | $0.7–1.1 \times 10^{-3}$ (sorbitol) $1.8 \times 10^{-4}$ (xylitol) | 7.9–8.1 | – | sorbitol, xylitol | [5.114] |
| Succinyl-CoA synthetase (succinate-CoA ligase, GDP-forming) E.C. 6.2.1.4 [9014-36-2] | porcine heart | 75 (42.5 + 34.5) | $4–8 \times 10^{-4}$ (succinate) $5–10 \times 10^{-6}$ (GTP) $5–20 \times 10^{-6}$ (CoA) | 8.3 | 5.8–6.4$^c$ | succinate | [5.115], [5.116] |
| Sulfite oxidase E.C. 1.8.3.1 [9029-38-3] | chicken liver | 110 (55) | $2.5 \times 10^{-5}$ (sulfite) | 8.5 | – | sulfite | [5.117] |
| Thrombin E.C. 3.4.21.5 [9002-04-4] | bovine plasma | 39 (α), 28 (β) | $1.3 \times 10^{-3}$ (B$_z^e$-Arg-p-nitroanilide) $5.9 \times 10^{-6}$ (tosyl-Gly-Pro-Arg-p-nitroanilide) | 9.0 | 5.3–5.75 | antithrombin III, fibrinogen, heparin, coagulation status | [5.118], [5.119] |
| Trypsin E.C. 3.4.21.4 [9002-07-7] | bovine pancreas | 23.3 | $9.4 \times 10^{-4}$ (B$_z^e$-Arg-p-nitroanilide) $4.3 \times 10^{-6}$ (B$_z^e$-Arg-ethyl ester) | 8.0 | 10.5–10.8 | $\alpha_1$-proteinase inhibitor, $\alpha_2$-macroglobulin | [5.120], [5.121] |
| Urate oxidase (uricase) E.C. 1.7.3.3 [9002-12-4] | Arthrobacter protophormiae | 170 (40) | $6.6 \times 10^{-5}$ (urate) | 9.0 | – | urate | [5.122] |
| Urease E.C. 3.5.1.5 [9002-13-5] | jack beans | 480 (75–83) | $1.05 \times 10^{-2}$ (urea) | 7.0 | 5.0–5.1 | urea; (marker enzyme) | [5.123], [5.124] |
| Xanthine oxidase E.C. 1.1.3.22 [9002-17-9] | cow milk | 283 (150) | $1.7 \times 10^{-6}$ (xanthine) $2.4 \times 10^{-5}$ (oxygen) | 8.5–9.0 | 6.2 | xanthine, hypoxanthine, phosphate | [5.125], [5.126] |

$^a$ Enzyme reaction does not obey Michaelis–Menten kinetics. $^b$ Depending on method (gel filtration or polyacrylamide gel electrophoresis in the presence of sodium dodecylsulfate). $^c$ Depending on method (ultracentrifugation or gel filtration). $^e$ Isoenzymes or multiple species. B$_z$ = benzoyl.

**Table 33.** Enzymes for therapeutic use

| Enzyme | E.C. number | CAS registry number | Source | $M_r$, $10^3$ dalton | Therapeutic use |
|---|---|---|---|---|---|
| $\alpha$-Amylase | 3.2.1.1 | [9000-90-2] | porcine pancreas | –[b] | digestive aid |
| $\beta$-Amylase | 3.2.1.2 | [9000-91-3] | | | digestive aid |
| Ancrod (*Agkistrodon rhodostoma* serine protease) | 3.4.21.28 | [9046-56-4] | *Agkistrodon rhodostoma* venom | 35.4 | peripheral artery diseases |
| Asparaginase | 3.5.1.1 | [9015-68-3] | *Escherichia coli* | 141 | treatment of lymphoblastic leukemia and lymphosarcomatosis |
| Batroxobin (*Bothrops atrox* serine protease) | 3.4.21.28 | [9039-61-6] | *Bothrop atrox* venom | | |
| Bromelain | 3.4.22.5 | [9001-00-7] | pineapple stem (*Ananas sativus*) | 28 | wound healing, digestive aid |
| Cellulase | 3.2.1.4 | [9012-54-8] | *Trichoderma viride* | 57 + 52 + 76 | digestive aid |
| Chymopapain | 3.4.22.6 | [9001-09-6] | papaya latex | 35 | treatment of prolapsed disk |
| Chymotrypsin | 3.4.4.5 | [9004-07-3] | bovine pancreas | 25 | wound healing, digestive aid, lens removal for cataract treatment |
| Collagenase | 3.4.24.3 | [9001-12-1] | *Clostridium histolyticum* | 72–81 | treatment of prolapsed disk |
| Deoxyribonuclease | 3.1.21.1 | [9003-98-9] | *Streptococcus* species | – | wound healing |
| Factor VII (proconvertin) | (3.4.21.21)[a] | [9070-16-0] | human plasma | 50 | bleeding, Factor VII deficiency |
| Factor IX (Christmas factor) | (3.4.21.27)[a] | [37316-87-3] | human plasma | 55.4 | hemophilia type B (Factor IX deficiency) |
| Factor Xa | 3.4.21.6 | [9002-05-5] | bovine or human plasma | 47.2 (Xa$\alpha$); 44.2 (Xa$\beta$) | coagulation disturbances, Factor X deficiency |
| Factor XIII (fibrin-stabilizing factor) | – | [9013-56-3] | human plasma | 340 | bleeding, Factor XIII deficiency |
| Hyaluronidase | 3.2.1.35 | [9001-54-1] | bovine testes | 61 | dispersion agent |
| Kallikrein (kininogenase) | 3.4.21.8 | [9001-01-8] | porcine pancreas | 27.1 (A), 28.9 (B) | peripheral vascular and coronary artery diseases, fertility disturbances |
| Lipase | 3.1.1.3 | [9001-62-1] | *Rhizopus arrhizus* | 43 | digestive aid |
| Lysozyme (muramidase) | 3.2.1.17 | [9001-63-2] | chicken egg white | 14.3 | infectious diseases |

| Enzyme | E.C. number[a] | CAS number | Source | Molecular mass | Application |
|---|---|---|---|---|---|
| Pancreatin | _[b] | [8049-47-6] | porcine pancreas | _[b] | pancreatic insufficiency, digestive aid |
| Papain | 3.4.22.2 | [9001-73-4] | papaya latex | 23 | digestive aid |
| Pepsin | 3.4.4.1 | [9001-75-6] | porcine stomach | 35 | digestive aid |
| Plasmin (fibrinolysin) | 3.4.21.7 | [9001-90-5] | human plasma | 87–91 | fibrinolytic agent, wound healing |
| Plasminogen activator, tissue | 3.4.21.99 | – | melanoma cell culture or recombinant mammalian cells | 67 | fibrinolytic agent |
| Ribonuclease | 3.1.27.5 | [9001-99-4] | bovine pancreas | 13.7 | wound healing |
| Subtilisin | 3.4.21.14 | [9014-01-1] | Bacillus species | 27.6 | wound cleaning |
| Superoxide dismutase | 1.15.1.1 | [9054-89-1] | bovine erythrocytes | 32.5 | inflammatory arthrosis, polyarthritis |
| Thrombin (fibrinogenase) | 3.4.21.5 | [9002-04-4] | human plasma | 39 ($\alpha$), 28 ($\beta$) | superficial bleeding |
| Trypsin | 3.4.4.4 | [9002-07-7] | bovine pancreas | 23.3 | wound cleaning, digestive aid |
| Urokinase/Pro-Urokinase | 3.4.21.31 | [9039-53-6] | human urine, cell culture recombinant bacteria | 49,54 (HMM), 31 (LMM)[c] | fibrinolytic agent |

[a] E.C. number of processed enzymes (Factor VII a, Factor IX a). [b] Mixture of pancreatic enzymes (proteases, amylase, lipase). [c] HMM = high molecular mass, LMM = low molecular mass.

enzymes produced from genetically engineered microorganisms or mammalian cell cultures.

Quality criteria for pharmaceutical enzymes depend on their mode of application and must be more stringent for enzymes destined to be administered intravenously than for those contained in ointments to be applied externally [5.13]. The practical aspects of enzyme therapy are described in Section 5.4. Table 33 is a survey of enzymes presently used therapeutically. With the exception of superoxide dismutase, all enzymes listed are hydrolases. Factor VIII (used for substitution treatment of hemophilia type A) and streptokinase (fibrinolytic agent) are not included in Table 33 (p. 160) because they are not enzymes but protein cofactors without enzymatic activity.

# 5.2. Enzymes in Diagnosis

This section describes the principles of enzyme use in diagnosis; the reader who wishes to get a deeper insight into this field is referred to special monographs, e.g., [5.127]. In clinical chemistry, enzymes are used as auxiliaries for determination of *substrate concentration* or *enzyme activity* and as labels for enzyme *immunoassays*. Reactions are monitored photometrically by absorption change, or they are measured by fluorescence. In addition, light–emitting reactions catalyzed by enzymes are of increasing practical importance, especially in immunoassays.

## 5.2.1. Determination of Substrate Concentration

Substrate concentrations are determined enzymatically in two ways: (1) end-point methods and (2) measurement of reaction rate [5.128], [5.129].

**End-Point Methods.** In the simplest case, the enzyme-catalyzed substrate conversion is practically complete, and either the decrease in substrate concentration or the increase in product concentration can be measured, e.g., by light absorption at a given wavelength. For substrate concentrations well below the value of the Michaelis constant ($K_m$), the reaction rate obeys the following equation [5.128]:

$$v = c_s \cdot V/K_m \tag{5.1}$$

Hence, the time required for completion (e.g., 99 % conversion) of the enzymatic reaction strongly depends on the maximum reaction rate ($V$) and the Michaelis constant ($K_m$) of the enzyme used.

*Determination of Glucose with Glucose Oxidase.* In the first reaction, glucose is oxidized with glucose oxidase (E.C. 1.1.3.4) [*9001-37-0*] at 10 U/mL and hydrogen peroxide is formed:

Glucose + $O_2$ + $H_2O$ $\longrightarrow$ Gluconate + $H_2O_2$

In the second reaction, hydrogen peroxide is used to oxidize a chromogen to form a dye with the aid of horse-radish peroxidase (E.C. 1.11.1.7) [9003-99-0] at 0.8 U/mL:

$$H_2O_2 + \text{chromogen} \longrightarrow \text{dye} + H_2O$$

2,2'-Azino-bis(3-ethyl-2,3-dihydrobenzthiazolsulfonate) [30931-67-0] (ABTS) is used as a chromogen [5.132].

*Determination of Urea.* Urea is hydrolyzed with urease (E.C. 3.5.1.5) [9002-13-5] at 0.7 U/mL and the ammonia formed is assayed by use of glutamate dehydrogenase (E.C. 1.4.1.3) [9029-12-3] at 6.2 U/mL:

$$\text{Urea} + H_2O \longrightarrow 2\,NH_3 + CO_2$$

$$\text{2-Ketoglutarate} + 2\,NH_4^+ + 2\,NADH \rightarrow 2\,\text{L-Glutamate} + 2\,NAD^+ + 2\,H_2O$$

The reaction is followed photometrically (decrease in NADH concentration).

**Kinetic Methods.** The drawback of end-point methods is that a considerable amount of time may be required before the reaction comes to completion; this limits the sample throughput. In addition, many time- and reagent-consuming sample blanks must be determined, to eliminate interference from light-absorbing serum constitutents. These problems can be overcome by assaying the substrate concentration via the rate of the enzyme-catalyzed reaction [5.129]. As shown in Equation (5.1), the kinetics is of pseudo-first order, and the rate is proportional to substrate concentration at concentrations well below $K_m$.

Such assays are most conveniently run on automatic analyzers using the so-called kinetic fixed-time approach, in which the absorbance is read at two constant points of time, and the slope of the absorbance change is calculated. Only one calibration is required.

These methods work only for substrate concentrations below the $K_m$ value, which sets an upper limit on the dynamic range. This upper concentration limit is increased in the presence of competitive inhibitors that increase the apparent $K_m$ of the enzyme. Kinetic substrate assays are described for many analytes.

*Urea* concentration is determined according to the reaction sequence mentioned previously.

*Determination of Glucose.*

$$\text{D-Glucose} + ATP \longrightarrow \text{D-Glucose 6-phosphate} + ADP$$

$$\text{D-Glucose 6-phosphate} + NADP^+ \longrightarrow \text{D-Glucono-$\delta$-lactone 6-phosphate} + NADPH + H^+$$

In the first reaction, glucose is phosphorylated by hexokinase (E.C. 2.7.1.1) [9001-51-8]; then glucose 6-phosphate is dehydrogenated by the action of glucose 6-phosphate dehydrogenase (E.C. 1.1.1.49) [9001-40-5]. The NADPH production is monitored photometrically.

*Determination of Triglycerides.* Triglycerides (fats) are hydrolyzed by lipase (E.C. 3.1.1.3) [9001-62-1] and carboxylesterase (E.C. 3.1.1.1) [9016-18-6]. Glycerol is then phosphorylated by glycerol kinase (E.C. 2.7.1.30) [9030-66-4]. The ADP formed in this reaction is rephosphorylated to ATP with phosphoenolpyruvate and pyruvate kinase (E.C. 2.7.1.40) [9001-59-6]. Finally, the pyruvate is hydrogenated by L-lactate dehydrogenase (E.C. 1.1.1.27) [9001-60-9] and the decrease in

NADH concentration is followed:

Triglyceride + 3 $H_2O$ ⟶ Glycerol + 3 Fatty acid

Glycerol + ATP ⟶ Glycerol 3-phosphate + ADP

ADP + Phosphoenolpyruvate ⟶ ATP + Pyruvate

Pyruvate + NADH + $H^+$ → L-Lactate + $NAD^+$

## 5.2.2. Determination of Enzyme Activity [5.130]

Ideally, the substrate concentration must be so high for an enzyme assay that the value of $K_m$ can be neglected. Then $v = V$, and the reaction rate is equal to the maximum rate defined by the Michaelis–Menten equation. The reaction obeys pseudo zero-order kinetics. In many cases, the enzyme to be assayed cannot be determined directly but must be coupled to a second enzyme-catalyzed reaction that serves as the indicator reaction. In this case, the conditions chosen must make the enzyme reaction which is to be determined the rate-limiting step.

*Activity of Alkaline Phosphatase* (E.C. 3.1.3.1) [9001-78-9].

4-Nitrophenylphosphate + $H_2O$ ⟶ Phosphate + 4-Nitrophenolate

This is the easiest case. A chromogenic substrate is cleaved by the enzyme whose activity is to be determined. At the optimum pH of this reaction, 9.8, the product is dissociated, and its rate of formation can be followed by the increase in light absorption at 405 nm.

*Activity of Creatine Kinase* (E.C. 2.7.3.2) [9001-15-4].

Creatine phosphate + ADP → Creatine + ATP.

This is an example of a reaction coupled with an auxiliary enzyme. The ATP formed in the reaction catalyzed by creatine kinase is used to phosphorylate glucose to glucose 6-phosphate, which is then dehydrogenated, as described previously. The rate of NADPH formation is a measure of the rate of the reaction catalyzed by creatine kinase.

## 5.2.3. Immunoassays

Enzyme immunoassays use enzymes as labels to determine the amount of immunocomplex (antigen–antibody) formed; for a review, see [5.131], [5.132]. Depending on the chemical reaction catalyzed by the respective enzyme, its activity can be measured photometrically, fluorometrically, or luminometrically. Enzymes can be bound to antibodies using such bifunctional coupling reagents as glutaraldehyde [111-30-8] [5.132], 3-maleinimidobenzoyl-*N*-hydroxysuccinimide [15209-14-0] [5.133], or bis(maleido)methyl ether [5.134], depending on the nature of the reactive groups on the enzyme. Horseradish peroxidase is bound directly to the antibody by reaction of carbonyl groups in the carbohydrate part of the enzyme with amino groups of the antibody by Schiff base formation and subsequent reduction with sodium borohydride [5.135].

*Alkaline phosphatase* (E.C. 3.1.3.1) [*9001-78-9*], used in enzyme immunoassays, should be homogeneous on electrophoresis, and its specific activity should be above 2500 U/mg when 4-nitrophenylphosphate is used as substrate at 37 °C. For conjugate synthesis, glutaraldehyde is recommended [5.132]. Alkaline phosphatase activity can also be assayed fluorometrically by using 4-methylumbelliferylphosphate [*3368-04-5*] as substrate. For use in immunohistochemistry, 5-bromo-4-chloro-3-indolylphosphate [*6578-06-9*] is especially suitable. A photometric amplification system for the immunoassay of alkaline phosphatase has been published recently [5.136].

*β-Galactosidase* (E.C. 3.2.1.23) [*9031-11-2*] should have a specific activity of more than 250 U/mg when 4-nitrophenyl-β-D-galactoside is used as substrate at 37 °C. The enzyme should not be present in the aggregated form. One enzyme molecule contains 12 mercapto groups; these are linked to antibodies by maleimide bifunctional reagents. The activity of β-galactosidase can be measured photometrically with 4-methylumbelliferyl-β-galactoside [*39940-54-0*]; in histochemistry, the substrate of choice is 5-bromo-4-chloro-3-indolyl-β-galactoside [*7240-90-6*].

*Horseradish peroxidase* (E.C. 1.11.1.7) [*9003-99-0*] contains 2–3 reactive amino groups per molecule and 12–14.5 wt% of carbohydrates, which can both be used for coupling to antibodies. The enzyme activity can be assayed photometrically in different ways. The chromogen 2,2'-azinobis[3-ethylbenzothiazoline-sulfonate] (ABTS) is especially suited as a substrate, and the peroxidase should have a specific activity of ca. 1000 U/L at 25 °C. 3,3',5,5'-Tetramethylbenzidine [*54827-17-7*] is recommended as a substrate for use in immunohistochemistry.

**Luminescence Reactions.** *Chemiluminescence* occurs when part of the free energy produced in a chemical reaction is emitted as light, because a reaction product is generated in its singlet excited state. *Bioluminescence* occurs in living organisms, where these reactions are catalyzed by enzymes. The corresponding enzymes are called *luciferases*, and the substrates being converted to light-emitting products are called *luciferins*; for a review see [5.137]; see also Section 2.2.5. Because light intensities can be measured with very high sensitivity, luminescence reactions are of increasing interest as indicator systems for immunoassays, giving detection limits comparable to those achieved with radioactive labels, e.g., see [5.138], [5.139].

**Firefly Bioluminescence.** For a long time, bioluminescence was considered to be something exotic. However, several groups have shown that these reactions follow the same rules as other enzyme-catalyzed reactions [5.137]. An example is the reaction catalyzed by luciferase of the American firefly *Photinus pyralis* (E.C. 1.13.12.7) [*61970-00-1*], as mentioned on p. 23. When Michaelis–Menten kinetics are applied to this reaction, the light intensity is

$$I = \frac{\mathrm{d}N}{\mathrm{d}t} = \Phi \cdot v = \Phi \frac{V \cdot c_{\mathrm{ATP}}}{K_{\mathrm{m}}}$$

where    $v$ = reaction rate
        $V$ = maximum reaction rate
        $I$ = light intensity
        $\Phi$ = quantum yield of the reaction
        $N$ = number of photons emitted

Hence, the light intensity is proportional to the reaction rate, and substrate assays using this reaction follow the rules of kinetic substrate determination. These considerations also apply to other enzyme-catalyzed chemiluminescence reactions. Bioluminescence catalyzed by firefly luciferase has been used in immunoassays in two different ways: (1) the luciferase can be used directly as a marker, e.g., [5.140]; or (2) pyruvate kinase can be used as a label, and the ATP formed in the rephosphorylation of ADP by phosphoenolpyruvate (see p. 164) is then assayed by the reaction catalyzed by firefly luciferase [5.141].

**Chemiluminescence Catalyzed by Horseradish Peroxidase.** The peroxidase from horseradish (E.C. 1.11.1.7) [9003-99-0] catalyzes the oxidation of luminol (5'-amino-2,3-dihydro-1,4-phthalazinedione [521-31-3] and luminol derivatives by hydrogen peroxide, with concomitant light emission:

Luminol + $H_2O_2$ + 2 OH$^-$ $\longrightarrow$ Aminophthalate + $N_2$ + 2 $H_2O$

Under normal conditions, this reaction is rather slow and, depending on the substrate concentrations, a lag phase of varying length is observed. Therefore, this reaction is not suitable for routine assay of peroxidase activity. However, several groups of compounds have been found recently that activate this reaction by eliminating the lag phase, which results in an increase in light intensity by a factor of 100–1000. The light signal is stable at a given set of concentrations. Firefly luciferin [2591-17-5] [5.142], 6-hydroxybenzothiazoles [5.143], and phenol derivatives such as 4-iodophenol [540-38-5] and 4-phenylphenol [92-69-3] are used as activators [5.144]. At pH 6, the phenol derivatives also act as activators of a photometric reaction catalyzed by peroxidase [5.145]. Hence, these compounds probably activate the peroxidase.

Immunoassays are described for the determination of several antibodies and metabolites of diagnostic use, based on this chemiluminescence reaction stimulated by peroxidase activator.

# 5.3. Enzymes for Food Analysis [5.1], [5.146]–[5.148]

Use of the catalytic activity of enzymes for analytical determination dates from the middle of the 19th century. Nevertheless, the use of enzymes for food analysis in its present form was established only about 20 or 25 years ago. The factors responsible for this achievement were the large-scale preparation of pure enzymes, the availability of suitable and inexpensive photometers, and the elaboration of both enzymatic procedures and methods for sample preparation.

In *food research*, the amounts of various food constitutents are determined, and changes occurring during technological processing and subsequent storage are monitored. Because of the great specificity manifest by enzymes, enzymatic methods are now used for this purpose.

In *industry*, raw materials are checked, production controls are carried out, and both the end products and the competing products are analyzed. Enzymatic methods were accepted quickly by industry because of the obvious advantage of using easy, rapid, and highly specific methods of analysis.

*Official inspections* involve regular analysis of samples and checking that laws, regulations, and guidelines are being observed. Checks are also made to determine if the food industry is fulfilling its obligation to make complete declarations according to the law. Enzymatic methods were used very early for these purposes not only because of the accurate results obtained from enzymatic analysis but also because of the great flexibility of the methods and their applicability to different types of sample materials. Because of tariffs on food, *customs officials* routinely analyze food by enzymatic methods. *Military inspectors* also use enzymes in the general control of food as well as in stability and storage investigations.

The use of enzymes for analytical determinations is recommended today by many national and international commissions; in fact, standardized enzymatic methods are widely available.

## 5.3.1. Carbohydrates

Carbohydrates are determined routinely in food analysis. The sugar components are an important part of the overall analytical picture: they constitute a large part of the caloric value of food, and the authenticity of food samples can be deduced from the relative amounts of different sugars. The presence of lactose indicates the use of milk. Starch serves as a raising agent in meat products and as a thickening agent.

**Glucose** [50-99-7] is by far the most abundant sugar and is usually determined by using the hexokinase method [5.1 a, vol. VI, pp. 163–172]:

$$\text{D-Glucose} + \text{ATP} \xrightarrow[\text{E.C. 2.7.1.1}]{\text{Hexokinase}} \text{ADP} + \text{Glucose 6-phosphate} \tag{5.2}$$

$$\text{Glucose 6-phosphate} + \text{NADP}^+ \xrightarrow[\text{E.C. 1.1.1.49}]{\text{Glucose 6-phosphate dehydrogenase}} \text{D-Gluconate 6-phosphate}$$
$$+ \text{ NADPH} + \text{H}^+ \tag{5.3}$$

Other enzymatic methods are not recommended for a variety of reasons: e.g., the enzymes are nonspecific (glucose dehydrogenase), or reducing substances in the sample interfere with the reaction (glucose oxidase–peroxidase method). In addition, other methods do not allow the simultaneous determination of fructose in the same cuvette. Glucose is usually determined together with other carbohydrates such as fructose, sucrose, maltose, and starch.

**Fructose** [*57-48-7*]. The enzyme hexokinase also acts on fructose, which is determined after glucose (Reactions 5.4, 5.5, and 5.3) [5.1 a, vol. VI, pp. 321–327]:

$$\text{D-Fructose} + \text{ATP} \xrightarrow[\text{E.C. 2.7.1.1}]{\text{Hexokinase}} \text{ADP} + \text{Fructose 6-phosphate} \tag{5.4}$$

$$\text{Fructose 6-phosphate} \xrightarrow[\text{E.C. 5.3.1.9}]{\text{Glucose phosphate isomerase}} \text{Glucose 6-phosphate} \tag{5.5}$$

Glucose and fructose are determined in wine and fruit juices, e.g., to detect the prohibited addition of sugar. Fructose and glucose are also determined together with sucrose.

**Galactose** [*59-23-4*] is easily determined by using the following reaction [5.1 a, vol. VI, pp. 104–112]:

$$\text{D-Galactose} + \text{NAD}^+ \xrightarrow[\text{E.C. 1.1.1.48}]{\text{Galactose dehydrogenase}} \text{D-Galactonic acid} + \text{NADH} + \text{H}^+ \tag{5.6}$$

This determination is of interest only for certain milk products (yogurt, soft cheese). The method can also be applied to acid hydrolysates of thickening agents like agar, guar, carrageenan, gum arabic, locust-bean gum, and tragacanth gum.

**Mannose** [*3458-28-4*]. The determination of free mannose is also of little interest. However, like galactose, it is an important component of thickening agents and can be estimated after subjecting these substances to acid hydrolysis according to Reactions (5.7), (5.8), (5.5), and (5.3) [5.1 a, vol. VI, pp. 262–267]:

$$\text{D-Mannose} + \text{ATP} \xrightarrow[\text{E.C. 2.7.1.1}]{\text{Hexokinase}} \text{ADP} + \text{Mannose 6-phosphate} \tag{5.7}$$

$$\text{Mannose 6-phosphate} \xrightarrow[\text{E.C. 5.3.1.8}]{\text{Phosphomannose isomerase}} \text{Fructose 6-phosphate} \tag{5.8}$$

**Sucrose** [*57-50-1*]. The disaccharide sucrose does not occur in animals but occurs in many plants in varying amounts. On hydrolysis, sucrose yields an equimolar mixture of glucose and fructose (Reaction 5.9) and is estimated via glucose (Reactions 5.2 and 5.3) [5.1 b, vol. 3, pp. 1176–1179]:

$$\text{Sucrose} + \text{H}_2\text{O} \xrightarrow[\text{E.C. 3.2.1.26}]{\beta\text{-Fructosidase}} \text{D-Glucose} + \text{D-Fructose} \tag{5.9}$$

In addition to measurement of glucose, fructose produced in Reaction (5.9) can also be measured according to Reactions (5.4), (5.5), and (5.3) to ensure accurate results or increase the sensitivity of the determination.

Sucrose, glucose, and fructose are determined in fruit and vegetable products (e.g., juices, jams, tomato puree, or potatoes), bread and biscuits, honey, ice cream, confectionery, desserts, diet food, sugar refinery products, or beverages. A large excess of glucose interferes with the accuracy of sucrose and fructose analysis. Therefore, excess glucose should be removed as completely as possible with a mixture of glucose oxidase and catalase.

**Maltose** *[69-79-4]*. Enzymatic hydrolysis of maltose yields two glucose molecules which are determined as described above [5.1 a, vol. VI, pp. 119–126]:

$$\text{Maltose} + \text{H}_2\text{O} \xrightarrow[\text{E.C. 3.2.1.20}]{\alpha\text{-Glucosidase}} 2 \text{ D-Glucose} \tag{5.10}$$

Sucrose, which also contains an $\alpha$-glucosidic bond, and maltotriose are hydrolyzed as well.

Maltose is determined in malt products, beer, baby food, and often together with partial hydrolysates of starch (glucose syrup, starch sugar; dextrins in beer).

**Lactose** *[63-42-3]*. The carbohydrate of milk, lactose, is an important constituent of infant diets. It is usually determined after hydrolysis (Reaction 5.11) via galactose and indicates the use of milk or milk products in the production of food [5.1 a, vol. VI, pp. 104–112]:

$$\text{Lactose} + \text{H}_2\text{O} \xrightarrow[\text{E.C. 3.2.1.23}]{\beta\text{-Galactosidase}} \text{D-Galactose} + \text{D-Glucose} \tag{5.11}$$

Analysis via galactose gives accurate results because very few samples contain large amounts of galactose. Lactose can also be estimated via glucose (Reactions 5.11, 5.2, and 5.3), but excess free glucose impairs the precision of the results and must be removed with a mixture of glucose oxidase and catalase (see lactulose). The amount of lactose is determined in milk (decreases with mastitis), milk products, and other food as a measure of the milk content, e.g., baby food, bread, biscuits, ice cream, chocolate, desserts, and sausages.

**Lactulose** *[4618-18-2]* is formed when lactose is heated. It is measured by means of Reactions (5.11), (5.4), (5.5), and (5.3) [5.1 a, vol. VI, p. 111] and is used to check the heating of milk.

**Raffinose** *[512-69-6]* occurs in relatively high concentrations in sugar beet. It accumulates in molasses during the production of sugar. Raffinose, which also occurs in soybean, is determined by using Reactions (5.12) and (5.6) [5.1 a, vol. VI, pp. 90–96]:

$$\text{Raffinose} + \text{H}_2\text{O} \xrightarrow[\text{E.C. 3.2.1.22}]{\alpha\text{-Galactosidase}} \text{D-Galactose} + \text{Sucrose} \tag{5.12}$$

Raffinose is measured at different stages in the production of beet sugar. The addition of soybean protein to food is detected indirectly by measuring raffinose, because soybean flour can contain up to 10% of this carbohydrate (addition of soybean is often subject to declaration).

**Starch** [*9005-25-8*]. The analysis of starch has played an important part in food analysis for over 100 years. It is determined via glucose after hydrolysis. The use of acid hydrolysis for this purpose is not advisable because other glucose polysaccharides and oligomers are also hydrolyzed by acids; in addition, undesired reactions can take place (e.g., conversion of glucose to fructose). Amylase is less suited for enzymatic hydrolysis because it does not break down starch to glucose completely. Amyloglucosidase is the enzyme of choice; however, in addition to starch, maltose and oligoglucosides are also hydrolyzed to glucose by this enzyme. Analysis is carried out by using Reactions (5.13), (5.2), and (5.3):

$$\text{Starch} + (n-1)\ \text{H}_2\text{O} \xrightarrow[\text{E.C. 3.2.1.3}]{\text{Amyloglucosidase}} n\ \text{D-Glucose} \tag{5.13}$$

Starch in the sample must be dissolved, which is achieved by heating in an autoclave or by treating with concentrated hydrochloric acid and dimethyl sulfoxide. All steps must be performed carefully according to instructions. Oligoglucosides (maltodextrins, glucose syrup, starch sugar) are separated by extraction with ethanol–water mixtures which do not dissolve starch.

Starch is determined in products such as flour, bread, biscuits, and meat products, which contain starch as a raising agent. Oligoglucosides are estimated in fruit juices, beverages, confectionery, jam, and ice cream. Dextrins (low molecular mass starch) are determined in beer; they indicate the presence of fermentable carbohydrates.

## 5.3.2. Organic Acids

Organic acids and their salts frequently appear in metabolism and are of varying importance. Their presence in food has a considerable effect on taste and may indicate fermentation.

**Acetic Acid** [*64-19-7*]. Acetic acid, a component of the "volatile acids," is determined specifically by using Reactions (5.14)–(5.16) [5.1 a, vol. VI, pp. 639–645]. The use of acetate kinase is not suitable because this enzyme also acts on propionate.

$$\text{Acetate} + \text{ATP} + \text{CoA} \xrightarrow[\text{E.C. 6.2.1.1}]{\text{Acetyl-CoA synthetase}} \text{Acetyl-CoA} + \text{AMP} + \text{Pyrophosphate} \tag{5.14}$$

$$\text{Acetyl-CoA} + \text{Oxaloacetate} + \text{H}_2\text{O} \xrightarrow[\text{E.C. 4.1.3.7}]{\text{Citrate synthetase}} \text{Citrate} + \text{CoA} \tag{5.15}$$

$$\text{L-Malate} + \text{NAD}^+ \underset{\text{E.C. 1.1.1.37}}{\overset{\text{L-MDH}}{\rightleftharpoons}} \text{Oxaloacetate} + \text{NADH} + \text{H}^+ \tag{5.16}$$

Reaction (5.16) is a preceding indicator reaction; the equilibrium must be considered when the calculations are being made.

Acetate is measured in wine, fruit and vegetable products, cheese, dressings, and vinegar as a check for fermentation.

**Ascorbic Acid** [*50-81-7*]. As a vitamin, ascorbic acid is of great biological importance in humans; it is also often used as a food additive in industry, e.g., for fruit and vegetable products. A chemical–enzymatic process (Reaction 5.17) is applied to the quantitative determination of ascorbic acid [5.1 a, vol. VI, pp. 376–385]:

$$\text{L-Ascorbic acid (XH}_2) + \text{MTT} \xrightarrow{\text{PMS}} \text{Dehydroascorbic acid (X)} + \text{Formazan}^- + \text{H}^+ \qquad (5.17)$$

where MTT = 3-(4,5-dimethylthiazolyl-2)-2,5-diphenyltetrazolium bromide
　　　PMS = 5-methylphenazinium methyl sulfate

Ascorbate is oxidatively removed in a sample blank reading in order to increase specificity:

$$\text{L-Ascorbic acid} + 1/2\,\text{O}_2 \xrightarrow[\text{E.C. 1.10.3.3}]{\text{Ascorbate oxidase}} \text{Dehydroascorbic acid} + \text{H}_2\text{O} \qquad (5.18)$$

Ascorbic acid is determined in fruit and vegetable products, meat products, milk, beer, wine, and flour. The vitamin content is often declared.

Dehydroascorbic acid [*490-83-5*] can be determined by using Reactions (5.17) and (5.18), after it has been chemically oxidized to ascorbic acid [5.1 a, vol. VI, pp. 376–385]:

$$\text{Dehydroascorbic acid} + \text{Dithiothreitol (reduced)} \longrightarrow \text{L-Ascorbic acid} + \text{Dithiothreitol (oxidized)}$$
$$(5.19)$$

**Aspartic Acid** [*56-84-8*]. Aspartic acid is estimated primarily in apple juice [5.1 a, vol. VIII, pp. 350–357]:

$$\text{L-Aspartate} + \alpha\text{-Oxoglutarate} \xrightarrow[\text{E.C. 2.6.1.1}]{\text{GOT}} \text{Oxaloacetate} + \text{L-Glutamate} \qquad (5.20)$$

$$\text{Oxaloacetate} + \text{NADH} + \text{H}^+ \xrightarrow[\text{E.C. 1.1.1.37}]{\text{L-MDH}} \text{L-Malate} + \text{NAD}^+ \qquad (5.21)$$

**Citric Acid** [*77-92-9*]. Citric acid is a key substance in metabolism. It occurs abundantly in plants and also in milk. It is determined by using Reactions (5.22) and (5.21) [5.1, vol. VII, pp. 2–12]:

$$\text{Citrate} \xrightarrow[\text{E.C. 4.1.3.6}]{\text{Citrate (pro-3S)-lyase}} \text{Oxaloacetate} + \text{Acetate} \qquad (5.22)$$

The decarboxylation of oxaloacetic acid to pyruvic acid can occur either chemically or enzymatically (oxaloacetate decarboxylase present as an impurity in citrate lyase). However, loss of oxaloacetate does not lead to inaccurate results because the pyruvate formed is also estimated by using the L-lactate dehydrogenase reaction (5.38).

Fruit and vegetable products, bread, cheese, meat products, beverages, wine, tea, and confectionery are all analyzed for citric acid.

**Formic Acid** [*64-18-6*]. An end product of the metabolism of bacteria and fungi, formic acid is employed as a preservative in a variety of foods. However, the use of this additive must comply with the law. Formic acid is determined according to [5.1a, vol. VI, pp. 668–672]:

$$\text{Formate} + \text{NAD}^+ + \text{H}_2\text{O} \xrightarrow[\text{E.C. 1.2.1.2}]{\text{Formate dehydrogenase}} \text{Hydrogen carbonate} + \text{NADH} + \text{H}^+ \tag{5.23}$$

**Gluconic Acid** [*526-95-4*]. Oxidation of glucose yields gluconic acid. $\delta$-Glucono-lactone is employed as a ripening agent in the production of sausages. The enzyme gluconate kinase is used to determine gluconic acid [5.1a, vol. VI, pp. 220–227]:

$$\text{D-Gluconate} + \text{ATP} \xrightarrow[\text{E.C. 2.7.1.12}]{\text{Gluconate kinase}} \text{D-Gluconate 6-phosphate} + \text{ADP} \tag{5.24}$$

$$\text{D-Gluconate 6-phosphate} + \text{NADP}^+ \xrightarrow[\text{E.C. 1.1.1.44}]{\text{6-PGDH}} \text{D-Ribulose 5-phosphate} + \text{NADPH} + \text{H}^+ + \text{CO}_2 \tag{5.25}$$

where 6-PGDH = 6-phosphogluconic acid dehydrogenase.

$\delta$-Gluconolactone is also determined by using Reactions (5.24) and (5.25) after conversion to gluconic acid by alkaline hydrolysis (pH 10–11):

$$\text{D-Glucono-}\delta\text{-lactone} + \text{H}_2\text{O} \longrightarrow \text{D-Gluconate} \tag{5.26}$$

**Glutamic Acid** [*6899-05-4*]. The enzyme glutamate dehydrogenase is applied to the quantitative determination of glutamic acid [5.1a, vol. VIII, pp. 369–376]:

$$\text{L-Glutamate} + \text{NAD}^+ + \text{H}_2\text{O} \xrightarrow[\text{E.C. 1.4.1.3}]{\text{Glutamate dehydrogenase}} \alpha\text{-Oxoglutarate} + \text{NADH} + \text{NH}_4^+ \tag{5.27}$$

The equilibrium of this reaction strongly favors the reductive synthesis of glutamate. Originally, hydrazine was used to trap $\alpha$-oxoglutarate to achieve quantitative conversion. This is more conveniently carried out today by using the enzyme diaphorase and iodonitrotetrazolium chloride (INT):

$$\text{NADH} + \text{H}^+ + \text{INT} \xrightarrow[\text{E.C. 1.8.1.4}]{\text{Diaphorase}} \text{NAD}^+ + \text{Formazan} \tag{5.28}$$

This color reaction is very sensitive; all reducing substances in the sample must be removed during its preparation to avoid interference.

**3-Hydroxybutyric Acid** [*300-85-6*]. The presence of hydroxybutyric acid in an egg indicates that fertilization has taken place and that the egg has been incubated for more than 6 d. Hydroxybutyric acid is estimated by using the following reaction:

$$\text{D-3-Hydroxybutyrate} + \text{NAD}^+ \xrightarrow[\text{E.C. 1.1.1.30}]{\text{3-HBDH}} \text{Acetoacetate} + \text{NADH} + \text{H}^+ \tag{5.29}$$

where 3-HBDH = 3-hydroxybutyrate dehydrogenase.

To ensure that this reaction is quantitative and to increase sensitivity, Reaction (5.29) is coupled to the color reaction (5.28).

**Isocitric Acid** [*6061-97-8*]. The determination of isocitric acid, an intermediate in the citric acid cycle, is of considerable importance for the analysis of fruit juice. The ratio of citrate to isocitrate is very constant; hence, analysis of isocitric acid can be used to show if citric acid has been added to the juice (adulteration). Isocitrate dehydrogenase is used to measure this acid quantitatively [5.1 a, vol, VII, pp. 13–19]:

$$\text{D-Isocitrate} + \text{NADP}^+ \xrightarrow[\text{E.C. 1.1.1.42}]{\text{Isocitrate dehydrogenase}} \alpha\text{-Oxoglutarate} + \text{NADPH} + \text{CO}_2 + \text{H}^+ \qquad (5.30)$$

The lactone or esters of isocitrate are first subjected to alkaline hydrolysis (Reaction 5.31) and then determined by using Reaction (5.30):

$$\text{D-Isocitric acid lactone (esters)} + \text{H}_2\text{O} \xrightarrow{\text{pH 9–10}} \text{Isocitric acid } (+\text{alcohol}) \qquad (5.31)$$

Isocitric acid is too expensive to be used to adulterate citrus fruit juices.

**Lactic Acid** [*598-82-3*]. One of the most important examples of the use of enzymes in food analysis is the enzymatic determination of the stereoisomeric forms of lactic acid. Only L-(+)-lactate is found in animals; it accumulates as the end product of glycolysis, e.g., in muscle. D-(−)-Lactate is formed by certain types of Lactobacilli. There are now no reservations about the intake of D-(−)-lactate in the diet. Indeed, the ADI value (acceptable daily intake) proposed by the WHO was relatively quickly withdrawn. The analysis is carried out by using the following reactions [5.1 a, vol. VI, pp. 582–592]:

$$\text{L-Lactate} + \text{NAD}^+ \xrightarrow[\text{E.C. 1.1.1.27}]{\text{L-LDH}} \text{Pyruvate} + \text{NADH} + \text{H}^+ \qquad (5.32)$$

$$\text{D-Lactate} + \text{NAD}^+ \xrightarrow[\text{E.C. 1.1.1.28}]{\text{D-LDH}} \text{Pyruvate} + \text{NADH} + \text{H}^+ \qquad (5.33)$$

The equilibria of the lactate dehydrogenase reactions favor formation of lactic acid. Originally, hydrazine was used to trap pyruvate in order to achieve quantitative conversion. Today, use of an enzymatic reaction, such as the following, is more convenient:

$$\text{Pyruvate} + \text{L-Glutamate} \xrightarrow[\text{E.C. 2.6.1.2}]{\text{GPT}} \text{L-Alanine} + \alpha\text{-Oxoglutarate} \qquad (5.34)$$

L-Lactate is determined in fruit and vegetable products, meat additives, and baking agents. Both D- and L-lactate are measured in such milk products as yogurt and cheese as a check for microbial activity, and in wine and beer where the lactate content correlates with sensory results.

**Malic Acid** [*6915-15-7*]. L-Malic acid, an intermediate in the tricarboxylic acid (TCA) cycle, is found in fruit (e.g., grapes) and vegetables. In analytical determination, it is oxidized in the presence of malate dehydrogenase and $\text{NAD}^+$ to yield oxaloacetic acid: the generated oxaloacetic acid is no longer trapped with hydrazine, but with an enzymatic reaction [5.1 a, vol. VII, pp. 39–47]:

$$\text{L-Malate} + \text{NAD}^+ \xrightarrow[\text{E.C. 1.1.1.37}]{\text{L-MDH}} \text{Oxaloacetate} + \text{NADH} + \text{H}^+ \tag{5.35}$$

$$\text{Oxaloacetate} + \text{L-Glutamate} \xrightarrow[\text{E.C. 2.6.1.1}]{\text{GOT}} \text{L-Aspartate} + \alpha\text{-Oxoglutarate} \tag{5.36}$$

L-Malic acid is analyzed in fruit products and wine. In comparison with chemical procedures which determine total malic acid content only, enzymatic methods can detect even small amounts of cheap D,L-malic acid added to the product. An enzymatic reaction for the determination of D-malate has been described in the literature, but the required enzyme is not available on the market.

**Oxalic Acid** [*144-62-7*]. More than half of all kidney stones are composed of calcium oxalate. Oxalic acid is also important because it influences the absorption of calcium in the intestine. It is determined by using Reactions (5.37) and (5.23) [5.1 b, vol. 3, pp. 1551–1555]:

$$\text{Oxalate} \xrightarrow[\text{E.C. 4.1.1.2}]{\text{Oxalate decarboxylase}} \text{Formate} + \text{CO}_2 \tag{5.37}$$

Oxalic acid is determined in beer (calcium oxalate is responsible for the Gushing effect), fruit, vegetables, and cocoa products. The oxalate oxidase method [5.1 a, vol. VI, pp. 649–656] is not recommended for food analysis because too many factors interfere with the reaction.

**Pyruvic Acid** [*127-17-3*] is a key intermediate in metabolism. It is determined by using the following reaction [5.1 a, vol. VI, pp. 570–577]:

$$\text{Pyruvate} + \text{NADH} + \text{H}^+ \xrightarrow[\text{E.C. 1.1.1.27}]{\text{L-LDH}} \text{L-Lactate} + \text{NAD}^+ \tag{5.38}$$

Wine is analyzed for pyruvic acid, which is an $\text{SO}_2$-binding component. In evaluating the quality of raw milk (and pasteurized milk), the determination of pyruvate is used as an alternative to microbial counting.

**Succinic Acid** [*110-15-6*]. Succinic acid is also an intermediate in the tricarboxylic acid cycle; it is determined by using Reactions (5.39), (5.40), and (5.38) [5.1 a, vol. VII, pp. 25–33]:

$$\text{Succinate} + \text{ITP} + \text{CoA} \xrightarrow[\text{E.C. 6.2.1.4}]{\text{Succinyl-CoA synthetase}} \text{IDP} + \text{Succinyl-CoA} + \text{p}_\text{i} \tag{5.39}$$

$$\text{IDP} + \text{PEP} \xrightarrow[\text{E.C. 2.7.1.40}]{\text{Pyruvate kinase}} \text{ITP} + \text{Pyruvate} \tag{5.40}$$

Fruits and fruit products are analyzed for succinic acid, which indicates the degree of ripeness. The presence of this acid in whole eggs is a sign of microbial activity following contamination. Other items sampled are cheese, soybean products, and wine.

## 5.3.3. Alcohols

Enzymes are also used for the analytical determination of alcohol, such as ethanol, glycerol, the sugar alcohols sorbitol and xylitol, or cholesterol, in food.

**Ethanol** [*64-17-5*]. Anaerobic metabolism by many microorganisms, notably yeast, results in the formation of ethanol. Reaction (5.41), employed in the determination of ethanol, was one of the first routinely used enzymatic analyses. The acetaldehyde generated was originally trapped with semicarbazide. However, its enzymatic oxidation (Reaction 5.50) is quicker and more efficient [5.1 a, vol. VI, pp. 598–606]:

$$\text{Ethanol} + \text{NAD}^+ \xrightarrow[\text{E.C. 1.1.1.1}]{\text{ADH}} \text{Acetaldehyde} + \text{NADH} + \text{H}^+ \qquad (5.41)$$

The ethanol concentration serves as an index of the quality of alcoholic drinks, such as wine, champagne, spirits, and beer. The ethanol content of products made from fruit indicates either the use of spoiled raw materials or the presence of yeast (e.g., in the case of kefir). The maximum ethanol concentration in drinks that are classified as "low in alcohol" or "nonalcoholic" has been established by law.

**Glycerol** [*56-81-5*] is widely distributed in nature, the bulk of it being bound as lipid. Glycerol is determined by using Reactions (5.42), (5.43), and (5.38) [5.1 b, vol. IV, pp. 1825–1831]:

$$\text{Glycerol} + \text{ATP} \xrightarrow[\text{E.C. 2.7.1.30}]{\text{Glycerol kinase}} \text{Glycerol 3-phosphate} + \text{ADP} \qquad (5.42)$$

$$\text{ADP} + \text{PEP} \xrightarrow[\text{E.C. 2.7.1.40}]{\text{Pyruvate kinase}} \text{ATP} + \text{Pyruvate} \qquad (5.43)$$

The determination of glycerol in wine is of considerable importance because the ratios of ethanol to glycerol and of glycerol to gluconic acid indicate immediately whether the wine has been adulterated by adding glycerol. Beer, spirits, and marzipan are also analyzed for glycerol.

**Sugar Alcohols.** *Sorbitol* [*50-70-4*] is obtained by reduction of fructose; it is of interest as a sugar substitute for people suffering from diabetes. The enzymatic determination of sorbitol can be carried out by using the following reaction, which is not specific, because the enzyme also acts on other polyols [5.1 a, vol. VI, pp. 356–362]:

$$\text{D-Sorbitol} + \text{NAD}^+ \xrightarrow[\text{E.C. 1.1.1.14}]{\text{Sorbitol dehydrogenase}} \text{D-Fructose} + \text{NADH} + \text{H}^+ \qquad (5.44)$$

This analysis can be made specific for sorbitol by quantitatively measuring the fructose produced, via Reactions (5.4), (5.5), and (5.3). Another possibility is to couple Reaction (5.44) to Reaction (5.28). Under these conditions, *xylitol* [*87-99-0*], which also takes part in the sorbitol dehydrogenase reaction, can be measured as well (5.1 a, vol. VI, pp. 484–490]:

$$\text{Xylitol} + \text{NAD}^+ \xrightarrow[\text{E.C. 1.1.1.14}]{\text{Sorbitol dehydrogenase}} \text{Xylulose} + \text{NADH} + \text{H}^+ \tag{5.45}$$

The difference between the quantitative determination of sorbitol via fructose and via INT–diaphorase gives the amount of xylitol present in the sample. Diet foods, pomaceous fruit products, ice cream, confectionery, and biscuits are analyzed for sorbitol. Xylitol is estimated in diet foods, chewing gum, and confectionery.

## 5.3.4. Other Food Ingredients

**Cholesterol** [57-88-5] is an important steroid with diverse physiological functions; e.g., it is a component of the plasma membrane. At the same time, excessive deposition of cholesterol in vascular tissue leads to atherosclerosis. Cholesterol is determined according to Reaction (5.46) coupled to Reactions (5.47) and (5.48) [5.1 a, vol. VIII, pp. 139–148]:

$$\text{Cholesterol} + \text{O}_2 \xrightarrow[\text{E.C. 1.1.3.6}]{\text{Cholesterol oxidase}} \Delta^4\text{-Cholestenone} + \text{H}_2\text{O}_2 \tag{5.46}$$

$$\text{H}_2\text{O}_2 + \text{Methanol} \xrightarrow[\text{E.C. 1.11.1.6}]{\text{Catalase}} \text{Formaldehyde} + 2\,\text{H}_2\text{O} \tag{5.47}$$

$$\text{Formaldehyde} + \text{NH}_4^+ + 2\,\text{Acetylacetone} \longrightarrow \text{Lutidine dye} + 3\,\text{H}_2\text{O} \tag{5.48}$$

Cholesterol is used as a measure of the egg content of food, e.g., noodles and eggnog. In general, the cholesterol content of food made from animal fats is very important. When food containing plant materials is analyzed for cholesterol, phytosterols with a 3$\beta$-hydroxyl group (other than lanosterol) interfere with the reaction.

**Triglycerides,** an important group of lipids containing glycerol, are normally determined in clinical chemistry. Low-fat food is also analyzed enzymatically for triglycerides, which are hydrolyzed with esterase or lipase to yield fatty acids and glycerol (Reaction 5.49). The glycerol generated is determined by Reactions (5.42), (5.43), and (5.38) [5.1 b, vol. IV, pp. 1831–1835]:

$$\text{Triglyceride} + 3\,\text{H}_2\text{O} \xrightarrow[\text{E.C. 3.1.1.1/E.C. 3.1.1.3}]{\text{Esterase and lipase}} \text{Glycerol} + 3\,\text{Fatty acids} \tag{5.49}$$

**Acetaldehyde** [75-07-0] is a flavor substance in beer, yogurt, and spirits. It is bound by SO$_2$ in wine. Acetaldehyde dehydrogenase linked to NAD is used to determine acetaldehyde [5.1 a, vol. VI, pp. 606–613]:

$$\text{Acetaldehyde} + \text{NAD}^+ + \text{H}_2\text{O} \xrightarrow[\text{E.C. 1.2.1.5}]{\text{Aldehyde dehydrogenase}} \text{Acetic acid} + \text{NADH} + \text{H}^+ \tag{5.50}$$

**Ammonia** [*7664-41-7*]. The simplest compound containing nitrogen and hydrogen is ammonia. Fruit juice, milk, biscuits, cheese, diet food, and meat products are analyzed for ammonia [5.1 a, vol. VIII, pp. 454–461]:

$$\alpha\text{-Oxoglutarate} + NADH + H^+ + NH_4^+ \xrightarrow[\text{E.C. 1.4.1.3}]{\text{Glutamate dehydrogenase}} \text{L-Glutamate} + NAD^+ + H_2O \quad (5.51)$$

**Nitrate.** The determination of nitrate by use of a stable enzyme was described for the first time in 1986:

$$\text{Nitrate} + NADPH + H^+ \xrightarrow[\text{E.C. 1.6.6.2}]{\text{Nitrate reductase}} \text{Nitrite} + NADP^+ + H_2O \quad (5.52)$$

The determination of nitrate is of great importance. This compound is the precursor of nitrite and nitrosamines; hence, it is considered hazardous. Samples of water, beverages, meat, milk products, fruits and vegetables, and baby food are among the items analyzed for nitrate. Indeed, fertilizers, which are primarily responsible for the nitrate in food, are also subject to analysis.

**Sulfite.** The enzymatic determination of sulfite was first described in 1983 [5.1 a, vol. VII, pp. 585–591]:

$$SO_3^{2-} + O_2 + H_2O \xrightarrow[\text{E.C. 1.8.3.1}]{\text{Sulfite oxidase}} SO_4^{2-} + H_2O_2 \quad (5.53)$$

$$H_2O_2 + NADH + H^+ \xrightarrow[\text{E.C. 1.11.1.1}]{\text{NADH peroxidase}} 2\,H_2O + NAD^+ \quad (5.54)$$

Sulfite is often used as a preservative in food technology and has diverse functions; e.g., it inactivates enzymes, prevents browning, and binds acetaldehyde in wine. Usually, too high concentrations are obtained when sulfur-containing samples, such as cabbage, leek, onion, garlic, and horseradish, are analyzed by either conventional distillation or the enzymatic procedure. However, use of the enzymatic method in wine analysis gives accurate results.

**Creatine and Creatinine.** Both creatine [*57-00-1*] and creatinine [*60-27-5*] are found in muscle. The amount of meat contained in soup, sauces, and meat extracts is determined by analyzing these foods for creatine and creatinine (Reactions 5.55, 5.56, 5.43, and 5.38) [5.1 a, vol. VIII, pp. 488–507]:

$$\text{Creatinine} + H_2O \xrightarrow[\text{E.C. 3.5.2.10}]{\text{Creatininase}} \text{Creatine} \quad (5.55)$$

$$\text{Creatine} + ATP \xrightarrow[\text{E.C. 2.7.3.2}]{\text{Creatine kinase}} \text{Creatine phosphate} + ADP \quad (5.56)$$

**Lecithin** [*8002-43-5*] (phosphatidylcholine) is the most important phospholipid occurring in plants and animals. The plural term lecithins refers to emulsifying agents, such as soybean preparations (soy lecithin contains 18–20 % phosphatidylcholine). Lecithin can be determined by using phospholipase D, which catalyzes its hydrolytic cleavage to form choline. The choline generated is subsequently measured as Reinecke salt. The method of choice, however, involves the use of phospholipase C from *Bacillus cereus* and of alkaline phosphatase to hydrolyze lecithin (Reactions 5.57 and 5.58), followed by heat inactivation of alkaline phosphatase and determination of choline (Reactions 5.59, 5.43, and 5.38) [5.1 a, vol. VIII, pp. 87–104]:

$$\text{Lecithin} + H_2O \xrightarrow[\text{E.C. 3.1.4.3}]{\text{Phospholipase C}} \text{1,2-Diglyceride} + \text{Phosphorylcholine} \tag{5.57}$$

$$\text{Phosphorylcholine} + H_2O \xrightarrow[\text{E.C. 3.1.3.1}]{\text{Alkaline phosphatase}} \text{Choline} + p_i \tag{5.58}$$

$$\text{Choline} + \text{ATP} \xrightarrow[\text{E.C. 2.7.1.32}]{\text{Choline kinase}} \text{Phosphorylcholine} + \text{ADP} \tag{5.59}$$

The sample can be prepared for analysis in various ways. An aqueous suspension can be made by ultrasonic treatment, or a sample solution can be prepared in *tert*-butyl alcohol and water (e.g., egg products, like eggnog). Alternatively, the sample can be dissolved in *tert*-butyl alcohol after extraction with organic solvents, or it can be prepared by alkaline hydrolysis with methanolic potassium hydroxide (e.g., cocoa products). Interpretation of the results can cause further problems, especially if acetone-soluble phosphorus compounds are used as comparison and if no difference is made between "lecithin" and "lecithins."

**Urea** [*57-13-6*] is the most important end product of protein metabolism, most of it being formed in the liver from ammonia. Urea is excreted via the kidneys. The enzyme urease is used for determination of urea (Reactions 5.60 and 5.51) [5.1 a, vol. VIII, pp. 444–449]:

$$\text{Urea} + H_2O \xrightarrow[\text{E.C. 3.5.1.5}]{\text{Urease}} 2\,NH_3 + CO_2 \tag{5.60}$$

Samples analyzed for urea are meat products and milk, where it serves as an index of protein in the diet of the cow. The urea content of wastewater and swimming pool water indicates the urine concentration in these waters.

# 5.4. Enzymes in Therapy

For many decades, attempts have been made to take advantage of the specifity and efficiency of enzymatic reactions for therapeutic use (see Table 33, p. 160). Some developments of enzymes for use as drugs did not adequately consider the pecularities inherent in their protein nature, and these enzymes no longer meet the requirements for modern drugs. In general, the following requirements must be met by enzyme therapy:

1) The enzyme must reliably reach its site of action in the tissue compartment.

2) The enzyme must function under the environmental conditions found at the presumed site of action, with respect to ionic milieu, substrate and cosubstrate supply, and presence of endogeneous inhibitors.
3) Therapeutic effects must be established in controlled trials and related to the particular activity of the enzyme applied.
4) The therapeutic benefit should outweigh the adverse reactions, particularly immunological complications.

Evidently, these requirements can be met positively by only a few enzymes approved up to now either as the main active principles or as additives in pharmaceutical formulations. Since a critical review of each individual enzyme with regard to dosage, route of administration, bioavailability, mode of action, and therapeutic value would be beyond the scope of this article, a selection of widely accepted and sufficiently documented enzymes is presented and complemented with topical trends.

## 5.4.1. Digestive Enzymes

Lytic enzyme deficiency syndromes are treated by oral substitution with enzyme preparations. Purity, source, and exact dosage of the digestive enzymes are not considered critical because of lack of absorption and intestinal proteolytic degradation. However, enzyme preparations must be adjusted to survive the pH changes during passage through the digestive tract and finally reach — in active form — the ionic environment adequate for its function.

Impaired gastric proteolysis can be treated with crude hog *pepsin* (E.C. 3.4.23.1) [*9001-75-6*] and an adequate dose of hydrochloric acid to optimize pepsin action (trade name: e.g., Enzynorm) [5.149].

Pancreatic insufficiency deserves more attention, since the exocrine function of the pancreas includes the degradation of alimentary protein, starch, and fat. Consequently, the complex pancreatic function is usually not replaced by isolated enzymes but by *pancreatin* [*8049-47-6*], a pancreatic extract containing, among other enzymes, trypsin (E.C. 3.4.21.4), chymotrypsin (E.C. 3.4.21.1), α-amylase (E.C. 3.2.1.1), and lipase (E.C. 3.1.1.3). Digestion of neutral fats, which is particularly impaired in pancreatic insufficiency [5.150], [5.151], is commonly promoted by the addition of bile acids [5.152]. Since pancreatic enzymes are denatured by the acidic pH of the stomach and are optimally active at alkaline duodenal pH, they are administered preferably as enterocoated formulation (trade names: e.g., Enzypan, Cotazym, Pancrex, Panzytrat, Gillazym, Pankreatan, Kreon).

Orally applied human or mammalian enzymes are not superior as digestive aids to lytic enzymes obtained from microbes or plants. Therefore, many substitutes for pancreatin such as *bromelain* (E.C. 3.4.22.4) [*9001-00-7*], *papain* (3.4.22.2) [*9001-73-4*], or extracts of *Aspergillus oryzae* (trade names: e.g., Luizym, Nortase) containing cellulases (E.C. 3.2.1.4), proteases, and α- and β-amylase (E.C. 3.2.1.1 and 3.2.1.2) have been developed and sometimes combined with pancreatin (trade name: e.g., Combizym).

The therapeutic efficacy of orally administered digestive enzymes has been established in controlled trials [5.153]–[5.155]. Adequate digestion of lipids by substitution therapy can be achieved even during complete pancreatic failure [5.150], [5.151]. However, minor digestive disorders probably do not require or benefit from treatment with digestive enzymes.

Orally administered lytic enzymes appear not to be toxic, although inhalation (e.g., of pancreatin dust) may cause allergic rhinitis [5.152]. Since many of these proteases degrade plasma proteins and cell membrane constituents in vitro, the obvious lack of systemic toxicity strongly argues against any significant absorption of the active enzymes from the gastrointestinal tract. In sharp contrast to this assumption, a variety of therapeutic effects of orally administered proteases such as fibrinolysis, prevention of tumor metastases, or improvement of spermatogenesis [5.156] have been tentatively attributed to partial absorption of the native enzymes. These discrepancies require clarification. However, up to this time absorption of any orally administered, intact protease has not been convincingly demonstrated in humans.

## 5.4.2. Debridement of Wounds

A variety of proteases, usually combined with broad-spectrum antibiotics in ointment, powder, or solution, are recommended for removal of fibrin layers from wounds to improve healing [5.157], [5.158]. Whether the beneficial effects of this kind of treatment can be attributed to the debridement itself or to the improved bioavailability of the antibiotics administered simultaneously is difficult to determine [5.159].

At present, only heterologous proteins such as bovine *plasmin* (E.C. 3.4.21.7 [*9001-90-5*]; trade names: Fibrolan, Lysofibrin, Actase, Elase), *trypsin* (E.C. 3.4.4.4 [*9002-07-7*]; trade name: Leukase), or a clostridial *collagenase* (E.C. 3.4.24.3 [*9001-12-1*]; trade names: Iruxol, Santyl, Biozyme C) are in use. This implies that despite topical application, the risk of allergic reactions must be considered.

## 5.4.3. Improvement of Blood Rheology

*Ancrod* (E.C. 3.4.21.28) [*9046-56-4*], a protease isolated from the venom of the Malayan pit viper, *Agkistrodon rhodostoma,* trade names: Venacil, Arwin) cleaves fibrinogen in a manner similar to thrombin, but does not promote cross-linking in the resulting fibrin gel [5.160]. Clinically, degradation of fibrinogen improves blood fluidity and is, therefore, believed to increase the supply of oxygen to the tissues in circulatory disorders [5.161]. In controlled trials, infusion of ancrod from several days up to four weeks yielded substantial defibrinogenation associated with marked and long-lasting clinical improvement in patients suffering from occlusive peripheral artery disease [5.162]. A similar mode of action and corresponding clinical results are reported for *batroxobin* (E.C. 3.4.21.29) [*9039-61-6*], a protease derived from the venom of *Bothrops atrox* (trade names: Defibrase, Botropase, Reptilase). Loss of activity after prolonged therapy, resulting from antibody formation, is observed with both snake venom proteases. Bleeding complications are the most common side effects [5.161].

*Kallikrein* isolated from hog pancreas (E.C. 3.4.21.8 [*9001-01-8*]; trade names: Padutin, Bioactin, Prokrein, Onokrein P) is also recommended for the treatment of peripheral arterial diseases. Its therapeutic effect, however, is attributed primarily to the limited proteolysis of kininogens, which generates the vasodilating mediator kallidin [5.163].

## 5.4.4. Thrombolysis

Therapeutic lysis of existing blood clots is the domain of plasminogen activators. Endogenous plasminogen is activated physiologically by the serine proteases urokinase (E.C. 3.4.21.31) and tissue plasminogen activator (E.C. 3.4.21.99). Plasmin (E.C. 3.4.21.7) generated in this manner degrades fibrin, the main protein constituent of blood clots.

*Streptokinase* (trade names: e.g., Streptase, Kabikinase, Awelysin, Kinalysin), the first plasminogen activator used in therapy, is not an enzyme as such, but directly activates plasminogen nonproteolytically by forming a stoichiometric 1:1 complex which, in turn, generates plasmin from residual plasminogen [5.164], [5.165]. Because of its complex mechanism of action, streptokinase is far from an ideal drug [5.166]. Being a protein derived from $\beta$-hemolytic streptococci, streptokinase acts as a strong antigen in humans. Since preexisting antibodies from streptococcal infections are frequently observed, pretreatment with corticosteroid is indicated when streptokinase is used for fibrinolysis [5.166]. Despite these disadvantages, the extensive clinical experience with streptokinase represents the basis of present knowledge of thrombolytic therapy. The principal indications for thrombolysis are occlusion of surgically inaccessible veins, lung embolism, and acute myocardial infarction [5.166]. The therapeutic benefit of recanalization of occluded coronary arteries, as far as cardiac function and long-term survival are concerned, has been debated until recently. A controlled prospective multicenter trial with 12 000 patients concluded that because of the prompt salvage of myocardial tissue, lytic therapy significantly improves the one-year survival rate if acute myocardial infarction is effectively treated within the first 3 h after the onset of symptoms. A long-term benefit is not observed, if thrombolysis is initiated more than 6 h after occlusion [5.167]. Experience with streptokinase, however, has also revealed the major problem of lytic therapy inherent in the principle of systemic plasminogen activation: plasmin attacks both fibrin and fibrinogen. Degradation of fibrinogen may lead to fatal bleeding complications.

*Urokinase* (E.C. 3.4.21.31 [*9039-53-6*]; trade names: Actosolv, Alphakinase, Abbokinase, Breokinase, Rheothromb, Ukidan), although less well-documented by clinical trials, is at present considered the drug of choice in thrombolytic therapy. It is a chemically and functionally defined enzyme [5.168]–[5.171], isolated from human urine or tissue culture. Thus, if pure, it does not cause any immunological complications. It specifically cleaves plasminogen to form plasmin with a straight dose-effect relationship. Therefore, it is more easily controlled than streptokinase. Urokinase is available in two biochemically related forms with different molecular masses (see Table 33, p. 160). Clinically relevant differences between both urokinase types have not yet been established. The major reason for the still restricted use of urokinase compared to streptokinase probably results from the comparatively high production cost of urokinase. However, a nonglycosylated human urokinase has recently been isolated from recombinant *Escherichia coli* and shown to be functionally equivalent to the human enzyme [5.172].

Recent trends in thrombolytic treatment center on the development of clot-specific plasminogen activators. *Tissue plasminogen activator* (tPA) effectively binds to fibrin. Thus, plasminogen activation preferably occurs at sites of fibrin deposition, and the therapeutic margin between dissolution of the thrombus and fibrinogen degradation is thereby broadened [5.173]. Human tissue plasminogen activator has been cloned in *E. coli* [5.174] and is now being produced in transformed mammalian cells. It was under intensive clinical investigation [5.175] and has been approved for the treatment of acute myocardial infarction (trade names: Activase, Actilyse).

The natural precursor of urokinase, the single-chain urokinase or *pro-urokinase*, also exhibits clot-specific lysis. Originally believed to be an inactive zymogen, pro-urokinase effectively activates plasminogen in the absence of fibrin but not in serum [5.176]. In a variety of animal models [5.177] and clinical investigations [5.178], pro-urokinase exhibited a therapeutic profile almost identical to that of tissue plasminogen activator despite obvious differences in the mechanism of action of the two compounds [5.179]. At present, two pro-urokinase species are available for clinical testing: a glycosylated form isolated from tissue culture [5.180] and a nonglycosylated form derived from recombinant *E. coli* [5.181], [5.182].

## 5.4.5. Support of Blood Clotting

Superficial bleeding resulting from trauma or surgical intervention has been treated for three decades with *thrombin* (E.C. 3.4.21.5) [9002-04-4] isolated from bovine blood (trade names: e.g., Thrombinar, Velyn, Topostatin). Thrombin represents the protease that cleaves fibrinogen, thereby initiating formation of insoluble fibrin. The fact that thrombin may not be applied intravascularly is due not to its heterologous nature but rather to its induction of intravascular coagulation.

Disturbances of the blood clotting system are treated by substitution with homologous proteins. The cascade of biochemical events leading to blood vessel occlusion is too complex to be discussed here in detail [5.183]. Congenital deficiencies of blood clotting factors, e.g., of *Factor VIII* (hemophilia A), *Factor IX* (hemophilia B), or *Factor XIII*, are associated with severe disturbance of blood coagulation and require substitution therapy. Patients with acquired relative deficiencies are also reported to benefit from the addition of coagulation factors, particularly before surgery. Depending on the type of deficiency, either such isolated factors as Factor VIII (trade names: e.g., Autoplex, Factorate, Profilate HS), Factor XIII (trade name: e.g., Fibrogammin HS), or Factor IX (trade name: Factor IX S-TIM 4) or blood fractions containing various factors (trade names: e.g., Profilnine HS 500 nonocomplex, Prothromplex S-TIM 4) are administered. Presently, all are prepared from pooled human blood. Therefore, the major problem in manufacturing such preparations lies in reliably eliminating viral contamination. Primarily for this reason, new sources are being developed by recombinant DNA technology [5.184], [5.185].

## 5.4.6. Therapy of Malignancies

Metabolic differences between malignant and normal cells are rare and restricted to special types of tumors. The various forms of lymphocytic leukemia depend on the availability of L-asparagine, an amino acid not essential to healthy human tissue. Based on this rationale, an *asparaginase* (E.C. 3.5.1.1) [*9015-68-3*] isolated from *Escherichia coli* (trade names: Crasnitin, Elspar, Kidrolase, Leunase) was developed for intravenous injection to starve floating and resident tumor cells by depriving the circulating blood of asparagine. This treatment, either alone or combined with polychemotherapy, induces remissions in acute childhood lymphoblastic leukemia, acute adult lymphocytic leukemia, non-Hodgkin's lymphoma, and also some non-lymphocytic forms of leukemia [5.186]–[5.191].

The main problems of this therapy are the rapid development of resistance by tumor cells, disturbances in the homeostatic system [5.192], and immunological complications resulting from the heterologous nature of the injected protein [5.193].

## 5.4.7. Chemonucleolysis in Intervertebral Disk Herniation

Herniation of the intervertebral disk is caused by a protrusion or extrusion of the nucleus pulposus into the vertebral channel, with concomitant painful pressure on nerve tissue. The purpose of chemonucleolysis is to dissolve the mucopolysaccharide protein complex of the protruding nucleus pulposus. This reduces the pressure on nerve roots without affecting the collagen and other structural proteins of the anulus fibrosus, adjacent bone, or nerve tissue. *Chymopapain* (E.C. 3.4.22.6) [*9001-73-4*], a plant protease isolated from *Carica papaya* (trade names: Discase, Chymodiactin), appears to approach the desired profile, if adequate dosages are injected directly into the nucleus pulposus [5.194]. The enzyme attacks the protein part of the insoluble mucopolysaccharide protein complex, thereby allowing further degradation of the polysaccharides. The therapeutic efficacy of chymopapain treatment, although controversial in the past, has now been established in a placebo-controlled, double-blind study performed on patients with unilateral sciatica resulting from verified intervertebral disk herniation [5.195]. Treatment, however, remains restricted to specialized neurosurgeons because of the difficult injection technique and the comparatively high complication rate. According to the manufacturers' instructions, myelitis associated with paresis or even paraplegia occurs rarely (1:18 000). These cases probably result from inexact injection. Anaphylactic reactions are more frequently observed (0.4%). Chymopapain treatment is therefore contraindicated in patients with a history of allergy and should never be repeated.

## 5.4.8. Treatment of Inflammation and Reperfusion Injury

During inflammation, the superoxide radical along with other mediators is released from phagocytes. This process, which represents an integral part of the host defense mechanism, is also considered to contribute to tissue destruction (e.g., depolymerization of hyaluronic acids, proteoglycans, and collagen) and to the lysis of cell membranes [5.196].

*Superoxide dismutase* (E.C. 1.15.1.1) [*9054-89-1*] effectively removes the superoxide radical by catalyzing its dismutation to yield molecular oxygen and hydrogen peroxide [5.197]. Therefore, injected superoxide dismutase, which spreads in the extracellular spaces, presumably prevents superoxide-dependent pathological events in noninfectious inflammatory disorders. Superoxide dismutase isolated from bovine tissue (trade names: Peroxinorm, Ontosein, Oxinorm) proved to be therapeutically effective in a series of placebo-controlled trials, if injected intra-articularly in patients suffering from osteoarthritis of the knee joint [5.198]. Impressive therapeutic results are also obtained in radiation cystitis and interstitial cystitis, if the enzyme is infiltrated into the mucosa of the urinary bladder [5.199]. Intrafocally administered superoxide dismutase substantially ameliorated the disabling symptoms in Peyronie's disease. Superoxide dismutase per se is not toxic. However, because of its bovine origin, the risk of allergic reactions in humans must be considered. With an overall incidence of 0.8% of allergic reactions and 0.04% of anaphylactic reactions [5.200], [5.201], the immunologic potential of superoxide dismutase is surprisingly low. An intravascular application of the heterologous protein is nevertheless considered to be too risky.

Systemic administration of high doses of superoxide dismutase would be required to prevent postischemic tissue damage. During ischemia, hypoxanthine accumulates because of degradation of high-energy nucleoside phosphates. Simultaneously, xanthine dehydrogenase is transformed into xanthine oxidase. Upon reoxygenation of the tissue, excess superoxide is formed by the xanthine oxidase reaction. Depending on the tissue, endothelial damage initiated by the xanthine oxidase reaction may be further complicated by invasion and activation of phagocytes. In animal experiments, the resulting tissue damage could be largely inhibited by pretreatment with xanthine oxidase inhibitors or by flooding the vascular bed with superoxide dismutase during reperfusion [5.202]–[5.204]. In all cases of unpredictable onset and duration of the ischemic event, prevention of reperfusion damage by superoxide dismutase is considered to be the alternative of choice. In this context, the salvage of myocardial tissue in patients subjected to fibrinolytic treatment (see Section 5.4.4) seems to be a most desirable goal. The recent development of a genuine human superoxide dismutase from recombinant yeast [5.205] offers the possibility of exploring this therapeutic approach.

# 6. Enzymes in Genetic Engineering

[6.1]–[6.21], [6.225]–[6.264]

Cloning techniques provide the basis for recombinant DNA technology and its applications in genetic engineering (Figs. 49 and 50). The term "genetic engineering" includes methods for the formation of new combinations of genetic material and for reintroducing and multiplying the recombinant nucleic acid molecules in a new cellular environment.

The availability of a large variety of class II restriction endonucleases and DNA modification methyltransferases with different sequence specificities is a prerequisite for recombinant DNA technology (Section 6.1). *Analytical applications* of these enzymes can be classified roughly as follows (see Fig. 50):

1) Molecular analysis of chromosome and genome structure (mapping, degree of methylation)
2) Characterization of phenotypically manifest or latent hereditary genetic defects on the DNA level (restriction fragment length polymorphism [RFLP] analysis)
3) Evidence of cell degeneration at an early stage by alteration of certain restriction sites in oncogene sequences
4) Taxonomy of viruses or other disease-causing organisms by the correlation of characteristic fragment patterns
5) Phylogenetic relationships by comparison of selected restriction sites in essential genes such as the hemoglobin gene.

In addition, DNA- and RNA-modifying enzymes are essential tools for *cloning technique*; these enzymes are discussed in Sections 6.2–6.6.

Cloning technique allows the identical multiplication of specific genes and fragments thereof in appropriate host cells to high copy numbers. From these overproducing cells — the clones — the coding DNA sequences and, after expression, the corresponding proteins can subsequently be isolated in large quantities and high purity.

The starting materials for gene cloning are first of all complementary mRNA molecules which are being retro-transcribed by the retroviral enzyme reverse transcriptase (E.C. 2.7.7.49) into complementary double-stranded DNA molecules (cDNA clones). The starting material may also be the structural gene itself which must first be isolated from the genome, i.e., the entire chromosomal DNA (genomic clones). Furthermore, in addition to the structural gene, essential *regulatory elements* must be available. Isolation of both specific gene fragments and regulatory sequences is achieved by fragmentation of the DNA with sequence-specific class II restriction endonucleases. This requires that the structural gene and the regulatory elements can be excised from the chromosome at exact nucleotide positions.

# Synthesis of Recombinant DNA

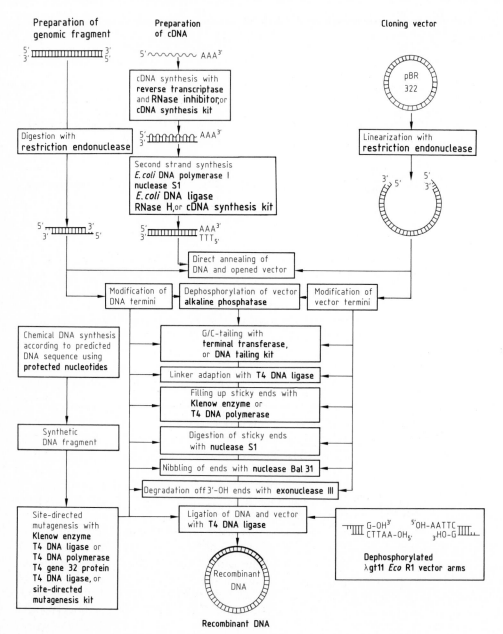

**Figure 49.** Use of DNA- and RNA-modifying enzymes in recombinant DNA synthesis
For abbreviations, see text

# Analysis of Recombinant DNA

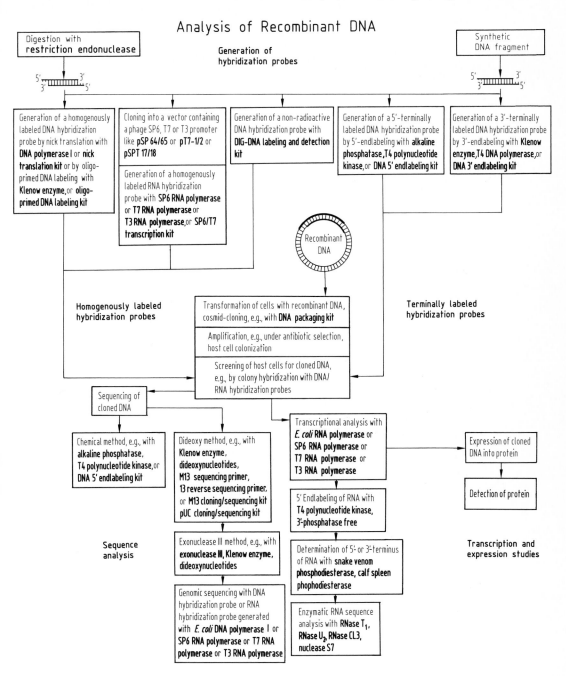

**Figure 50.** Use of RNA- and DNA-modifying enzymes in recombinant DNA analysis
For abbreviations, see text.

The cDNA or the genomic fragments obtained by restriction endonucleases can then be modified further by DNA-modifying enzymes such as exonucleases, kinases, or phosphatases. The fragment ends can also be altered in their sequence specificity by attachment of synthetic oligonucleotides, i.e., *linkers* or *adapters*.

With the help of T4 DNA ligase, the isolated fragments are inserted into special cloning vehicles, the *vectors*. These vectors may, for example, be extrachromosomal circular plasmids (e.g., pBR 322), appropriate derivatives of specific bacteriophages (e.g., λ, M 13), or eukaryotic viruses (e.g., SV 40).

The chimeric molecules obtained in this way, i.e., the recombinant DNA, are then inserted into new host cells, such as prokaryotic (e.g., *Escherichia coli*), yeast (e.g., *Saccharomyces cerevisiae*), or eukaryotic (e.g., CHO) cells which are easy to cultivate. All vectors carry the *origin element ori*, a DNA sequence for the autonomous replication of the DNA hybrid. The recombinant DNA is thus able to amplify in transformed cells irrespective of the regular cell cycle. Growth of the host cells under selective conditions, e.g., on culture media containing specific antibiotics, ensures that only transformed bacteria replicate. Bacterial cells that do not take up recombinant DNA molecules die after a short period in such selective media. In this way, only transformed cells are able to grow, and the introduced vectors, and thus the inserted DNA fragments, are identically amplified in a selective way to different copy numbers, depending on the nature of the vector used. This is referred to as *cloning* because under selective growth conditions, all bacteria can be traced back to a homogenous population of identically transformed cells, the cell clone.

Large quantities of identical recombinant DNA molecules can be purified easily from the cultivated bacteria. Subsequently, the inserted DNA fragment can again be excised from the vector with the help of the initially used class II restriction endonucleases. This DNA fragment that contains the gene to be amplified can be characterized, e.g., by physical mapping and sequencing.

In the final stage, the identically amplified DNA fragment is inserted into efficient expression vectors which have been provided with regulatory control elements. After renewed introduction into appropriate host cells, the regulatory elements are triggered by growth conditions in such a way that primarily the cloned genes are transcribed into complementary mRNA. In prokaryotes, the synthesized mRNA is directly translated into colinear proteins by the ribosomes. In eukaryotes, the mRNA is first synthesized as a precursor mRNA which is processed into mature mRNA by capping, splicing, and polyadenylation. The structure and sequence of the various RNA species can be analyzed by using RNA-modifying enzymes. As in prokaryotes, the mature mRNA is then translated into colinear protein but often further processed by selective cleavage and/or glycosylation. Often the overproduced protein is synthesized in such high concentrations that it precipitates in the cell (refractile [RF] bodies) and can, therefore, be isolated from the cell mass in high yield and with high purity under denaturing conditions.

# 6.1. Restriction Endonucleases and DNA Modification Methyltransferases

[6.22]–[6.29]

## 6.1.1. Classification

In the search for enzymes that are the basis of restriction–modification systems [6.33], [6.140], class I enzymes were discovered in 1968 [6.137], [6.151]. The first member of a class II restriction endonuclease, *Hind*II, was found in 1970 [6.115], [6.215]. Later, enzymes of class III were also isolated [6.113], which have properties between those of class I and class II. A selection of important class II restriction endonucleases (ENases) and DNA modification methyltransferases (MTases) is given in Table 34 (pp. 190–197).

**Class I** enzymes are coded by the three gene segments *hsd*S, *hsd*R, and *hsd*M, expressing the functions for sequence specificity, restriction, and DNA modification methyl transfer. The corresponding restriction endonucleases and DNA modification methyltransferases are enzyme complexes of high molecular mass. These enzyme complexes require magnesium ions, ATP, and SAM as essential cofactors. Class I enzymes bind to and methylate specific sequences, but cut unspecifically after translocation into the 3'-direction [6.34], [6.68], [6.156], [6.251]. An example of a class I system is the *Eco*B restriction–modification system. The early steps of SAM-binding, i.e., enzyme activation and formation of the initial enzyme–DNA complex, are common to both restriction and modification; however, these reaction pathways diverge after the stable complex has been formed. In the case of recognition, ATP leads to the release of the enzyme from modified DNA; it stimulates translocation in unmodified recognition complexes. This reaction is coupled to ATP hydrolysis. Sequential cleavage of both DNA strands occurs in a limited sequence stretch some 100 to 1000 base pairs away from the binding site.

**Class II** enzymes are expressed as two single proteins from two different genes acting as separate enzymes. Class II restriction endonucleases hydrolyze both DNA strands at specific phosphodiester bonds within, or very near, the recognition sequence. Their activity depends only on magnesium ions. The corresponding class II DNA modification methyltransferases methylate both DNA strands, independent of the presence of the respective nuclease in an SAM-dependent reaction. An example of a class II system is the *Eco*RI restriction–modification system. Class II restriction endonucleases are by far the most important tools for the synthesis of recombinant DNA because of their absolute sequence specificity for both binding and cleavage [6.73], [6.157], [6.163], [6.164], [6.190], [6.212], [6.242].

**Class III** enzymes are also expressed from two genes. However, like class I enzymes, the corresponding restriction endonucleases and DNA modification methyltransferases form enzyme complexes in which both activities act simultaneously.

**Table 34.** Recognition sequences of important class II restriction endonucleases (ENases) and DNA modification methyltransferases (MTases)

**A. Class II enzymes with palindromic recognition sequences**
**A.1. Tetra-, hexa-, or octanucleotide recognition sequences (subclass IIP enzymes)**
**A.1.1. Interal AT palindromes**

| Position no. | Recognition sequence | Enzyme | Number of recognition sites | | | | | | | Microorganism | References |
|---|---|---|---|---|---|---|---|---|---|---|---|
| | | | λ | Ad2 | SV40 | ΦX174 | M13mp7 | pBR322 | pBR328 | | |
| 1 | AATT | TspEI | 188 | 0 | 39 | 25 | 65 | 8 | 13 | *Thermus* species, strain EI | [6.23] |
| 2 | AAATTT | — | 16 | 13 | 4 | 5 | 3 | 0 | 0 | — | — |
| 3 | $\overset{0\ +}{\mathrm{G/AATTC}}$ | EcoRI | 5 | 5 | 1 | 0 | 2 | 1 | 1 | *Escherichia coli* RY13 | [6.88], [6.100], [6.227] |
| | $\overset{M}{\mathrm{GAATTC}}$ | M·EcoRI | 5 | 5 | 1 | 0 | 2 | 1 | 1 | *Escherichia coli* RY13 | [6.64], [6.88] |
| 4 | CAATTG | MfeI | 8 | 4 | 4 | 1 | 0 | 0 | 0 | *Mycoplasma fermentans* | [6.23] |
| 5 | TAATTA | — | 8 | 4 | 2 | 1 | 3 | 0 | 0 | — | — |
| 6 | $\overset{m}{\mathrm{GA}}/\mathrm{TC}$ | DpnI | 116 | 87 | 8 | 0 | 8 | 22 | 27 | *Diplococcus pneumoniae* | [6.130], [6.131] |
| | $/\overset{0+}{\mathrm{GATC}}$ | MboI | 116 | 87 | 8 | 0 | 8 | 22 | 27 | *Moraxella bovis* | [6.63], [6.77] |
| | $\overset{+}{\mathrm{GATC}}$ | NdeII | 116 | 87 | 8 | 0 | 8 | 22 | 27 | *Neisseria denitrificans* | [6.236], [6.241] |
| | $/\overset{0+}{\mathrm{GATC}}$ | Sau3AI | 116 | 87 | 8 | 0 | 8 | 22 | 27 | *Staphylococcus aureus* | [6.63], [6.220] |
| | $\overset{M}{\mathrm{GATC}}$ | M·EcodamI | 116 | 87 | 8 | 0 | 8 | 22 | 27 | *Escherichia coli* HB101 | [6.48], [6.98], [6.214] |
| 7 | $\mathrm{A}/\overset{0}{\mathrm{GATCT}}$ | BglII | 6 | 11 | 0 | 0 | 1 | 0 | 0 | *Bacillus globigii* | [6.65], [6.179] |
| 8 | $\overset{0+0}{\mathrm{G/GATCC}}$ | BamHI | 5 | 3 | 1 | 0 | 2 | 1 | 1 | *Bacillus amyloliquefaciens* H | [6.194], [6.246] |
| | $\overset{M}{\mathrm{GGATCC}}$ | M·BamHI | 5 | 3 | 1 | 0 | 2 | 1 | 1 | *Bacillus amyloliquefaciens* RUB 500 | [6.99], [6.162] |
| | $(^A_G)/\mathrm{GATC}(^T_C)$ | XhoII | 21 | 22 | 3 | 0 | 4 | 8 | 7 | *Xanthomonas holcicola* | [6.173] |
| 9 | $\overset{0+}{\mathrm{CGAT}}/\mathrm{CG}$ | PvuI | 3 | 7 | 0 | 0 | 1 | 1 | 1 | *Proteus vulgaris* | [6.81] |
| 10 | $\mathrm{T}/\overset{+0}{\mathrm{GATCA}}$ | BclI | 8 | 5 | 1 | 0 | 0 | 0 | 1 | *Bacillus caldolyticus* | [6.37] |
| 11 | $\mathrm{CATG}/$ | NlaIII | 181 | 183 | 16 | 22 | 15 | 26 | 27 | *Neisseria lactamica* | [6.186] |
| 12 | ACATGT | — | 2 | 9 | 0 | 0 | 3 | 1 | 0 | — | — |

| No. | Sequence | Enzyme | | | | | | | | Organism | References |
|---|---|---|---|---|---|---|---|---|---|---|---|
| 13 | GCATG/C (°,+) | SphI | 6 | 8 | 2 | 0 | 0 | 1 | 1 | *Streptomyces phaeochromogenes* | [6.74], [6.118] |
| 14 | C/CATGG (+) | NcoI | 4 | 20 | 3 | 0 | 0 | 0 | 1 | *Nocardia corallina* | [6.132] |
|  | C/C(A/T)(A/T)GG (+) | StyI | 10 | 44 | 8 | 0 | 1 | 1 | 2 | *Escherichia coli* KM 201 [pST27 *hsd*, S-a] | [6.155] |
| 15 | TCATGA | BspHI | 8 | 3 | 2 | 1 | 4 | 4 | 4 | *Bacillus* species H | [6.23] |
| 16 | TATA | – | 113 | 65 | 17 | 16 | 16 | 8 | 13 |  | – |
| 17 | ATATAT | – | 11 | 3 | 0 | 4 | 4 | 1 | 3 |  | – |
| 18 | GTATAC | SnaI | 3 | 3 | 0 | 0 | 0 | 1 | 0 | *Sphaerotilus natans* C | [6.183] |
| 19 | CTATAG | – | 0 | 4 | 2 | 0 | 0 | 0 | 0 |  | – |
| 20 | TTATAA | – | 12 | 4 | 3 | 2 | 2 | 0 | 0 |  | – |

### A.1.2. Internal GC palindromes

| No. | Sequence | Enzyme | | | | | | | | Organism | References |
|---|---|---|---|---|---|---|---|---|---|---|---|
| 21 | AG/CT (+) | AluI | 143 | 158 | 34 | 24 | 24 | 16 | 14 | *Arthrobacter luteus* | [6.192] |
|  | AGCT (M) | M·AluI | 143 | 158 | 34 | 24 | 24 | 16 | 14 | *Arthrobacter luteus* | [6.127] |
| 22 | A/AGCTT (+) | HindIII | 6 | 12 | 6 | 0 | 0 | 1 | 1 | *Haemophilus influenzae* R$_d$ | [6.102], [6.172], [6.178] |
| 23 | GAGCT/C (°) | SacI | 2 | 16 | 0 | 0 | 0 | 0 | 0 | *Streptomyces achromogenes* | [6.32] |
| 24 | CAG/CTG (+) | PvuII | 15 | 24 | 3 | 3 | 3 | 1 | 1 | *Proteus vulgaris* | [6.81] |
| 25 | TAGCTA | – | 2 | 4 | 2 | 0 | 3 | 0 | 0 |  | – |
| 26 | GG/CC (+,°) | HaeIII | 149 | 216 | 18 | 11 | 15 | 22 | 30 | *Haemophilus aegyptius* | [6.47], [6.152] |
|  | GGCC (M) | M·HaeIII | 149 | 216 | 18 | 11 | 15 | 22 | 30 | *Haemophilus aegyptius* | [6.145] |
| 27 | AGG/CCT (+) | StuI | 6 | 11 | 7 | 1 | 0 | 0 | 0 | *Streptomyces tubercidicus* | [6.208] |
| 28 | GGGCC/C (+) | ApaI | 1 | 12 | 1 | 0 | 0 | 0 | 0 | *Acetobacter pasteurianus* sub. *pasteurianus* | [6.206] |
|  | G(A)GC(C)/C | BanII | 7 | 57 | 2 | 2 | 1 | 2 | 2 | *Bacillus aneurinolyticus* | [6.219] |
| 29 | C/GGCCG (+) | EclXI | 2 | 19 | 0 | 0 | 0 | 1 | 2 | *Enterobacter cloacae* 590 | [6.23] |
|  | GC/GGCCGC (°,+) | NotI | 0 | 7 | 0 | 0 | 0 | 0 | 0 | *Nocardia otitidis-caviarum* | [6.202] |
|  | (C)/GGCC(A) (+) | EaeI | 39 | 70 | 0 | 2 | 3 | 6 | 7 | *Enterobacter aerogenes* | [6.107], [6.253] |

Table 34. (continued)

| Position no. | Recognition sequence | Enzyme | Number of recognition sites | | | | | | | Microorganism | References |
|---|---|---|---|---|---|---|---|---|---|---|---|
| | | | λ | Ad2 | SV40 | ΦX174 | M13mp7 | pBR322 | pBR328 | | |
| 30 | TGG/CCA | BalI | 18 | 17 | 0 | 0 | 1 | 1 | 1 | *Brevibacterium albidum* | [6.78] |
| 31 | CG/CG | FnuDII | 157 | 303 | 0 | 14 | 18 | 23 | 24 | *Fusobacterium nucleatum D* | [6.138] |
| 32 | A/CGCGT | MluI | 7 | 5 | 0 | 2 | 0 | 0 | 0 | *Micrococcus luteus* | [6.118] |
| 33 | A/C(A/G)(C/T)GT | AflIII | 20 | 25 | 0 | 2 | 3 | 1 | 0 | *Anabaena flos-aquae* | [6.244] |
| | G/CGCGC | BssHII | 6 | 52 | 0 | 1 | 0 | 0 | 0 | *Bacillus stearothermophilus* H3 | [6.132] |
| 34 | CCGC/GC | KspI | 4 | 33 | 0 | 1 | 0 | 0 | 0 | *Kluyvera* species 631 | [6.23] |
| | C/C(A/T)(C/T)GG | DsaI | 46 | 82 | 3 | 3 | 2 | 2 | 3 | *Dactylococcopsis salina* | [6.23] |
| 35 | TCG/CGA | NruI | 5 | 5 | 0 | 2 | 0 | 1 | 2 | *Nocardia rubra* | [6.165] |
| 36 | TGCA | | 272 | 206 | 36 | 18 | 15 | 21 | 19 | | |
| 37 | ATGCA/T | NsiI | 14 | 9 | 3 | 0 | 0 | 0 | 0 | *Neisseria sicca* | [6.55] |
| 38 | G/TGCAC | ApaLI | 4 | 7 | 0 | 1 | 1 | 3 | 2 | *Acetobacter pasteurianus* | [6.248] |
| | G(A/T)GC(A/T)/C | AspI | 28 | 38 | 0 | 3 | 2 | 8 | 7 | *Achromobacter* species | [6.23] |
| | G(A/G)GC(T/C/A)C | Bsp1286I | 38 | 105 | 4 | 3 | 4 | 10 | 10 | *Bacillus sphaericus* | [6.161] |
| 39 | CTGCA/G | PstI | 28 | 30 | 2 | 1 | 1 | 1 | 1 | *Providencia stuartii* 164 | [6.51], [6.211] |
| | CTGCAG | M · PstI | 28 | 30 | 2 | 1 | 1 | 1 | 1 | *Providencia stuartii* 164 | [6.238] |
| 40 | TTGCAA | – | 13 | 23 | 3 | 2 | 2 | 1 | 1 | | – |
| **A.1.3. Internal CG palindromes** | | | | | | | | | | | |
| 41 | A/CGT | MaeII | 143 | 83 | 0 | 19 | 22 | 10 | 12 | *Methanococcus aeolicus* PL-15/H | [6.203] |
| 42 | AACGTT | – | 7 | 3 | 0 | 3 | 2 | 4 | 5 | | – |
| 43 | GACGT/C | AatII | 10 | 3 | 0 | 1 | 0 | 1 | 1 | *Acetobacter aceti* | [6.219] |
| 44 | CACGTG | PmaCI | 3 | 10 | 0 | 0 | 0 | 0 | 0 | *Pseudomonas maltophilia* CB50P | [6.23] |
| 45 | TAC/GTA | SnaBI | 1 | 0 | 0 | 0 | 1 | 0 | 0 | *Sphaerotilus natans* | [6.42] |

| No. | Sequence | Enzyme | | | | | | | | Organism | Reference |
|---|---|---|---|---|---|---|---|---|---|---|---|
| 46 | GCG/C | CfoI | 215 | 375 | 2 | 18 | 25 | 31 | 25 | *Clostridium formicoaceticum* | [6.142] |
| | GCG/C | HhaI | 215 | 375 | 2 | 18 | 25 | 31 | 25 | *Haemophilus haemolyticus* | [6.193] |
| | GCGC (M) | M·HhaI | 215 | 375 | 2 | 18 | 25 | 31 | 25 | *Haemophilus haemolyticus* | [6.144] |
| 47 | AGC/GCT | Eco47III | 2 | 13 | 1 | 0 | 2 | 4 | 3 | *Escherichia coli* RFL47 | [6.108] |
| 48 | GG/CGCC | NarI | 1 | 20 | 0 | 2 | 1 | 4 | 5 | *Nocardia argentiensis* | [6.57] |
| | (A)GCGC/(T/C) | HaeII | 48 | 76 | 1 | 8 | 6 | 11 | 9 | *Haemophilus aegyptius* | [6.191], [6.231] |
| | G(A)/CG(T)C | AhaII | 40 | 44 | 0 | 7 | 1 | 6 | 7 | *Aphanothece halophytica* | [6.244] |
| | G/G(T)(A)CC | BanI | 25 | 57 | 1 | 3 | 6 | 9 | 12 | *Bacillus aneurinolyticus* | [6.219] |
| 49 | CGCGCG | — | 1 | 48 | 0 | 0 | 1 | 1 | 0 | *Fischerella* species | – |
| 50 | TGCGCA | FspI | 15 | 17 | 0 | 1 | 0 | 4 | 3 | *Fischerella* species | [6.223] |
| 51 | C/CGG | HpaII | 328 | 171 | 1 | 5 | 19 | 26 | 33 | *Haemophilus parainfluenzae* | [6.145], [6.207] |
| | CCGG (M) | M·HpaII | 328 | 171 | 1 | 5 | 19 | 26 | 33 | *Haemophilus parainfluenzae* | [6.145] |
| | C/CGC | MspI | 328 | 171 | 1 | 5 | 19 | 26 | 33 | *Moraxella* species | [6.234] |
| | CCGG (M) | M·MspI | 328 | 171 | 1 | 5 | 19 | 26 | 33 | *Moraxella* species | [6.111] |
| 52 | ACCGGT | — | 13 | 5 | 0 | 0 | 0 | 0 | 1 | | – |
| 53 | GCC/GGC | NaeI | 1 | 13 | 1 | 0 | 1 | 4 | 6 | *Nocardia aerocolonigenes* | [6.58] |
| | (A)CCGG(C) | Cfr10I | 61 | 40 | 1 | 0 | 1 | 7 | 10 | *Citrobacter freundii* | [6.109] |
| 54 | CCC/GGG | SmaI | 3 | 12 | 0 | 0 | 0 | 0 | 0 | *Serratia marcescens* S$_b$ | [6.70] |
| | C/CCGGG | XmaI | 3 | 12 | 0 | 0 | 0 | 0 | 0 | *Xanthomonas malvacearum* | [6.70] |
| 55 | C/(T)CG(A)G | AvaI | 8 | 40 | 0 | 1 | 1 | 1 | 1 | *Anabaena variabilis* | [6.105], [6.160] |
| | T/CCGGA | BspMII | 24 | 8 | 0 | 0 | 0 | 1 | 1 | *Bacillus* species M | [6.159] |
| 56 | T/CGA | TaqI | 121 | 50 | 1 | 10 | 14 | 7 | 13 | *Thermus aquaticus* YTI | [6.208] |
| | TCGA (M) | M·TaqI | 121 | 50 | 1 | 10 | 14 | 7 | 13 | *Thermus aquaticus* YTI | [6.199] |
| 57 | AT/CGAT | ClaI | 15 | 2 | 0 | 0 | 2 | 1 | 1 | *Caryophanon latum* L | [6.147] |
| | ATCGAT (M) | M·ClaI | 15 | 2 | 0 | 0 | 2 | 1 | 1 | *Caryophanon latum* L | [6.148] |

**Table 34.** (continued)

| Position no. | Recognition sequence | Enzyme | Number of recognition sites | | | | | | | Microorganism | References |
|---|---|---|---|---|---|---|---|---|---|---|---|
| | | | λ | Ad2 | SV40 | ΦX174 | M13mp7 | pBR 322 | pBR328 | | |
| 58 | G/TCGAC | SalI | 2 | 3 | 0 | 0 | 2 | 1 | 1 | Streptomyces albus G | [6.31] |
| | GT(C/T)/(A/G)AC | HincII | 35 | 25 | 7 | 13 | 2 | 2 | 2 | Haemophilus influenzae Rc | [6.173] |
| | GT(C/T)/(A/G)AC | HindII | 35 | 25 | 7 | 13 | 2 | 2 | 2 | Haemophilus influenzae Rd | [6.114], [6.115], [6.215] |
| | GT/(A/G)AC | AccI | 9 | 17 | 1 | 2 | 2 | 2 | 1 | Acinetobacter calcoaceticus | [6.253] |
| 59 | C/TCGAG | XhoI | 1 | 6 | 0 | 1 | 0 | 0 | 0 | Xanthomonas holcicola | [6.83] |
| 60 | TT/CGAA | FspII | 7 | 1 | 0 | 0 | 0 | 0 | 2 | Fischerella species | [6.223] |

**A.1.4. Internal TA palindromes**

| Position no. | Recognition sequence | Enzyme | Number of recognition sites | | | | | | | Microorganism | References |
|---|---|---|---|---|---|---|---|---|---|---|---|
| | | | λ | Ad2 | SV40 | ΦX174 | M13mp7 | pBR 322 | pBR328 | | |
| 61 | ATAT | — | 230 | 81 | 21 | 20 | 42 | 9 | 19 | | — |
| 62 | AAT/ATT | SspI | 20 | 5 | 6 | 1 | 6 | 1 | 2 | Sphaerotilus species | [6.86] |
| 63 | GAT/ATC | EcoRV | 21 | 9 | 1 | 0 | 0 | 1 | 1 | Escherichia coli J62[pLG74] | [6.119], [6.200] |
| 64 | CA/TATG | NdeI | 7 | 2 | 2 | 0 | 3 | 1 | 0 | Neisseria denitrificans | [6.240] |
| 65 | TATATA | — | 4 | 5 | 0 | 0 | 1 | 1 | 1 | | — |
| 66 | GT/AC | RsaI | 113 | 83 | 12 | 11 | 18 | 3 | 4 | Rhodopseudomonas sphaeroides | [6.141] |
| 67 | AGT/ACT | ScaI | 5 | 5 | 0 | 0 | 0 | 1 | 2 | Streptomyces caespitosus | [6.91], [6.121], [6.225] |
| 68 | G/GTACC | Asp718I | 2 | 8 | 1 | 0 | 0 | 0 | 0 | Achromobacter species 718 | [6.41] |
| | GGTAC/C | KpnI | 2 | 8 | 1 | 0 | 0 | 0 | 0 | Klebsiella pneumoniae OK8 | [6.211], [6.230] |
| 69 | CGTACG | SplI | 1 | 4 | 0 | 2 | 0 | 0 | 0 | Spirulina platensis PL15/H | [6.23] |
| 70 | TGTACA | — | 5 | 5 | 2 | 0 | 1 | 0 | 0 | | — |
| 71 | C/TAG | MaeI | 13 | 54 | 12 | 3 | 4 | 5 | 4 | Methanococcus aeolicus PL-15/H | [6.203] |
| 72 | A/CTAGT | SpeI | 0 | 3 | 0 | 0 | 0 | 0 | 0 | Sphaerotilus species | [6.56] |
| 73 | G/CTAGC | NheI | 1 | 4 | 0 | 0 | 0 | 1 | 1 | Neisseria mucosa | [6.56] |
| 74 | C/CTAGG | AvrII | 2 | 2 | 2 | 0 | 0 | 0 | 0 | Anabaena variabilis uw | [6.196] |

| No. | Recognition sequence | Enzyme | | | | | | | | Organism | References |
|---|---|---|---|---|---|---|---|---|---|---|---|
| 75 | +T/CTAGA | XbaI | 1 | 1 | 5 | 8 | 0 | 0 | 0 | *Xanthomonas badrii* | [6.254] |
| 76 | TTAA | MseI | 195 | 115 | 47 | 35 | 61 | 15 | 17 | *Micrococcus species* | [6.23] |
| 77 | ATTAAT | AsnI | 17 | 3 | 3 | 2 | 6 | 1 | 1 | *Arthrobacter species N-CM* | [6.23] |
| 78 | GTT/AAC | HpaI | 14 | 6 | 4 | 3 | 0 | 0 | 0 | *Haemophilus parainfluenzae* | [6.76], [6.207] |
| 79 | C/TTAAG | AflII | 3 | 4 | 1 | 2 | 0 | 0 | 0 | *Anabaena flos-aquae* | [6.244] |
| 80 | TTT/AAA | DraI | 13 | 12 | 12 | 2 | 5 | 3 | 5 | *Deinococcus radiophilus* | [6.184] |

**A.2. Class II enzymes with penta- or heptanucleotide recognition sequences (subclass IIW enzymes)**

| No. | Recognition sequence | Enzyme | | | | | | | | Organism | References |
|---|---|---|---|---|---|---|---|---|---|---|---|
| 81 | G/G(A/T)CC | AvaII | 35 | 73 | 6 | 1 | 1 | 8 | 7 | *Anabaena variabilis* | [6.105], [6.160] |
| | (A/G)G/G(A/T)CC(C/T) | PpuMI | 3 | 23 | 1 | 0 | 0 | 2 | 0 | *Pseudomonas putida M* | [6.159] |
| | CG/G(A/T)CCG | RsrII | 5 | 2 | 0 | 0 | 0 | 0 | 0 | *Rhodopseudomonas sphaeroides* G30 | [6.170] |
| 82 | CC(A/T)GG | ApyI | 71 | 136 | 17 | 2 | 7 | 6 | 10 | *Arthrobacter pyridinolis* | [6.93] |
| | CC(A/T)GG | BstNI | 71 | 136 | 17 | 2 | 7 | 6 | 10 | *Bacillus stearothermophilus* | [6.201] |
| | /CC(A/T)GG | EcoRII | 71 | 136 | 17 | 2 | 7 | 6 | 10 | *Escherichia coli* R245 | [6.39], [6.46], [6.125] |
| 83 | CC(C/G)GG | NciI | 114 | 97 | 0 | 1 | 4 | 10 | 10 | *Neisseria cinerea* | [6.241] |
| 84 | G/ANTC | HinfI | 148 | 72 | 10 | 21 | 26 | 10 | 10 | *Haemophilus influenzae* R$_f$ | [6.153] |
| 85 | G/GNCC | Cfr13I | 74 | 164 | 11 | 2 | 4 | 15 | 16 | *Citrobacter freundii* RFL13 | [6.110] |
| | G/GNCC | Sau96I | 74 | 164 | 11 | 2 | 4 | 15 | 16 | *Staphylococcus aureus* PS96 | [6.221] |
| | (A/G)G/GNCC(C/T) | DraII | 3 | 44 | 3 | 0 | 0 | 4 | 2 | *Deinococcus radiophilus* | [6.61], [6.92] |
| 86 | GC/NGC | Fnu4HI | 380 | 411 | 24 | 31 | 15 | 42 | 37 | *Fusobacterium nucleatum* 4H | [6.135] |

**Table 34.** (continued)

| Position no. | Recognition sequence | Enzyme | Number of recognition sites | | | | | | | Microorganism | References |
|---|---|---|---|---|---|---|---|---|---|---|---|
| | | | $\lambda$ | Ad2 | SV40 | $\Phi$X174 | M13mp7 | pBR322 | pBR328 | | |
| 87 | $\overset{0\,+}{\text{CC/NGG}}$ | *Scr*FI | 185 | 233 | 17 | 3 | 11 | 16 | 20 | *Streptococcus cremonis* F | [6.71] |
| | $\overset{+}{\text{/CCNGG}}$ | *Dsa*V | 185 | 223 | 17 | 3 | 11 | 16 | 20 | *Dactylococcopsis salina* | [6.23] |
| 88 | /GTNAC | *Mae*III | 156 | 118 | 14 | 17 | 25 | 17 | 18 | *Methanococcus aeolicus* PL-15/H | [6.203] |
| | G/GTNACC | *Bst*EII | 13 | 10 | 0 | 0 | 0 | 0 | 0 | *Bacillus stearothermophilus* ET | [6.133] |
| 89 | $\overset{+}{\text{C/TNAG}}$ | *Dde*I | 104 | 97 | 20 | 14 | 29 | 8 | 9 | *Desulfovibrio desulfuricans* Norway | [6.142] |
| | $\overset{0\ \ 0}{\text{CC/TNAGG}}$ | *Sau*I | 2 | 7 | 0 | 0 | 1 | 0 | 1 | *Streptomyces aureofaciens* IKA18/4 | [6.229] |

**A.3. Class II enzymes with recognition sequences containing internal $(N)_x$ sequences (subclass IIN enzymes)**

| Position no. | Recognition sequence | Enzyme | $\lambda$ | Ad2 | SV40 | $\Phi$X174 | M13mp7 | pBR322 | pBR328 | Microorganism | References |
|---|---|---|---|---|---|---|---|---|---|---|---|
| 90 | GGN/NCC | *Nla*IV | 82 | 178 | 16 | 6 | 6 | 24 | 26 | *Neisseria lactamica* | [6.186] |
| 91 | CACNN/GTG | *Dra*III | 10 | 10 | 1 | 1 | 1 | 0 | 0 | *Deinococcus radiophilus* | [6.61], [6.92] |
| 92 | GAANN/NNTTC | *Asp*700 | 24 | 5 | 0 | 3 | 2 | 2 | 1 | *Achromobacter* species 700 | [6.41], [6.116], [6.117] |
| 93 | $\overset{+\ \ 0}{\text{GAANN/NNTTC}}$ | *Xmn*I | 24 | 5 | 0 | 3 | 2 | 2 | 1 | *Xanthomonas manihotis* 7AS1 | [6.136] |
| 94 | $\overset{+\ \ 0}{\text{GCC(N)}_4\text{/NGGC}}$ | *Bgl*I | 29 | 20 | 1 | 0 | 1 | 3 | 5 | *Bacillus globigii* | [6.35], [6.65], [6.233] |
| 95 | $\overset{+}{\text{CCA(N)}_5\text{/NTGG}}$ | *Bst*XI | 13 | 10 | 1 | 3 | 0 | 0 | 0 | *Bacillus stearothermophilus* X1 | [6.132] |
| 96 | $\overset{0\ \ 0}{\text{GGCC(N)}_4\text{/NGGCC}}$ | *Sfi*I | 0 | 3 | 1 | 1 | 0 | 0 | 0 | *Streptomyces fimbriatus* | [6.185] |

Restriction Endonucleases and DNA Modification Methyltransferases 197

**B. Class II enzymes with nonpalindromic recognition sequences (subclass IIT and IIS enzymes)**

**B. 1. Subclass IIT enzymes**

| No. | Recognition sequence | Enzyme | | | | | | | | Organism | Reference |
|---|---|---|---|---|---|---|---|---|---|---|---|
| 97 | GAATG-CN/N<br>CTTAC/GN-N | BsmI | 46 | 10 | 4 | 3 | 1 | 1 | 3 | Bacillus stearothermophilus NUB36 | [6.161] |
| 98 | TGACN/N-NGTCN<br>ACTGN-N/NCAGN | AspI | 1 | 6 | 0 | 0 | 0 | 2 | 0 | Achromobacter species 699 | [6.23] |

**B. 2. Subclass IIS enzymes**

| No. | Recognition sequence | Enzyme | | | | | | | | Organism | Reference |
|---|---|---|---|---|---|---|---|---|---|---|---|
| 99 | $\overset{+}{G}CAGC(N)_8$<br>$CGTCG(N)_{12}$ | BbvI | 199 | 179 | 22 | 14 | 8 | 21 | 11 | Bacillus brevis | [6.82] |
| 100 | $ACCTGC(N)_4$<br>$TGGACG(N)_8$ | BspMI | 41 | 39 | 0 | 3 | 4 | 1 | 2 | Bacillus species M | [6.159] |
| 101 | $\overset{(+)}{G}GATG(N)_9$<br>$CCTAC(N)_{13}$ | FokI | 150 | 78 | 11 | 8 | 4 | 12 | 11 | Flavobacterium okeanokoites | [6.118] |
| 102 | $GACGC(N)_5$<br>$CTGCG(N)_{10}$ | HgaI | 102 | 87 | 0 | 14 | 7 | 11 | 12 | Haemophilus gallinarum | [6.117] |
| 103 | $\overset{(+)+}{GGTGA}(N)_8$<br>$CCACT(N)_7$ | HphI | 168 | 99 | 4 | 9 | 18 | 12 | 18 | Haemophilus parahaemolyticus | [6.122] |
| 104 | $\overset{+}{GAAGA}(N)_8$<br>$CTTCN(N)_7$ | MboII | 130 | 113 | 16 | 11 | 12 | 11 | 12 | Moraxella bovis | [6.49], [6.69], [6.77] |
| 105 | $CCTC(N)_7$<br>$GGAG(N)_7$ | MnlI | 262 | 397 | 51 | 34 | 61 | 26 | 30 | Moraxella nonliquefaciens | [6.252] |
| 106 | $\overset{o}{G}CATC(N)_5$<br>$CGTAG(N)_9$ | SfaNI | 169 | 84 | 6 | 12 | 7 | 22 | 17 | Streptococcus faecalis | [6.205] |

Class III enzymes recognize specific sequences, but cleave rather specifically 25 to 27 base pairs downstream from this region. The endonuclease activity depends on magnesium ions and ATP, whereas methylation occurs in the presence of SAM [6.34], [6.95], [6.106], [6.113], [6.156].

Class II restriction endonucleases and DNA modification methyltransferases are ubiquitous in prokaryotic cells [6.23], [6.25], [6.27]. In accordance with the proposals made by SMITH and NATHANS, the various enzymes are abbreviated by taking the first letters of the bacterial name [6.214]. The various restriction endonu-cleases and DNA modification methyltransferases occur in almost all genera of the two kingdoms of eu- and archaebacteria. In particular, class II enzymes are known in both gram-positive and gram-negative eubacteria, in cyanobacteria, as well as in thermophilic, halophilic, methanogenic, and sulfur-dependent archaebacteria. In contrast, eukaryotic restriction–modification systems have not been described yet. Specifically cleaving eukaryotic endonucleases, such as the HO-nuclease in yeast or the retroviral endonucleases, have other functions and are characterized by more complex recognition sequences [6.66], [6.124].

In addition to class II restriction–modification systems of single specificity, a whole series of multiple component systems were discovered to contain several restriction–modification systems of different specificity. Particularly complex systems, which contain up to six different specificities, are found in certain species of *Nostoc* (*Nsp*7524I–V) and *Dactylococcopsis salina* (*Dsa*I–VI); *Neisseria gonorrhoeae* contains eight DNA modification methyltransferases (M · *Ngo*I–VIII) [6.23], [6.123], [6.188].

Methods for separating the various enzyme species include size fractionation, column chromatography with anion and cation exchangers, and hydrophobic or affinity materials with DNA or DNA-like heparin as ligands [6.90], [6.180]. At the end of 1989 as many as 1024 different class II restriction endonucleases with at least 163 different specificities were known, which have been isolated from 860 different bacterial strains (Table 34, p. 190–197). The corresponding DNA modification methyltransferases have been characterized only for the most important systems. Currently known enzymes are described completely in [6.23], [6.25], [6.27].

The active enzymes of the class II restriction endonucleases consist of protein dimers of identical subunits, whereas the corresponding DNA modification methyl-transferases act as monomeric proteins. The molecular masses under denaturing conditions range from $20 \times 10^3$ to $70 \times 10^3$ dalton for both class II restriction endonucleases and DNA modification methyltransferases. The only protein of higher molecular mass known so far is *Bst*I which is about $400 \times 10^3$ dalton [6.157].

As an example, the *Eco*RI endonuclease monomer is a 277 amino acid protein with a molecular mass of 31 063 dalton, which is processed after translation by removing the terminal *N*-formyl-methionine. The corresponding methyltransferase protein consists of 326 amino acids with a mo-lecular mass of 38 048 dalton. This protein is also processed by removing the N-terminal dipeptide fMet-Ala [6.89], [6.166].

Examples of cloned restriction–modification systems are *Bsu*RI from *Bacillus subtilis* [6.120], [6.231]; *Eco*RI [6.43], [6.54], [6.89], [6.96], [6.139], [6.166]; *Eco*RII [6.125]; and *Eco*RV [6.44], [6.45], [6.126] from *Escherichia coli*; *Hha*II from *Haemo-*

*philus parainfluenzae* [6.144], [6.204]; *Pae*R7 from *Pseudomonas aeruginosa* [6.80], [6.226]; *Pst*I from *Providencia stuartii* [6.134], [6.239]; and *Pvu*II from *Proteus vulgaris* [6.38]. A complete list is given in [6.23]. According to gene mapping of the various systems, the proteins for both restriction endonuclease and DNA modification methyltransferase may be plasmid or chromosomally coded. Further analysis of the two genes shows that they are either arranged in tandem to form transcription units or expressed as independent genes [6.89], [6.120], [6.166], [6.231]. When the known protein and genetic sequences of corresponding restriction endonucleases and DNA modification methyltransferases were compared, no statistically significant sequence homology could be detected [6.120], [6.166]. Circular dichroism studies also indicate extensive differences in secondary structure [6.166]. According to these observations, the two proteins have different evolutionary origins and interact with DNA in different ways.

## 6.1.2. Activity of Class II Restriction Endonucleases

An enzyme unit, U, is normally defined as the amount of protein that completely digests 1 μg of the DNA from bacteriophage $\lambda$ in 1 h at 37 °C. To analyze the completeness of the reaction, the fragment mixture is separated in 0.5–2% agarose gels by horizontal or vertical electrophoresis (see p. 26). The use of $\lambda$ DNA as the standard test substrate has two advantages. This DNA has 48 502 base pairs and contains cleavage sites for almost all known restriction endonucleases, which allows a comparison of activities [6.23], [6.59], [6.60], [6.197]. Because the complete sequence of $\lambda$ DNA is known, any cleavage pattern can be calculated in advance by computer.

### 6.1.2.1. Reaction Parameters

The activity of *Eco*RI and other class II restriction endonucleases not only depends on the presence of specific recognition sites and essential divalent cations, but also very strongly on such reaction parameters as temperature, ionic strength, and pH [6.242]. The activity of class II restriction endonucleases is also frequently stimulated by the presence of reducing agents such as 2-mercaptoethanol or dithiothreitol.

On the plasmid pBR322, *Eco*RI has a temperature optimum of 40 °C. The apparent activation energy for the overall enzymatic reaction is 5.1 kJ/mol. The enzyme is inactivated irreversibly above 45–50 °C [6.182]. The optimal pH for a specific *Eco*RI cleavage is 7.5; the optimal NaCl concentration, 150 mmol/L [6.247].

The optimal temperature for enzymatic activity often corresponds to the optimal growth temperature of the organism from which the enzyme is isolated [6.242]. Accordingly, enzymes such as *Bcl*I, *Bst*EII, *Mae*I–III, or *Taq*I with high temperature optima are isolated from thermophilic organisms [6.37], [6.133], [6.198], [6.203]. Enzymes are also known which require higher ionic strength (e.g., *Mae*III with 350 mmol/L NaCl) or higher pH values (e.g., *Bgl*I pH 9.5) for maximal activity [6.36], [6.203].

### 6.1.2.2.  Additional Structural Requirements Influencing Activity

The structure of the substrate also strongly influences the activity of class II restriction endonucleases at individual cleavage sites. Important parameters are

1) G/C content and base distribution in natural DNA
2) Size of the DNA and cleavage frequencies
3) Length and base composition of the flanking sequences
4) Position of the cleavage sites with respect to each other
5) DNA tertiary structure
6) Modification of DNA and protein attachment.

**Guanosine/Cytidine Content.** In natural DNA, the G/C content varies broadly. The DNA from *Micrococcus lysodeikticus* contains 72% G/C residues, but the DNA from *Euglena gracilis* chloroplasts is composed of only 25% G/C residues [6.87], [6.235]. Although this second DNA consists of about $130 \times 10^3$ base pairs, there is not a single cleavage site for *Sma*I recognizing only G/C base pairs [CCC/GGG]. Furthermore, the base distribution in natural DNA is nonstatistical. In eukaryotic SV40 DNA, the dinucleotide CpG occurs very rarely [6.169]. As a result, the cleavage frequency for such class II restriction endonucleases as *Mae*II, *Cfo*I, *Hpa*I, *Taq*I, or *Fnu*DII, which contain the CpG dinucleotide in their tetranucleotide recognition sequences, is extremely low [6.23], [6.53]. By contrast, the number of cleavage sites on prokaryotic pBR322 DNA of similar size is about ten times higher, with values like those obtained from enzymes such as *Alu*I or *Hae*III, which have no CpG dinucleotides in their tetranucleotide recognition sequences [6.23].

**Length of DNA.** The influence of DNA length can be demonstrated by the *Sal*I cleavage of pBR322 DNA [6.31]. The length of this plasmid is only about one-tenth the length of the bacteriophage $\lambda$ DNA [6.175], [6.197], [6.222]. The $\lambda$ DNA has only two cleavage sites, but the much smaller pBR322 DNA still has one cleavage site. Because the number of cleavage sites per microgram of DNA increases fivefold with pBR322 DNA, digestion of the same amount of $\lambda$ DNA with *Sal*I requires about five times more enzyme as compared with the plasmid DNA.

**Flanking Sequences.** The influence of flanking sequences is demonstrated by the observation that short oligonucleotides consisting, for example, only of the hexanucleotide *Eco*RI recognition sequence GAATTC are not cleaved. Even if several hexanucleotides were ligated to longer DNA molecules to stabilize the double strand — which would be completely denatured in the short hexanucleotide at 37 °C — they could not be digested with *Eco*RI. Only if the recognition sequence is extended on both ends by one or more flanking nucleotides the enzyme can exert its activity. However, in these cases, higher quantities of enzyme and longer reaction times are required. Normal reaction rates can be achieved only by further increasing the length of the flanking sequences [6.154].

**Base Composition of Flanking Sequences.** The base composition within the flanking sequence distinctly influences the actual cleavage rate at the individual recognition site. This effect was first observed for *Eco*RI which digests the various sites on λ or Ad2 DNA with rates differing by a factor of ten [6.72], [6.163]. A similar preferential digestion has also been observed with *Hin*dIII on λ, *Hga*I on ΦX174, and *Pst*I on plasmid pSM1 DNA [6.30], [6.52]. In the latter case, significant resistance to cleavage was conferred on G/C-rich flanking regions.

More significant differences in individual cleavage rates were observed on pBR322 DNA with *Nae*I which recognizes the G/C-rich sequence GCC/GGC [6.58]. The enzyme *Nae*I cuts pBR322 DNA at four different sites, two of which are digested rapidly and the third 5–10 times more slowly. Complete cleavage at the fourth site, however, proceeds 50 times slower [6.71]. Similar slow cleavage sites were also observed for the enzymes *Nar*I [GG/CGCC], *Sac*II [CCG/CGG], and *Xma*III [C/GGCCG] which recognize sequences with alternating G and C residues [6.23], [6.67], [6.128]. In all these examples, the slower reaction rates at particular sites are attributed to the sequences within the flanking regions. However, whether a particular sequence element or an altered DNA conformation is responsible for the strong decrease in cleavage rate is still unknown.

**Relative Position of Cleavage Sites.** A similar decrease in cleavage rate is observed if two or more recognition sites that would be digested readily alone are only a few nucleotides apart from each other. After digestion at one of the two sites, the essential flanking sequences for efficient cleavage at the neighboring position are missing. Therefore, after the first cleavage, the DNA can only be hydrolyzed very slowly at the second site. This is the case for the multiple cloning sites of the M13mp derivatives that are used frequently for sequencing studies. In these polylinker regions, the recognition sequences for several restriction endonucleases are located directly next to each other [6.249].

## 6.1.3. Specificity of Class II Restriction Endonucleases

The specificity of the presently known class II restriction endonucleases is defined by three criteria:

1) recognition sequence,
2) position of cleavage site, and
3) influence of methylation.

However, all three characteristics of enzyme specificity are known only for a limited number of enzymes. For many class II restriction endonucleases, only the recognition sequences have so far been determined [6.23], [6.27].

### 6.1.3.1. Palindromic Recognition Sequences

For the majority of class II enzymes, the recognition sequence on the double-stranded DNA is characterized by a two-fold axis of rotational symmetry. Such symmetric recognition sequences are called palindromes. In a palindromic sequence, a horizontal complementary arrangement exists in addition to the normal base pairing between A:T and G:C. For example, the *Eco*RI recognition sequence is

$$^{5'}\text{G/AATT-C}^{3'}$$
$$_{3'}\text{C-TTAA/G}_{5'}$$

Most palindromic recognition sequences are tetra-, penta-, hexanucleotides, but defined octanucleotides also exist [6.202]. In addition, palindromic recognition sequences of extended length exist, e.g., for *Bst*XI, in which only the flanking nucleotides are specifically recognized. In most of these cases, the flanking trinucleotides are unambiguously defined. However, one case of nonequivocally defined flanking trinucleotides is known with *Dra*II [6.61], [6.92]:

$$(^\text{A}_\text{G})\text{G/GNCC}(^\text{T}_\text{C})$$

An example of an enzyme that recognizes flanking tetranucleotides is *Sfi*I [6.185]:

$$\text{GGCCNNNN/NGGCC}$$

Palindromic recognition sequences can be arranged according to homologies within their recognition sequences (Table 35). In this approach, the sequences are ordered by starting from the shortest central palindromes AT, GC, CG, and TA. Tetranucleotide palindromes are created for each dinucleotide sequence by adding one nucleotide on the left side and its complementary nucleotide on the right side. Repeated application of this cycle generates hexa- and octanucleotide palindromes. Out of the 16 possible tetranucleotides, 13 sequences are currently covered by corresponding enzymes. For the 64 possible hexanucleotides, 51 enzyme specificities are known [6.23].

The cleavage sites for class II restriction endonucleases lie within or directly next to the recognition sequence; they are marked by a slash (/) in Table 36 (p. 204). As in recognition, the positions in the two complementary strands have a two-fold rotational symmetry. Either *blunt* double-stranded ends or *protruding* single-stranded 5'- or 3'-ends are formed. The lengths of the protruding terminal single-stranded regions depend on the position of the cleavage sites within the recognition sequence. Independent of cleavage positions, all class II restriction endonucleases form fragments that have 5'-phosphate and 3'-hydroxyl fragment ends. The only exception known so far is *Nci*I, which produces 3'-phosphorylated fragment termini [6.103].

Some class II restriction endonucleases are able to digest DNA·RNA hybrids. The enzymes *Alu*I, *Hae*III, *Hha*I, and *Taq*I cut the DNA strand of the hybrid at the correct recognition sequences [6.158].

Sequence-specific digestion of *single-stranded* DNA has also been shown for a number of enzymes. However, like double-stranded DNA, base-paired recognition sequences are necessary for cleavage of single-stranded DNA. Therefore, refolding effects between different recognition sites, which result in transiently formed secondary canonical structures, play a decisive role. The cleavage rates depend mainly on the stability of these double-stranded structures [6.101], [6.168], [6.250].

**Table 35.** Palindromic recognition sequences of restriction endonucleases

|   | AT | GC | CG | TA |   |
|---|----|----|----|----|---|
| A | *Tsp*EI | *Alu*I | *Mae*II | | T |
| G | *Sau*3AI | *Hae*III | *Cfo*I | *Rsa*I | C |
| C | *Nla*III | *Fnu*DII | *Hpa*II | *Mae*I | G |
| T | | | *Taq*I | *Mse*I | A |
| AA | | *Hind*III | | *Ssp*I | TT |
| GA | *Eco*RI | *Sac*I | *Aat*II | *Eco*RV | TC |
| CA | *Mfe*I | *Pvu*II | *Pma*Cl | *Nde*I | TG |
| TA | | | *Sna*Bl | | TA |
| AG | *Bgl*II | *Stu*I | *Eco*47III | *Sca*I | CT |
| GG | *Bam*HI | *Apa*I | *Nar*I | *Asp*718I | CC |
| CG | *Pvu*I | *Xma*III | | *Spl*I | CG |
| TG | *Bcl*I | *Bal*I | *Mst*I | | CA |
| AC | | *Mlu*I | | *Spe*I | GT |
| GC | *Sph*I | *Bss*HII | *Nae*I | *Nhe*I | GC |
| CC | *Nco*I | *Sac*II | *Sma*I | *Avr*II | GG |
| TC | *Bsp*HI | *Nru*I | *Bsp*MII | *Xba*I | GA |
| AT | | *Nsi*I | *Cla*I | *Asn*I | AT |
| GT | *Sna*I | *Apa*LI | *Sal*I | *Hpa*I | AC |
| CT | | *Pst*I | *Xho*I | *Afl*II | AG |
| TT | | | *Mla*I | *Dra*I | AA |
|   | AT | GC | CG | TA |   |

## 6.1.3.2. Nonpalindromic Recognition Sequences

In addition to most class II enzymes that act on palindromic sequences (subclasses IIP, IIW, and IIN), beside subclass IIT and IIU enzymes of subclass IIS like *Fok*I, *Hga*I, *Ksp* 632I, *Mbo*II, or *Mnl*I have been found, which recognize nonsymmetric sequences. These enzymes digest the double-stranded DNA at precise distances downstream from their recognition sites [6.23], [6.49], [6.217], [6.218], [6.252].

The spatial separation between recognition and cleavage has been utilized to digest any predetermined sequence in single-stranded DNAs like those of the various M13mp derivatives [6.181], [6.224]. Universal cleavage in the case of subclass IIS enzymes is achieved by applying a specifically designed *bivalent adapter*. This synthetic DNA element is constructed from a constant double-stranded domain harboring (1) the recognition sequence of the subclass IIS enzymes and (2) a variable single-stranded domain complementary to the sequence to be cleaved. The linearized single-stranded target DNA can be converted to double-stranded DNA by a Klenow-catalyzed polymerization reaction using the synthetic deoxyoligonucleotide as a primer. By this approach, double-stranded DNA molecules are created that can be cleaved precisely at any position. Thus, the new method permits the design of

**Table 36.** Cleavage positions of restriction endonucleases and generated fragment ends

| Enzyme | Recognition sequence | End produced |
|--------|---------------------|--------------|
| *Hae*III | GG/CC<br>CC/GG | blunt end |
| *Eco*RI | G/AATT-C<br>C-TTAA/G | 5′-protruding |
| *Pst*I | C-TGCA/G<br>G/ACGT-C | 3′-protruding |
| *Not*I | GC/GGCC-GC<br>CG-CCGG/CG | 5′-protruding |
| *Eco*RII | /CC($^A_T$)GG<br>GG($^A_T$)CC/ | 5′-protruding |
| *Sau*I | CC/TNA-GG<br>GG-ANT/CC | 5′-protruding |
| *Xmn*I | GAANN/NNTTC<br>CTTNN/NNAAG | blunt end |
| *Sfi*I | GGCCN-NNN/NGGCC<br>CCGGN/NNN-NCCGG | 3′-protruding |
| *Bst*XI | CCAN-NNNN/NTGG<br>GGTN/NNNN-NACC | 3′-protruding |
| *Mnl*I | CCTCNNNNNNN/N<br>GGAGNNNNNNN/N | blunt end |
| *Hga*I | GACGCNNNNN/NNNNN-N<br>CTGCGNNNNN-NNNNN/N | 5′-protruding |
| *Mbo*II | GAAGANNNNNNN-N/N<br>CTTCTNNNNNNN/N-N | 3′-protruding |

novel enzyme specificities by combining subclass IIS restriction endonucleases with bivalent DNA adapter oligonucleotides mediating the novel sequence specificities.

### 6.1.3.3. Isomers

Isomers are class II restriction endonucleases that are isolated from various organisms but recognize identical sequences. Isomeric enzymes, however, may have identical or different cleavage positions within the common recognition sequence (iso-/hetero-schizomers) or may show identical or different sensitivity to particular modified bases (iso-/hetero-hypekomers). The greatest variability of isomers is observed with 61 different enzymes known so far all recognizing the sequence GATC [6.23].

In this collection of isomers, well-known examples for heterohypekomers, i.e., isomeric enzymes with different sensitivities to particular modified bases are *Dpn*I, *Mbo*I, and *Sau*3AI [6.63], [6.77],

[6.131]. Whereas *Sau*3AI is insensitive to methylation of the internal adenine residue, *Mbo*I is completely inhibited by this methylation. In contrast, *Dpn*I is fully dependent on this methylation because the enzyme does not cleave the unmethylated GATC sequence. The DNA from pBR322, isolated from wild-type *Escherichia coli* cells, is highly methylated at adenosine residues within GATC sequences by *E. coli*dam methyltransferase that specifically recognizes GATC [6.98], [6.232]. Consequently, *Sau*3AI and *Dpn*I will cleave this DNA, whereas *Mbo*I is unable to digest this substrate.

Another important pair of heterohypekomers, *Hpa*II and *Msp*I, shows a different sensitivity with respect to cytosine methylation. Methylation of the central cytosine residue within their common tetranucleotide recognition sequence CCGG renders the cleavage site resistant to digestion with *Hpa*II but sensitive to cleavage with *Msp*I [6.111], [6.145]. This ability to discriminate between methylated and unmethylated cytosine residues is used to detect methylated CpG dinucleotides which are thought to be involved in eukaryotic gene regulation [6.94], [6.187], [6.245].

An example of heteroschizomers, i.e. isomeric enzymes with different cleavage positions, is *Asp*718I and *Kpn*I. Both enzymes recognize the sequence GGTACC but cleave between either 5'-terminal guanosine residues or 3'-terminal cytosine residues [6.41], [6.230]. The 5'-terminal protruding ends of *Asp*718I fragments can be efficiently labeled with T4 polynucleotide kinase, whereas the 3'-ends are suitable substrates for 3'-endlabeling reactions with Klenow enzyme. Labeling of the 3'-protruding ends of *Kpn*I fragments can be obtained via the tailing reaction with terminal transferase.

The most commonly used cloning vehicles often have no restriction sites for particular enzymes. In such cases, cloning can be achieved by the use of enzymes from an enzyme family. Members of an enzyme family all produce identical single-stranded fragment ends, whereas the complete recognition sequences of individual enzymes are different in the flanking nucleotides. The best known enzyme family is the GATC family, whose members are *Sau*3AI, *Bgl*II, *Bam*HI, *Bcl*I, and *Xho*II [6.23]. All these enzymes produce single-stranded ends of the sequence GATC. The respective hexanucleotide recognition sequences are lost after recombination of fragments produced by different enzymes of this family, but the common internal nucleotides remains the same in all possible combinations. In this way, for example, *Bgl*II fragments which are cloned into the *Bam*HI site of pBR322 DNA can be recovered from the vector by cutting with *Sau*3AI. Another enzyme family exists for the sequence CTAG with the enzymes *Mae*I, *Spe*I, *Nhe*I, *Avr*II, and *Xba*I [6.23].

## 6.1.4. Changes in Sequence Specificity

**Relaxation.** For many class II restriction endonucleases such as *Eco*RI, relaxation of the specificity occurs when the solvent is changed [6.73], [6.79], [6.143], [6.209], [6.228], [6.247]. This altered activity of *Eco*RI is known as *Eco*RI* activity. Relaxed specificity markedly increases the number of small fragments, whereas the large fragments disappear. According to sequencing data, the subcanonical sequences differ from the canonical recognition sites by at least one base pair [6.75].

Relaxation of both flanking positions yields the tetranucleotide AATT as the predominant *Eco*RI* specificity. However, relaxation of internal positions has also been observed [6.195].

Relaxation of the specificity of class II restriction endonucleases has been attributed generally to the effects of altered reaction environment on the nature of the complexes formed between the DNA helix and the *Eco*RI dimer. Because *Eco*RI* is an inherent enzyme property, it cannot be eliminated by more extensive purification. However, relaxation specificity can be suppressed by using suitable buffers; thus relaxation can be prevented by the following incubation conditions [6.143], [6.228], [6.247]:

high ionic strength,
decreased pH,
use of $Mg^{2+}$ ions rather than $Mn^{2+}$, $Co^{2+}$, or $Zn^{2+}$ ions,
low enzyme concentration,
low concentration of glycerol or other organic solvents such as dimethyl sulfoxide,
    which destabilize the double-stranded DNA helix, or
short incubation periods.

In addition, incubation *temperature* strongly affects relaxation of specificity. As shown for *Pst*I, an increase in incubation temperature from 37 to 42 °C suppresses the relaxed specificity almost completely even at higher concentrations and longer incubation periods. In contrast, lowering the incubation temperature to 30 °C strongly favors *Pst*I* activity. When the incubation temperature is further lowered to 25 or 20 °C, the relaxed activity becomes predominant even under normal incubation conditions. Analogous observations have been made for *Bam*HI, *Bst*EII, *Eco*RI, *Sph*I, *Sal*I, and *Taq*I [6.189].

**Modification of DNA by Methylation.** Additional changes in the activity and specificity of class II restriction endonucleases are attributed to DNA modifications such as methyl transfer [6.25]. Certain enzymes are inhibited not only by the corresponding methyltransferase counterpart but also by DNA methylation mediated by *Escherichia coli* *dam*I [GATC] or *dcm*I methyltransferase [CC($^A_T$)GG] [6.97], [6.146], [6.232]. These two methyltransferases are not part of the restriction-modification systems but have other biological functions such as the regulatory role of *dam*I methyltransferase in mismatch repair and DNA replication [6.150], [6.216], [6.237].

When the recognition sequences of *dam*I- or *dcm*I-specific DNA methyltransferase (M · *Ecodam*I; M · *Ecodcm*I) and the particular class II restriction endonuclease overlap totally, the activity on DNA isolated from *E. coli* is completely inhibited. Well-known examples of different M · *Ecodam*I influences at $N^6$-methylated adenine residues in GATC sequences are the isoschizomers *Dpn*I, *Mbo*I, and *Sau*3AI mentioned previously [6.63], [6.77], [6.181]. Similar considerations apply to the M · *Ecodcm*I that inhibits *Eco*RII but activates *Apy*I by 5C-methylation of the internal cytosine residue [6.46], [6.93]; however, *Bst*NI is not affected by this particular methylation [6.93], [6.176]. The inhibitory effects of M · *Ecodam*I or M · *Ecodcm*I can be eliminated by isolation of the DNA from *E. coli* cells in which both methyltransferase genes have been inactivated by mutation [6.97], [6.98], [6.146].

Partial overlapping with the M · *Ecodam*I recognition sequence is observed for a variety of enzymes and frequently inhibits their activity [6.149], [6.165]. As an example, after M · *Ecodam*I-specific methylation, *Cla*I digestion is inhibited specifically at the heptanucleotide sequences **GATC**GAT or ATC**GATC** [6.147]. All the other *Cla*I sites are unaffected. Therefore, sequential action of isolated M · *Ecodam*I methyltransferase and *Cla*I endonuclease generates the following new sequence specificity:

$$\begin{pmatrix} A \\ G \\ T \end{pmatrix} ATCGAT \begin{pmatrix} T \\ G \\ A \end{pmatrix}$$

In contrast, the heptanucleotides mentioned above are cleaved specifically with endonuclease *Dpn*I after methylation of DNA with *Cla*I methyltransferase [6.148]. In this approach, the other *Dpn*I cleavage sites remain stable, and both enzyme pairs complement each other in creating new sequence specificities.

**Permutation of Fragment Ends.** Another approach for creating novel cleavage sites applies the permutation of generated fragment ends with 5′-protruding termini [6.174]. Sites generated by such enzymes as *Eco*RI, *Bam*HI, or *Sal*I can be rendered blunt-ended by filling in the protruding ends with deoxynucleoside triphosphates and Klenow enzyme. This ligation creates a new symmetry center which corresponds to the ligation point of the two filled-in ends. If, for example, a *Sal*I site [G/TCGAC] is treated in this way, a 10 base pair palindromic sequence GT**CGATCG**AC with a novel *Pvu*I cleavage site [(CGAT/CG)] is created. After *Pvu*I cleavage, a third restriction site may be generated by treatment with nuclease S1. After ligation, another novel cleavage site is created because the resulting octanucleotide GT**CGCGAC** contains a new *Nru*I cleavage site [TCG/CGA]. Analogous conversion of restriction sites can be accomplished for all the other enzymes that generate fragments with 5′-protruding ends. Interconversion of restriction sites could generate unique novel cloning sites without the need of synthetic linkers. This should improve the flexibility of genetic engineering experiments.

**Modification by Bacteriophages.** Bacteriophage-induced DNA modification is also known to strongly influence the activity of class II restriction endonucleases. The respective phage genomes contain all or most of the cytosine or thymine residues substituted at the 5-position [6.104], [6.112]. Substitution of cytosine by 5-methyl-cytosine (*Xanthomonas oryzae* phage XP12) or glycosylated 5-hydroxymethyl-cytosine (*Escherichia coli* phage T4), or of thymine by phosphoglucuronated and glycosylated 5-(4′,5′-dihydroxypentyl)uracil (*Bacillus subtilis* phage SP15) renders the DNA resistant to almost all known class II restriction endonucleases. The only exception is *Taq*I, which cleaves the modified DNAs extensively. Substitution of 5-hydroxyuracil or uracil for thymine (*B. subtilis* phages SPO1 or PBS1) results only in a decreased cleavage rate but not in complete inhibition of the various enzyme activities.

**Endonuclease Inhibitors.** Other inhibitors of class II restriction endonucleases are DNA-binding agents such as ethidium bromide, actinomycin D, proflavine, distamycin A, or neotrypsin [6.84], [6.167], [6.171]. Spermine or spermidine also have an inhibitory effect at high concentrations, whereas low concentrations of these substances have a stimulatory effect. This stimulation is similar to that observed after addition of proteins such as *E. coli* RNA polymerase or T4 gene 32 protein which bind tightly to the single-stranded termini generated by many class II restriction endonucleases [6.62], [6.129], [6.177], [6.178].

## 6.1.5.  Novel Class II Restriction Endonucleases

New sequence specificities are also obtained by screening bacteria for novel class II restriction endonucleases [6.28]. The first method for finding new enzymes consists of examining those bacterial strains that are resistant to certain bacteriophages. As an alternative procedure, representatives of previously unexamined families from the wide range of eu- and archaebacteria are screened systematically for the presence of new class II restriction endonucleases. For example, in a total of 348 microorganisms screened, 105 novel enzymes were found, predominantly in gram-negative bacteria [6.117]. To detect the novel specificities, aliquots of bacterial extracts are incubated with a DNA of known sequence such as $\lambda$ DNA for various periods under standard incubation conditions. The fragment mixtures are analyzed by electrophoresis on agarose gels.

The new sequence specificity is determined with computer assistance by correlating experimentally observed and predicted fragment patterns. In a following step, the corresponding fragments obtained with other DNAs, e.g., from pBR322 or SV40, are analyzed. In a third step, the precise cleavage positions with the recognition sequence are fixed in sequencing gels by accurate fragment length determination of both strands of a particular DNA fragment created with the novel enzyme.

The three steps in characterization of the specificity and cleavage positions of a new enzyme may be exemplified by means of the recently discovered class II restriction endonuclease *Mae*I. This new enzyme can be isolated from the archaebacterium *Methanococcus aeolicus* PL15/H, along with *Mae*II and *Mae*III which are also new enzymes with novel specificities [6.203]. The enzyme *Mae*I recognizes the tetranucleotide sequence C/TAG and cleaves both DNA strands specifically between the 5'-flanking cytosine and thymine nucleotides. The enzyme *Mae*II also recognizes a tetranucleotide sequence [A/CGT], whereas *Mae*III acts on the pentanucleotide sequence /GTNAC.

Novel class II restriction endonucleases with more complex recognition sequences have also been found. Examples are *Dra*II and *Dra*III which were isolated recently as minor species from *Deinococcus radiophilus* in addition to the already known main activity *Dra*I, an isoschizomer of *Aha*III [6.243]. The enzyme *Dra*II recognizes a novel type of heptanucleotide with ambiguities in the flanking nucleotides:

$$\left(^G_A\right)G/GNCC\left(^G_T\right)$$

The enzyme *Dra*III is characterized by the nonanucleotide CACNNN/GTG with three internal undefined nucleotides [6.61], [6.92].

# 6.2. DNA Polymerases

Table 38 lists the properties of important DNA polymerases under A (p. 236).

## 6.2.1. *Escherichia coli* DNA Polymerase I

The enzyme *E. coli* polymerase I, also called deoxynucleoside triphosphate: DNA deoxynucleotidyltransferase (DNA-directed) (E.C. 2.7.7.7) [*9012-90-2*], is obtained from *Escherichia coli*.

**Properties.** *Escherichia coli* DNA poylmerase I consists of a single polypeptide with a molecular mass of $109 \times 10^3$ dalton [6.447] and contains one $Zn^{2+}$ ion per molecule [6.654]. The enzyme consists of two domains joined by a protease-sensitive peptide linker [6.296], [6.461], although the complete protein is approximately spherical with a diameter of 6500 pm [6.360], [6.403]. *Escherichia coli* DNA polymerase I contains a single free mercapto group able to form a dissociable complex with $Hg^{2+}$. The protein can be carboxymethylated without loss of activity [6.448].

One unit of *E. coli* DNA polymerase I is defined as the enzyme activity that incorporates 10 nmol of total nucleotide into acid-precipitable material in 30 min at 37 °C [6.605].

**Uses.** *Escherichia coli* DNA polymerase I is used predominantly for the in vitro labeling of DNA by *nick translation* [6.262]. Several published procedures all follow a common principle (Fig. 51). Trace amounts of DNase I (see Section 6.4.1) introduce nicks into unlabeled DNA, thus exposing internal 3'-hydroxyl groups. These nicks move toward the 3'-terminus by the subsequent sequential action of *E. coli* DNA polymerase I. First, the enzyme removes the nucleotide on the 5'-side of the nick. In the coupled reaction that follows, the excised nucleotide is replaced with a labeled nucleotide at the 3'-terminus of the nick. Repeated action of *E. coli* polymerase I replaces all nucleotides with labeled nucleotides downstream from the nick. Below 22 °C, the reaction will not proceed further than one complete exchange of the existing unlabeled DNA strand for a labeled one, as shown in Figure 51. At the right ratio of DNA, DNase I, and *E. coli* DNA polymerase I, about 20–40% of the labeled nucleotides are incorporated into the product, with a yield in label density of $10^8–10^9$ dpm per microgram of DNA. Only 20–30% of this incorporated label is commonly found in foldback DNA [6.639]. The amount of foldback DNA is often higher if the unlabeled DNA fragment is relatively small (<1000 nucleotides) [6.608]. In this case, more 3'-ends are probably used for DNA synthesis than internal 3'-termini at nicks.

## 6.2.2. Klenow Enzyme

Klenow enzyme is the largest fragment obtained by proteolysis of *Escherichia coli* DNA polymerase I with subtilisin; it is also called deoxynucleoside triphosphate: DNA deoxynucleotidyltransferase (DNA-directed) (E.C. 2.7.7.7) [*9012-90-2*].

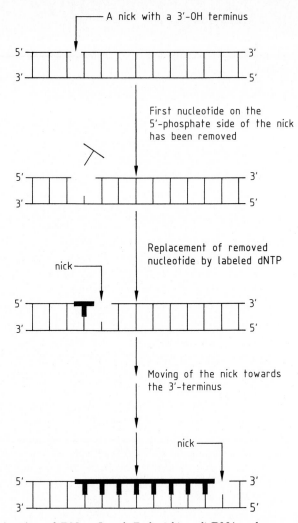

A nick with a 3′-OH terminus
5′
3′
3′
5′
First nucleotide on the
5′-phosphate side of the nick
has been removed
5′
3′
3′
5′
Replacement of removed
nucleotide by labeled dNTP
nick
5′
3′
3′
5′
Moving of the nick towards
the 3′-terminus
nick
5′
3′
3′
5′

**Figure 51.** Sequential action of DNase I and *Escherichia coli* DNA polymerase I during nick translation

**Figure 52.** DNA sequencing with the dideoxy chain-termination method

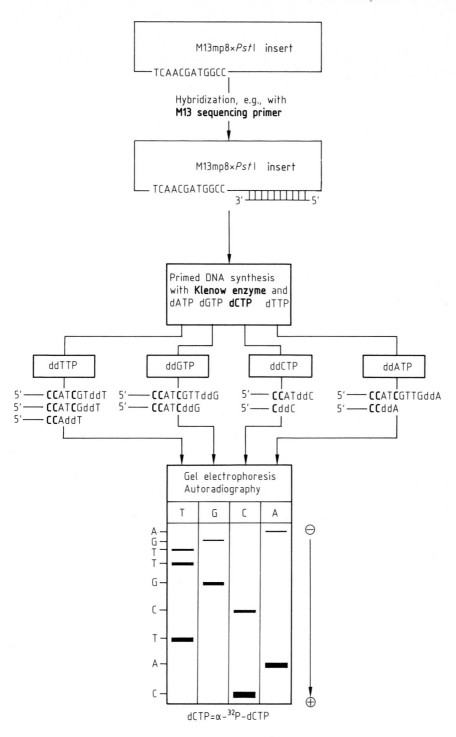

**Properties.** Klenow enzyme has a molecular mass of $75 \times 10^3$ dalton. It can be obtained by subtilisin treatment, which digests the protease-sensitive peptide linker between the two functional domains of the native enzyme [6.439]. The large fragment carries the $5'-3'$ polymerase and the $3'-5'$ exonuclease activities of intact DNA polymerase I, but lacks the $5'-3'$ exonuclease activity located on the small fragment of $36 \times 10^3$ dalton. Klenow enzyme catalyzes the addition of deoxynucleotides to the 3'-hydroxyl terminus of a primer annealed to the template DNA.

One unit of Klenow enzyme is defined as the enzyme activity that incorporates 10 nmol of total nucleotide into an acid-precipitable fraction in 30 min at 37 °C [6.605].

**Uses.** Klenow enzyme is used in a wide variety of techniques, such as the following:

1) Conversion of 3'-recessed ends of restricted DNA fragments to blunt ends by the fill-in reaction [6.688]
2) 3'-Endlabeling of DNA fragments by use of $\alpha$-$^{32}$P-deoxynucleotides [6.688]
3) Second-strand synthesis of cDNA [6.615]
4) Homogeneous labeling of DNA fragments with random oligodeoxynucleotides as primers (oligolabeling) [6.365]
5) Sequencing of DNA by the dideoxy chain termination method [6.622] (Fig. 52)
6) Elongation of oligonucleotides mediating mismatch formation for site-directed mutagenesis [6.661]

## 6.2.3. T4 DNA Polymerase

The enzyme T4 DNA polymerase, also called deoxynucleoside triphosphate: DNA deoxynucleotidyltransferase (DNA-directed) (E.C. 2.7.7.7) [9012-90-2], is obtained from phage T4-infected *Escherichia coli*.

**Properties.** The enzyme T4 DNA polymerase consists of a single polypeptide with a molecular mass of $114 \times 10^3$ dalton [6.396]. This polymerase displays a relatively broad pH optimum ranging from 8 to 9; at pH values of 7.5 and 9.7, ca. 50 % of the optimal activity is observed. Maximal polymerase activity requires 6 mmol/L of $Mg^{2+}$; this activity is decreased to one-fourth if $Mg^{2+}$ is replaced by $Mn^{2+}$ at its optimal concentration of 0.1 mmol/L.

One unit of T4 DNA polymerase is defined as the enzyme activity that incorporates 10 nmol of $^3$H-dTTP into acid-precipitable DNA products in 30 min at 37 °C [6.396].

The T4 DNA polymerase catalyzes the addition of mononucleotides onto the 3'-hydroxyl terminus of a primer annealed to a single-stranded region of a DNA template. Because fully duplex DNA cannot serve as a template primer, DNA must first be made partially single-stranded by digestion of the 3'-termini with *E. coli* exonuclease III (see Section 6.4.2) [6.605].

Single-stranded DNA can also serve as a template primer for T4 DNA polymerase. This polymerase is unable to use a DNA duplex that contains a nick as template primer. However, addition of the T4 gene 32 protein facilitates strand displacement, which allows the T4 DNA polymerase to replicate even the nicked duplex DNA in vivo and in vitro [6.266], [6.560]. Although T4 DNA polymerase lacks $5'-3'$ exonuclease activity, it contains an extremely active $3'-5'$ exonuclease that shows a strong specificity for single-stranded DNA. Its turnover number is about 250 times greater than that of the $3'-5'$ exonuclease associated with *E. coli* DNA polymerase I and about 3 times greater than the turnover number of its own $5'-3'$ polymerase activity [6.436].

**Uses.** The enzyme T4 DNA polymerase can act as either a $5'-3'$ polymerase or a $3'-5'$ exonuclease. Polymerization occurs when all three of the following substrates are available to the enzyme: a polynucleotide template, a 3'-hydroxyl primer at least one residue shorter than the template, and the appropriate dNTPs or NTPs complementary to the template. In the absence of any one of these three components, the enzyme functions as an $3'-5'$ exonuclease.

The T4 DNA polymerase can be used to label blunt or recessed 3'-termini of DNA by its $5'-3'$ polymerase activity. In the case of blunt ends, the $3'-5'$ exonuclease activity of the enzyme is responsible for digestion of duplex DNA to produce molecules with recessed 3'-termini. On subsequent addition of labeled dNTPs, the partially digested DNA molecules serve as primer–templates that are regenerated by the polymerase into intact, double-stranded DNA. Molecules labeled to high specific activity by this technique may be used as sensitive hybridization probes. They have two advantages over probes prepared by nick translation. First, they lack the artifactural hairpin structures that may be produced during nick translation. Second, they can be converted easily into strand-specific probes by cleavage with suitable restriction endonucleases in the unlabeled central region of the DNA molecule [6.330]. In combination with T4 gene 32 protein, the T4 DNA polymerase is used for gap filling in site-directed mutagenesis experiments in which short oligonucleotides are employed for mismatch formation [6.339].

## 6.2.4. Reverse Transcriptase

The enzyme AMV reverse transcriptase, also called deoxynucleoside triphosphate:DNA deoxynucleotidyltransferase (DNA-directed) (E.C. 2.7.7.49), is obtained from avian myeloblastosis virus.

**Properties.** The enzyme AMV reverse transcriptase is one of the gene products of the RNA genome of avian myeloblastosis virus existing in two copies within the core structure of the retrovirus particle. Retroviruses which are capable of replicating without helper viruses have at least three genes arranged in the following order

1) *gag* (encoding structural proteins of the inner coat and core),

2) *pol* (encoding reverse transcriptase and DNA endonuclease), and
3) *env* (encoding glycoproteins of the envelope).

A *gag – pol* precursor protein, p180$^{gag-pol}$, is one of the primary protein products; it is processed into a smaller protein, p130$^{gag-pol}$ [6.359], [6.387]. Further proteolysis releases two polypeptides designated as $\alpha$ and $\beta$, the latter being phosphorylated [6.627]. The enzymatically active forms of reverse transcriptase exist as $\alpha$, $\beta\beta$, and $\alpha\beta$. The molecular mass of the $\alpha$-subunit is $68 \times 10^3$ dalton and of the $\beta$-subunit $92 \times 10^3$ dalton.

One unit of reverse transcriptase is defined as the enzyme activity that incorporates 1 nmol of $^3$H-dTMP into acid-precipitable products in 10 min at 37 °C with poly(dA) · p(T)$_{15}$ as template primer [6.402].

Because the amino acid sequence of $\alpha$ is a subset of the $\beta$-subunit, the mature form $\alpha\beta$ is considered to be formed by proteolytic cleavage of the $\beta\beta$ dimer into $\alpha\beta$ and p32, containing DNA endonuclease activity [6.499]. Enzymatic activities associated with $\alpha\beta$ are (1) RNA-dependent DNA polymerase, (2) DNA-dependent DNA polymerase, (3) RNase H, and (4) an unwinding activity normally attributed to single-strand binding proteins [6.703]. The $\alpha$-subunit contains both the polymerase and the RNase H activity which has the unique ability to degrade the RNA in DNA · RNA hybrids in an exonucleolytic manner [6.315], [6.455]. This exonucleolytic activity of RNase H can proceed from either the 5'- or the 3'-terminus. However, it cannot act on RNA linkages in covalently closed duplex circles. Reverse transcriptase was shown to synthesize poly(dT) transcripts in vitro in a progressive way with poly(rA)$_{1100}$ · oligo(dT)$_{12-18}$ as template and primer [6.402]. In addition, the capacity of reverse transcriptase to use poly(rC) · oligo(dG)$_{12}$ as substrate is a useful feature for discrimination between the viral polymerase activity and the various host DNA polymerases [6.458], [6.696]. Effective priming is observed if the primer is more than eight nucleotides in length [6.379].

**Uses.** The enzyme AMV reserve transcriptase is widely used for synthesis of cDNA transcripts of specific RNA sequences in vitro (Fig. 53). Synthesis of the primary DNA strand is catalyzed by the RNA-dependent DNA polymerase activity. The DNA-dependent DNA polymerase activity is responsible for second-strand synthesis in cDNA formation when reverse transcriptase is used as polymerizing agent. The DNA-dependent DNA polymerase activity can be inhibited by addition of actinomycin D [6.378]. Transcripts of cDNA are employed for analysis of structure, organization, and expression of eukaryotic genes. Comparison between cDNA and genomic DNA sequences elucidates genomic rearrangements, the existence of intervening sequences, and details of the splicing events [6.292], [6.293], [6.689]. Synthesis of cDNA is also a powerful tool for the isolation of functional DNA sequences and the generation of corresponding hybridization probes. More recently, cloning of cDNA copies of the RNA virus genomes has opened another way to study details of their genomic structure [6.460], [6.614], [6.616]. Several protocols for cDNA synthesis have been described [6.354], and typical procedures have been reviewed recently [6.394], [6.428], [6.620]. In addition, new methods have been developed for highly efficient cloning of full-length cDNA and sequencing of RNA [6.391] by using AMV reverse transcriptase [6.565].

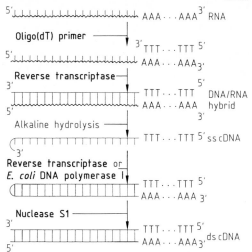

**Figure 53.** Use of AMV reverse transcriptase for in vitro synthesis of cDNA

## 6.2.5. Terminal Transferase

Terminal transferase, also called nucleoside triphosphate: DNA deoxynucleo-tidylexotransferase (E.C. 2.7.7.31) [*9027-67-2*], is obtained from calf thymus.

**Properties.** Terminal transferase from calf thymus catalyzes the addition of de-oxynucleotides to the 3'-hydroxyl terminus of a DNA primer [6.291].

One unit of terminal transferase is defined as the enzyme activity that incorporates 1 nmol of dAMP into acid-insoluble products in 60 min at 37 °C with $d(pT)_6$ as primer [6.318].

Terminal transferase from calf thymus has a molecular mass of $32 \times 10^3$ dalton. Under denaturing conditions, two nonidentical subunit structures are observed with molecular masses of $26.5 \times 10^3$ and $8 \times 10^3$ dalton. The isoelectric point is at pH 8.6, and the pH optimum at 7.2. At 35 °C, terminal transferase has an absolute requirement for an oligodeoxynucleotide primer containing at least three deoxynucleotides and a free 3'-hydroxyl end. However, the reaction is not template-dependent. Any dNTP, including substituted deoxynucleotides, can be polymerized by the enzyme [6.704]. The highest polymerization rate for purine deoxyribonucleoside triphosphates is obtained when cacodylate buffer and $Mg^{2+}$ are used, whereas the use of $Co^{2+}$ instead of $Mg^{2+}$ increases the polymerization of pyrimidine bases [6.591].

**Uses.** Because terminal transferase requires single-stranded DNA as primer, the incorporation efficiency is highest for dsDNA with 3'-protruding ends. However, blunt ends and 3'-recessive ends are also tailed by reverse transcriptase, with reduced efficiency [6.342], [6.616]. Terminal transferase is used mainly to add homopolymer tails to DNA fragments for the construction of recombinant DNA. By use of this tailing method, any double-stranded DNA fragment can be joined to a cloning vehicle. Optimal lengths for G/C tails are about 20–25 nucleotides, for A/T tails about 100 nucleotides [6.262]. In addition to the tailing reaction, terminal transferase is used for 3'-end-labeling of DNA fragments with $\alpha$-$^{32}$P-labeled dNTPs or ddNTPs [6.285], [6.694] and for the addition of a single nucleotide to the 3'-ends of DNA for in vitro mutagenesis experiments [6.342].

# 6.3. RNA Polymerases

Table 38 lists important RNA polymerases under B (p. 239).

## 6.3.1. SP6 RNA Polymerase

The enzyme SP6 RNA polymerase, also called nucleoside triphosphate:RNA nucleotidyltransferase (DNA-directed) (E.C. 2.7.7.6) [*9014-24-8*], is obtained from phage SP6-infected *Salmonella typhimurium* LT2.

**Properties.** The enzyme SP6 RNA polymerase consists of a single polypetide chain of $96 \times 10^3$ dalton [6.299].

One unit of SP6 RNA polymerase is defined as the enzyme activity that catalyzes the incorporation of 1 nmol of GTP into acid-precipitable RNA products in 60 min at 37 °C [6.299].

The SP6 RNA polymerase requires a DNA template and $Mg^{2+}$ as a cofactor for RNA synthesis, and is strongly stimulated by bovine serum albumin or spermidine [6.299]. The enzyme is not inhibited by the antibiotic rifampicin. This inhibitor affects bacterial RNA polymerases but not bacteriophage-coded RNA polymerases such as SP6 RNA polymerase [6.299], [6.421]. SP6 RNA polymerase possesses a very stringent promoter specificity which is distinct from that of analogous enzymes induced by bacteriophages T7, T3, gh1, or other T7-like bacteriophages [6.309], [6.314], [6.692]. Thus SP6 RNA polymerase is able to transcribe only bacteriophage SP6 DNA [6.452] or DNA cloned downstream from SP6 promoter as, for example, in the plasmids pSP64 or pSP65 [6.537].

**Uses.** The enzyme SP6 RNA polymerase is used for specific in vitro synthesis of RNA transcripts obtained from distinct DNA sequences cloned into the multiple cloning site of the vectors pSP64 or pSP65 [6.537]. In these vectors, the SP6 promoter located in front of the multiple cloning site directs specific initiation of the transcription reaction. Restriction endonuclease cleavage of the vector at a unique restriction

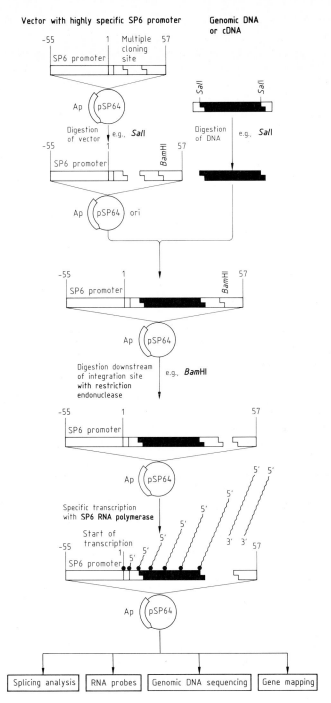

**Figure 54.** Use of SP6 RNA polymerase for in vitro synthesis of mRNA

site downstream from the inserted DNA sequences provides a discrete terminus for the run-off RNA transcript. Sense or anti-sense RNA transcripts of unique length can be generated, depending on the orientation of the integrated DNA (Fig. 54).

*Splicing Analysis.* When this vector system is applied, SP6 RNA polymerase can be used to synthesize large amounts of precursor RNA for in vitro and in vivo studies of the splicing reaction [6.401], [6.473], [6.537].

*RNA Probes.* The enzyme SP6 RNA polymerase is also useful for the generation of homogeneously labeled, single-stranded RNA probes of highest specific radioactivity. These probes may be applied in Northern- and Southern-blot hybridization techniques [6.744]. After DNase treatment and removal of the nonincorporated nucleotides by ethanol precipitation, the RNA probes generated with the pSP64/65 vectors are ready for hybridization without prior purification by gel electrophoresis. A tenfold increase in sensitivity may be observed when RNA probes are used instead of nick-translated DNA probes [6.537].

*Genomic DNA Sequencing.* Radioactively labeled RNA probes derived from SP6 RNA polymerase are also applied in genomic DNA sequencing [6.326] and in situ hybridization experiments.

*Gene Mapping.* M13 libraries may be screened with RNA probes synthesized by SP6 RNA polymerase in a one-step procedure that identifies the orientation of cloned DNA fragments.

*Synthesis of Specific RNA.* An additional application of SP6 RNA polymerase is the synthesis of specific RNA strands for use in the RNase protection technique. After hybridization to total cell RNA, nonhomologous sequences are digested with a mixture of RNase A and RNase $T_1$. Only the homologous double-stranded regions will give stable hybrids of characteristic length [6.744].

## 6.3.2. T7 RNA Polymerase

The enzyme T7 RNA polymerase, also called nucleoside triphosphate:RNA nucleotidyltransferase (DNA-directed) (E.C. 2.7.7.6) [*9014-24-8*], is obtained from phage T7-infected *Escherichia coli*.

**Properties.** The enzyme T7 RNA polymerase consists of a single polypeptide chain of $98 \times 10^3$ dalton [6.673].

One unit of T7 RNA polymerase is defined as the enzyme activity that incorporates 1 nmol of GMP into acid-precipitable RNA products in 60 min at 37 °C [6.338].

The enzyme T7 RNA polymerase is highly specific for a T7 promoter (17 base pairs) including the transcription start site [6.353]. The promoter sequence differs from that of bacteriophage SP6-coded RNA polymerase (Section 6.2.1) by only two nucleotides. Although the sequences are similar, T7 RNA polymerase specifically transcribes only T7 DNA or DNA cloned downstream of the T7 promoter. Appropriate cloning vectors are pT7-1 and pT7-2 containing the strong T7 promoter $\Phi10$ in front of a polylinker region with eight unique restriction sites [6.686]. In pT7-1 and pT7-2, the polylinker regions are integrated in opposite orientation. In addition to a DNA template with a T7 promoter, T7 RNA polymerase requires $Mg^{2+}$ as cofactor and the four NTPs as substrates. Spermidine or bovine serum albumin stimulates the activity of the enzyme. In contrast to bacterial RNA polymerases, T7 RNA polymerase is not inhibited by rifampicin [6.299], [6.421].

**Uses.** The T7 transcription system consisting of T7 RNA polymerase and vectors pT7-1 and pT7-2 is used for in vitro synthesis of specific RNA. For this purpose, the template DNAs are cloned into the polylinker regions of these vectors downstream from the T7 promoter. By cutting the cloned DNA at the 3′-end with an appropriate restriction endonuclease before the transcription reaction has started, run-off RNA transcripts of defined size are obtained. By using the T7 vectors pT7-1 and pT7-2 with oppositely oriented polylinker regions, DNA can be transcribed into sense and anti-sense mRNA.

The T7 system allows in vitro synthesis of either homogeneously labeled (in the presence of radioactive or biotinylated NTPs) or unlabeled RNA molecules. Since the ratio of the produced RNA to template DNA reaches 30:1, elimination of the DNA is usually not necessary but can be achieved by treatment with RNase-free DNase I. The RNA probes are highly specific for the complementary DNA strand; therefore, they are especially useful for different RNA or DNA blotting techniques [6.657], [6.666], in situ hybridization [6.745], genomic sequencing [6.326], or nuclease S1 studies. This property also allows the screening of M13 gene banks for clones containing DNA with a defined orientation. With unlabeled primary mRNA transcripts, splicing of RNA may be studied in vivo or in vitro. With the aid of in vitro synthesized anti-sense mRNA, expression of the corresponding genes can be suppressed after introduction of this RNA into eukaryotic cells [6.536]. The T7 system also allows the synthesis of pure RNA transcripts for in vitro translation studies.

# 6.4. DNA Nucleases

Table 38 lists important DNA nucleases under C (pp. 241–243).

## 6.4.1. DNase I

The enzyme DNase I, also called deoxyribonuclease I (E.C. 3.1.21.1), is obtained from bovine pancreas.

**Properties.** DNase I from bovine pancreas is a glycoprotein with a molecular mass of $31 \times 10^3$ dalton. The pH optimum of the enzyme is 7.0. One unit of DNase I is defined as the enzyme activity that produces after incubation of calf thymus DNA an absorbance increase of 0.001 in one minute at 25 °C [6.484], [6.488].

DNase I is normally isolated as a mixture of four isoenzymes: A, B, C, and D. The complete amino acid sequence of isoenzyme A, which is the main component, has been determined [6.504]. Isoenzymes A and B differ only in the composition of the carbohydrate side chain attached to asparagine 18, but isoenzymes C and D have the histidine at position 118 replaced by a proline [6.619]. All DNase I isoenzymes contain two disulfide bridges, one of which can be reduced in the presence of $Ca^{2+}$ without loss of activity [6.582]. DNase I degrades double-stranded DNA in an endonucleolytic way to yield 5′-oligonucleotides [6.489], [6.550]. DNase I requires divalent metal ions for DNA hydrolysis, with maximal enzymatic activity in the presence of $Ca^{2+}$ and $Mg^{2+}$ or $Ca^{2+}$ and $Mn^{2+}$ [6.581]. The mode of action and the specificity of the enzyme depend on the type of divalent cations present [6.305], [6.449]. DNase I appears to be primarily a digestive enzyme. However, DNase I-like activities have also been found in nondigestive tissues, which suggests that the enzyme might have other functions [6.550]. It may play a role in the regulation of actin polymerization, because it forms a very stable complex with monomeric actin and causes depolymerization of filamentous actin [6.430], [6.492], [6.524].

**Uses.** DNase I is used for labeling dsDNA by nick translation [6.262]. RNase-free DNase I is also applied to template digestion after in vitro synthesis of RNA with bacteriophage SP6-, T7-, or T3-coded RNA polymerases [6.299].

## 6.4.2. Exonuclease III

Exonuclease III, also called exodeoxyribonuclease III (E.C. 3.1.11.2), is obtained from *Escherichia coli*.

**Properties.** Exonuclease III is a monomeric protein with a molecular mass of $28 \times 10^3$ dalton [6.716].

One unit of exonuclease III is defined as the enzyme activity that releases 1 nmol of acid-soluble nucleotides from calf thymus DNA treated with ultrasonic waves in 30 min at 37 °C [6.604].

The enzyme is a multifunctional protein which catalyzes the hydrolysis of several types of phosphoester bonds in dsDNA.

In addition, the following minor activities are found [6.590], [6.612], [6.613]:

an endonuclease activity, specific for apurinic and apyrimidinic sites,
a 3′-phosphatase activity, and
an RNase H activity, which degrades the RNA strand in DNA · RNA hybrids.

The pH optima for the various endo- and exonucleolytic activities are between 7.6 and 8.5. A lower pH optimum between 6.8 and 7.4 has been reported for phosphatase activity. For optimal exonucleolytic activity, $Mg^{2+}$ or $Mn^{2+}$ is required, whereas $Zn^{2+}$ inhibits exonuclease III [6.605], [6.613]. In addition, the activity depends strongly on temperature, salt concentration, and ratio of DNA to enzyme concentration [6.415], [6.655]. Also, the rate at which mononucleotides are released decreases in a base-dependent mode in the following order $C \gg A \sim T \gg G$ [6.513].

**Uses.** Exonuclease III is used mainly for the generation of single-stranded terminal regions from dsDNA by acting as a 3′−5′ exonuclease. The DNA modified in this way is used as a substrate for labeling DNA with Klenow enzyme or for sequencing studies [6.414], [6.415], [6.655]. Another application of exonuclease III is to shorten progressively both strands of dsDNA by additional digestion of exonuclease III-generated single-stranded tails with nuclease S1 [6.416]. Exonuclease is also used to determine stable lesions in defined sequences of DNA [6.617].

## 6.4.3. Nuclease S1

Nuclease S1, also called *Aspergillus* nuclease S1 (E.C. 3.1.30.1), is obtained from *Aspergillus oryzae*.

**Properties.** Nuclease S1 is a monomeric protein with a molecular mass of $38 \times 10^3$ dalton [6.566], [6.705]. The enzyme is a glycoprotein containing 18% carbohydrate residues [6.566]. Its isoelectric point is at pH 4.3–4.4 [6.618].

One unit of nuclease S1 is defined as the enzyme activity that releases 1 μg of acid-soluble deoxynucleotides from denatured DNA in 1 min at 37 °C [6.705].

The pH optimum is 4.0–4.3, with half-maximal rates at pH 3.3 and 4.9. Higher pH values (4.6–5.0) have been used to avoid possible nicking of DNA substrates by depurination. The enzyme requires $Zn^{2+}$ for optimal activity. The ions $Co^{2+}$ and $Hg^{2+}$ can replace $Zn^{2+}$, but they are less effective [6.705]. Nuclease S1 is optimally active at NaCl concentrations of 0.1 mol/L and is relatively unaffected by ionic strengths between 0.01 and 0.2 mol/L of NaCl [6.435]. The enzyme is remarkably stable to such denaturing agents as urea, sodium dodecyl sulfate, or formamide [6.742]. When the enzyme is used in high concentrations, the salt concentration must be high enough (>0.2 mol/L) to suppress nonspecific nicking [6.706].

**Uses.** Under optimal conditions, the rates of nuclease S1-catalyzed hydrolysis of single- and double-stranded nucleic acids have been estimated to differ by a factor of 75 000 [6.725]. Thus, the main application of nuclease S1 is to trim single-stranded protruding ends of DNA or RNA without significant nibbling of the duplex ends [6.279], [6.418]. Mapping of transcripts is performed by analyzing S1-resistant RNA · DNA hybrids in alkaline agarose or denaturing polyacrylamide gels [6.280], [6.281]. High-resolution mapping is obtained after end-labeling the DNA with [32]P by analyzing the S1-resistant DNA fragments on sequencing gels [6.712]. Use of this method allows mapping of the 5′- and 3′-ends of mRNA [6.408]. In contrast to exonuclease VII, single-stranded loops in dsDNA or RNA · DNA hybrids are digested by nuclease S1 [6.321], [6.382]. Thus, mapping of introns in eukaryotic genes is possible by digesting the hybrid of mature mRNA and the [32]P-labeled codogenic DNA strand with nuclease S1 (Fig. 55).

A further application that takes advantage of this property is digestion of hairpin structures during cDNA synthesis [6.394]. An S1 mapping based on analysis by two-dimensional gel electrophoresis is also described [6.364]. The methodology of S1 mapping has been used successfully to study the regulation of cloned genes reintroduced into eukaryotic cells [6.713].

## 6.4.4. Nuclease Bal 31

Nuclease Bal 31, also called exodeoxyribonuclease (E.C. 3.1.11), is obtained from *Alteromonas espejiani*.

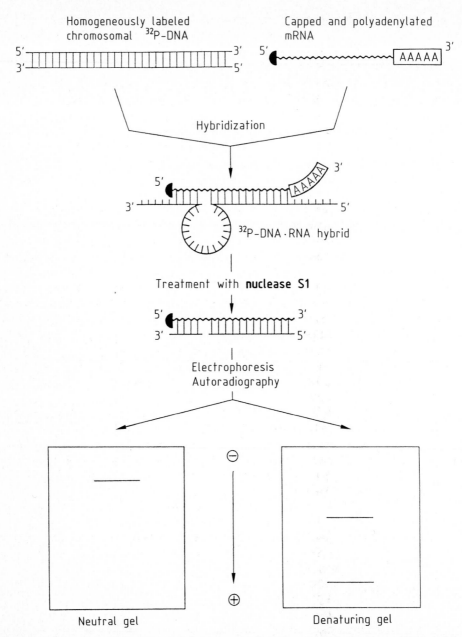

**Figure 55.** Mapping of introns in eukaryotic genes by digesting a hybrid of mature mRNA and the coding strand of DNA with nuclease S1

**Properties.** Nuclease Bal 31 progressively degrades both strands of dsDNA from the ends inward. The enzyme also has a ssDNA-specific endonuclease activity which cleaves supercoiled DNA to yield, for example, the linear form. No dsDNA-specific endonuclease activity is associated with the enzyme.

One unit of nuclease Bal 31 is defined as the enzyme activity that releases 600 base pairs in 10 min at 30 °C from 2 µg of linearized pUR222 DNA [6.399].

Enzyme activity requires $Ca^{2+}$ and $Mg^{2+}$, and is optimal at an NaCl concentration of 0.6 mol/L. Nuclease Bal 31 is not inhibited by sodium dodecyl sulfate or urea. Fragments of DNA shortened with nuclease Bal 31 can be ligated readily with T4 DNA ligase either directly or after filling in any single-stranded end with Klenow enzyme.

**Uses.** In molecular cloning experiments, various specific manipulations of DNA sequences must be accomplished: for example, coding sequences must be joined to a promoter in the proper reading frame, deletions of predetermined DNA segments are required to map functional regions, or transcription units must be localized by trimming to the minimum essential size. If suitable restriction sites do not exist at the proper location on the DNA sequence, nuclease Bal 31 can be applied.

Nuclease Bal 31 can also be used for mapping DNA with restriction endonucleases [6.495]. The DNA segment to be mapped is shortened to various lengths with the nuclease. Each sample is then digested with the corresponding restriction endonuclease, and the resulting fragments are analyzed by gel electrophoresis. Along with shortening of the original DNA segment, restriction fragments disappear from the gel in exactly the same order in which the restriction sites are arranged along the DNA strand.

Another use of nuclease Bal 31 is to transfer any restriction site located adjacent to a preselected position [6.573]. A linearized plasmid containing a cloned DNA sequence is digested with nuclease Bal 31 to a predetermined length, cut with a single cutting restriction enzyme, circularized with T4 DNA ligase, and transformed into *Escherichia coli* cells. Clones are selected which are sensitive to digestion with the restriction enzyme used before.

The ssDNA-specific endonuclease activity of nuclease Bal 31 has also been exploited to investigate the secondary structure in supercoiled DNAs or an altered helical structure in nonsupercoiled DNA produced by carcinogenic or mutagenic agents [6.491].

# 6.5. RNA Nucleases

Table 38 lists important RNA nucleases under D (p. 244).

## 6.5.1. RNase H

The enzyme RNase H, also called endoribonuclease (E.C. 3.1.26.4), is obtained from *Escherichia coli*.

**Properties.** The enzyme activity found in *E. coli* which hydrolyzes the RNA of RNA · DNA hybrids is designated RNase H, whereas the activity against the RNA in RNA · RNA duplexes is called RNase III [6.610].

One unit of RNase H is the enzyme activity that produces 1 nmol of acid-soluble ribonucleotides from $^3$H-poly(A) · poly(dT) in 20 min at 37 °C [6.427].

RNase H acts as an endoribonuclease and degrades the RNA strand of RNA · DNA hybrids of natural origin, and of synthetic complexes like poly(A) · poly(dT). RNase H produces ribonucleotides with 5'-phosphate and 3'-hydroxyl termini. Almost no activity is detected with polyribonucleotides alone or polymers annealed to their complementary ribopolymer. For optimal activity, RNase H requires $Mg^{2+}$ which can be replaced only partially by $Mn^{2+}$. The enzyme has its maximum activity in the presence of compounds containing mercapto groups and is inhibited by *N*-ethylmaleimide. The pH optimum is between 7.5 and 9.1. RNase H activity is relatively insensitive to salt; 50 % of its activity is retained in the presence of NaCl at 0.3 mol/L [6.283].

**Uses.** In addition to the use of *E. coli* RNase H for the study of in vivo functions such as RNA-primed initiation of DNA synthesis [6.427], RNase H is applied in the synthesis of cDNA by combination of the classical first-strand synthesis with the novel second-strand synthesis mediated by *E. coli* DNA polymerase I, *E. coli* RNase H, and *E. coli* DNA ligase [6.412], [6.565]. Furthermore, RNase H is used to detect RNA · DNA regions in dsDNA of natural origin [6.407], [6.455]. A further application of the enzyme is the removal of poly(A) sequences of mRNA, which leads to increased electrophoretic homogeneity of mRNA in gel electrophoresis [6.707]. RNase H may also be used for the site-specific enzymatic cleavage of RNA. With this method, a synthetic DNA oligomer will hybridize only to complementary single-stranded regions of an RNA molecule, which are therefore digested by RNase H in a site-specific manner [6.347].

## 6.5.2. Site-Specific RNases

### 6.5.2.1. RNase A

The enzyme RNase A, also called pancreatic ribonuclease, (E.C. 3.1.27.5), is obtained from bovine pancreas.

**Properties.** RNase A is an endonuclease that cleaves RNA but not DNA. The enzyme specifically attacks pyrimidine nucleotides by cleaving the 3'-adjacent phosphodiester bond Py/pN. The molecular mass is $13.7 \times 10^3$ dalton and the pH optimum is 7.0–7.5 [6.294], [6.429]. Cyclic 2',3'-pyrimidine nucleotides are obtained as intermediates. Pyrimidine 3'-phosphates and oligonucleotides with terminal pyrimidine 3'-phosphate groups are the final products.

One unit of RNase A is defined as the enzyme activity that produces total conversion of RNA as substrate in 1 min at 25 °C. The decrease in absorbance $A_0$ to $A_1$ corresponds to total conversion, $A_1$ being the final absorbance [6.483].

**Uses.** RNase A is used as A- and G-specific RNase in RNA sequencing [6.288], [6.479].

### 6.5.2.2. RNase CL3

**Properties.** The enzyme RNase CL3, also called endoribonuclease (E.C. 3.1.27.1), is obtained from chicken liver.

RNase CL3 cleaves RNA predominantly at Cp/N bonds and produces fragments with 3'-terminal cytidine phosphate. Bonds of the type Ap/N and Gp/N are cleaved less frequently, and Up/N bonds are cleaved to a very small extent [6.288]. The molecular mass is $16.85 \times 10^3$ dalton; the pH optimum is at 6.5–7.5 [6.502].

One unit of RNase CL3 is defined as the enzyme activity that releases a sufficient quantity of acid-soluble oligonucleotides to produce an absorbance increase of 1.0 at 260 nm in 15 min at 37 °C [6.502]. One unit corresponds to the decomposition of ca. 40 µg of poly(C).

**Uses.** RNase CL3 is used as minus-U-specific RNase in RNA sequencing [6.288], [6.479]. The Cp/N cleavage is more efficient in the presence of urea and at 50 °C, i.e., under conditions in which the secondary structure of RNA is destroyed [6.479], [6.480]. Under these denaturing conditions, approximately five to ten times as much enzyme must be used, but predominantly C-specific cleavage is obtained.

### 6.5.2.3. RNase $T_1$

Ribonuclease $T_1$ (E.C. 3.1.27.3), also called endoribonuclease, is obtained from *Aspergillus oryzae*.

**Properties.** RNase $T_1$ is an endonuclease that cleaves RNA but not DNA. The enzyme specifically attacks the 3'-adjacent phosphodiester bond Gp/N. RNase $T_1$ is therefore used as G-specific RNase in RNA sequencing [6.288], [6.479]. The molecular mass is $11.1 \times 10^3$ dalton, the pH optimum is at 7.4 [6.356], [6.625]. Cyclic 2',3'-guanosine nucleotides are obtained as intermediates, and guanosine 3'-phosphates and oligonucleotides with guanosine 3'-phosphate terminal groups are the final products.

One unit of RNase $T_1$ is defined as the enzyme activity that releases a sufficient quantity of acid-soluble oligonucleotides to produce an absorbance increase of 1.0 at 260 nm with denatured DNA in 15 min at 37 °C [6.356].

**Uses.** RNase $T_1$ is used especially for RNA sequencing and RNA fingerprinting.

### 6.5.2.4. RNase U$_2$

Ribonuclease U$_2$ (E.C. 3.1.27.4), also called endoribunuclease, is obtained from *Ustilago sphaerogena*.

**Properties.** RNase U$_2$ is an endonuclease that cleaves RNA but not DNA. The enzyme specifically attacks purine nucleotides by cleaving the 3′-adjacent phosphodiester bond Pu/pN [6.348]. The molecular mass is $10 \times 10^3$ dalton, and the pH optimum is 3.5. Cyclic 2′,3′-purine nucleotides are obtained as intermediates. Purine 3′-phosphates and oligonucleotides with purine 3′-phosphate terminal groups are the final products. Reversal of the final step can be used for the synthesis of ApN or GpN [6.687].

One unit of RNase U$_2$ is defined as the enzyme activity that releases a sufficient quantity of oligonucleotides to produce an absorbance increase of 1.0 at 260 nm with denatured DNA in 15 min at 37 °C [6.687].

**Uses.** RNase U$_2$ is used as A- and G-specific RNAse in RNA sequencing [6.288], [6.479].

### 6.5.2.5. Nuclease S7

Nuclease S7, also called endonuclease (E.C. 3.1.31.1), is obtained from *Staphylococcus aureus*.

**Properties.** Nuclease S7 is an endonuclease that cleaves both DNA and RNA. The products obtained after cleavage of RNA or DNA with nuclease S7 are oligo- and mononucleotides with terminal 3′-phosphate groups. The enzyme manifests A or U specificity in RNA digestion at pH 7.5 [6.422].

One unit of nuclease S7 is defined as the enzyme activity that releases a sufficient amount of acid-soluble oligonucleotides to produce an absorbance increase of 1.0 at 260 nm with denatured DNA in 30 min at 37 °C [6.422].

**Uses.** The enzyme is used widely to optimize the standard rabbit reticulocyte lysate. The mRNA-dependent protein synthesis system with only low amounts of endogenous RNA is obtained by incubation of the lysate with the enzyme in the presence of CaCl$_2$ [6.576]. Because of the strict dependence of the nuclease on Ca$^{2+}$ ions, it can easily be inactivated by a specific chelating agent such as ethylenediaminetetraacetic acid. During incubation at 50 °C and pH 7.5 in the presence of CaCl$_2$ and urea, NpA and NpU bonds are cleaved. Nuclease S7 is therefore also used as A- and U-specific RNase in RNA sequencing [6.479]. This reaction generates RNA fragments with 3′-phosphate termini. Thus, in sequencing gels, the generated RNA fragments are shifted downstream by one nucleotide compared with RNase T$_1$ or RNase U$_2$ fragments of identical length.

### 6.5.2.6. Site-Specific RNases in RNA Sequence Analysis

Base-specific RNases and nucleases with RNA and DNA specificity can be used for enzymatic sequencing of RNA molecules [6.348], [6.587], [6.648], [6.650] (Fig. 56). Because the sequence of RNA molecules larger than some hundred nucleotides cannot be determined directly, two alternative approaches are possible:

1) Fingerprint analysis, for example, of RNase $T_1$-resistant fragments [6.730], [6.732]
2) Sequencing RNA after partial digestion of large RNA chains (e.g., with RNase $T_1$) to produce a set of fragments covering the whole molecule [6.672].

In both approaches, the RNA fragments must be terminally labeled with $^{32}P$ in vitro prior to separation. Endlabeling of RNA can be achieved in two ways:

1) Labeling the 5'-terminus by using T4 polynucleotide kinase and $\gamma$-$^{32}P$-ATP as substrate (left branch, Fig 56) [6.355], [6.531], [6.664]
2) Labeling the 3'-terminus by using T4 RNA ligase and 5'-$^{32}P$-Cp as substrate (right branch, Fig. 56) [6.361].

In addition, 3'-endlabeling of tRNA may be obtained with tRNA nucleotidyltransferase. Any RNA molecules that fail to be polyadenylated can be labeled at the 3'-terminus with *E. coli* poly(A) polymerase [6.624], [6.652], [6.720]. After isolation of the various RNase $T_1$ subfragments, RNA sequencing consists of three analytical steps, as shown in Figure 56, bottom part.

**Determination of 5'- and 3'-Terminal Nucleotides.** Information about the nature of the 5'-terminally labeled nucleotide is obtained by complete digestion of the fragment with nuclease P1, resulting in 5'-$^{32}P$-nucleotides [6.489], [6.650], [6.735]. Excessive digestion of 3'-terminally labeled fragments with RNase $T_2$ permits the identification of the 3'-termini by analysis of the liberated 3'-$^{32}P$-nucleotides.

**Sequence Determination of the Terminal Nucleotides by Mobility-Shift Analysis.** Sequence information concerning limited regions of the 5'- and 3'-termini is obtained by partial enzymatic digestion of alternatively 5'- or 3'-labeled oligonucleotides with randomly acting endonuclease P1 or snake venom phosphodiesterase, acting as 3'-exonucleases; by randomly acting endonuclease S1 or calf spleen phosphodiesterase, acting as 5'-exonucleases; [6.320], [6.489], [6.647], [6.649], or by chemical treatment with $NaHCO_3$ or ethylenediaminetetraacetic acid [6.720].

**Complete Sequence Analysis by Partial Digestion with Base-Specific Nucleases.** In addition to chemical RNA sequencing [6.575], RNA sequencing by reverse transcription in cDNA [6.391], enzymatic RNA sequencing by partial digestion with the base-specific RNases mentioned in the previous sections (RNase $T_1$, RNase $U_2$, nuclease S7, and RNase CL3) is a well-established procedure [6.288], [6.348], [6.479], [6.650]. Reaction details are given in Table 37.

Identification of purine bases within the RNA chain is achieved by digestion of RNA with RNase $T_1$ and RNase $U_2$ [6.348], [6.587], [6.648], [6.650]. Digestion with RNase $T_1$ (G-specific) results in fragments with 3'-terminal Gp residues. Digestion

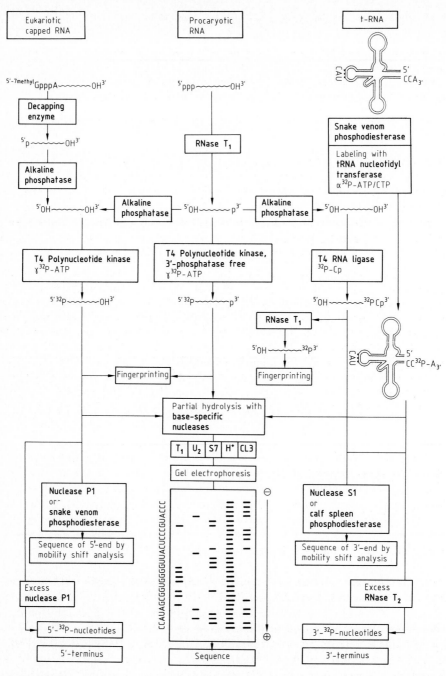

**Figure 56.** Use of nucleases with RNA and RNA:DNA specificity in sequencing RNA
$H^+$ = acid hydrolysis

**Table 37.** Partial digestion of RNA with base-specific RNases

| Enzyme | Conditions | Specificity |
|--------|-----------|-------------|
| RNase T$_1$ | urea, 8 mol/L, pH 3.5; 50 °C | $^{5'}$Gp/N$^{3'}$ |
| RNase U$_2$ | urea, 8 mol/L, pH 3.5; 50 °C | $^{5'}$Ap/N$^{3'}$ and ($^{5'}$Gp/N$^{3'}$) |
| RNase S7 | urea, 8 mol/L, pH 7.5; 50 °C | $^{5'}$Np/A$^{3'}$ and $^{5'}$Np/U$^{3'}$ |
| RNase CL3 | urea, 8 mol/L, pH 8.0; 50 °C | $^{5'}$Cp/N$^{3'}$ and ($^{5'}$Ap/N$^{3'}$ + $^{5'}$Gp/N$^{3'}$) |

with RNase U$_2$ (A- and G-specific) under appropriate conditions results in fragments with mainly 3′-terminal Ap residues (A-specific). Pyrimidine residues are localized by digestion of RNA with nuclease S7 at pH 7.5 and RNase CL3 at pH 8.0. Digestion with RNase CL3 results mainly in fragments with 3′-terminal Cp residues, less frequently in fragments with 3′-terminal Ap and Gp residues (minus-U-specific = C-, A-, and G-specific). Digestion with nuclease S7 at pH 7.5 results in fragments with 5′-terminal A and U residues (A- and U-specific) [6.479].

Care must be taken in reading the sequence ladder, because — compared with T$_1$ or U$_2$ fragments of corresponding length — the electrophoretic mobility of 5′-endlabeled S7 fragments is shifted downward by exactly one nucleotide; 3′-endlabeled S7 fragments are shifted upward by one nucleotide. This difference in eletrophoretic mobility compared with T$_1$ and U$_2$ fragments is caused by the cleavage site of the enzyme upstream from A or U residues. Normal electrophoretic mobility is observed with RNA fragments obtained with RNase CL3 [6.288], [6.502]. Digestion with RNase CL3 results mainly in fragments with 3′-terminal Cp residues. The formation of fragments with 3′-terminal Ap and 3′-terminal Gp residues is much less frequent. 3′-Terminal Up residues are generated to only a very small extent. Thus the information obtained by digestion of RNA with nuclease S7 and RNase CL3 is useful for unambiguous discrimination between both pyrimidine residues.

# 6.6. Modifying Enzymes

Table 38 lists important DNA-modifying enzymes under E (p. 245).

## 6.6.1. Alkaline Phosphatase

Alkaline phosphatase, also called orthophosphoric monoester phosphohydrolase (alkaline optimum) (E.C. 3.1.3.1) [*9001-78-9*], is obtained from calf intestine.

**Properties.** The molecular mass of alkaline phosphatase from calf intestine is $140 \times 10^3$ dalton. The enzyme is a dimeric glycoprotein composed of two identical or similar subunits with a molecular mass of $69 \times 10^3$ dalton each. Alkaline phosphatase contains four Zn$^{2+}$ ions per molecule [6.355].

One unit of alkaline phosphatase is defined as the enzyme activity that hydrolyzes 1 μmol of 4-nitrophenyl phosphate in 1 min at 37 °C in a diethanolamine buffer

(1 mol/L); 4-nitrophenyl phosphate concentration: 10 mmol/L; $MgCl_2$ concentration: 0.25 mmol/L; pH 9.8 [6.549].

The purified enzyme is most stable in the pH range 7.5–9.5 and is inactivated rapidly at lower pH. Protection against acidic denaturation is observed in the presence of inorganic phosphate [6.369]. Alkaline phosphatase can also be inactivated either by treatment with NaOH or by heating for 45 min at 65 °C in the presence of nitrilotriacetic acid or ethylenediaminetetraacetic acid to chelate the essential $Zn^{2+}$ ions [6.651]. Alternatively, alkaline phosphatase can be inhibited by inorganic phosphate. In the first two methods, RNA or DNA may be damaged. However, treatment with the chelating agents completely inactivates alkaline phosphatase without damage to the polynucleotide.

**Uses.** Alkaline phosphatase catalyzes the hydrolysis of numerous phosphate esters, such as esters of primary and secondary alcohols, sugar alcohols, cyclic alcohols, phenols, and aminoalcohols. Phosphodiesters do not react. The enzyme is used preferentially to selectively cleave terminal phosphate groups from oligonucleotides and monophosphate esters. In molecular biology, alkaline phosphatase is used primarily for dephosphorylation of 5′-phosphorylated DNA or RNA ends. These 5′-hydroxylated substrates can be effectively 5′-endlabeled with T4 polynucleotide kinase and $\gamma$-$^{32}$P-ATP as substrate [6.355], [6.394], [6.507], [6.530], [6.531], [6.588], [6.602]. 3′-Phosphorylated ends of RNA are also dephosphorylated with alkaline phosphatase [6.588], [6.648]. Either DNA or RNA that is $^{32}$P-labeled at the 5′-terminus is most frequently used in chemical DNA sequencing [6.530], [6.531] or in RNA sequencing by degradation of end-labeled RNA with base-specific RNases [6.588], [6.648]. However, alkaline phosphatase is also involved in the labeling of DNA and RNA fragments used for mapping and fingerprinting studies [6.451], [6.695]. The enzyme is further used in the construction of recombinant DNA molecules. Self-annealing of the vector DNA can be suppressed considerably by dephosphorylation of the linearized vector molecule prior to insertion of the DNA fragment to be cloned [6.290], [6.438], [6.699].

## 6.6.2. T4 DNA Ligase

The enzyme T4 DNA ligase, also called poly(deoxyribonucleotide):poly(deoxyribonucleo-tide) ligase (AMP-forming) (E.C. 6.5.1.1) [*9015-85-4*], is obtained from bacteriophage T4-infected *Escherichia coli*.

**Properties.** The molecular mass of T4 DNA ligase is $65 \times 10^3$ dalton [6.555], [6.729]. The enzyme is composed of a single subunit [6.574]. The T4 DNA ligase depends on ATP as a cofactor in the joining reaction [6.663]. The energy of ATP hydrolysis to yield AMP and pyrophosphate is used to form a phosphodiester linkage between polynucleotide chains. The enzyme also catalyzes an exchange reaction between pyrophosphate and ATP [6.719].

One unit of T4 DNA ligase is defined as the enzyme activity that converts 1 nmol of $^{32}$P from pyrophosphate into material that can be absorbed by Norit in 20 min at 37 °C [6.719]. The optimal pH range for T4 DNA ligase is 7.2–7.8; at pH 6.9 and 8.0, the enzyme has 46 and 56%, respectively, of its activity at pH 7.6 [6.717]. The enzyme requires $Mg^{2+}$ for activity; $Mn^{2+}$ is only 25% as effective as $Mg^{2+}$. Low concentrations of $NH_4^+$ ions have no effect on the T4 DNA ligase reaction. Higher levels of monovalent cations inhibit the enzyme completely [6.584].

**Uses.** The T4 DNA ligase is used to ligate DNA fragments with either 5′- or 3′-protruding or blunt ends. This enzyme is the only DNA ligase known that can catalyze blunt-end joining [6.635]. An important use of T4 DNA ligase is the preparation of recombinant DNA molecules for cloning experiments. Hydrogen-bonded recombinant DNA molecules can be generated by annealing two DNA fragments containing cohesive ends; T4 DNA ligase is the enzyme of choice for joining such cohesive ends since it requires a smaller overlapping sequence than *E. coli* DNA ligase.

Cohesive ends can be generated by cleavage of DNA with class II restriction endonucleases. An alternative approach is the addition of complementary homopolymer tails to the appropriate fragments with terminal transferase (G/C tailing). Cohesive ends can also be generated by blunt-end ligation with a synthetic DNA linker that contains the recognition sequence for a restriction endonuclease producing cohesive termini.

After labeling internal 5′-ends with T4 polynucleotide kinase, T4 DNA ligase can also be used to identify 3′- and 5′-end groups at single-stranded interruptions by nearest neighbor analysis [6.717]. Further, T4 DNA ligase can be applied to determine the ability of other enzymes to act at nicks and gaps in duplex DNA molecules. In addition, T4 DNA ligase can be used to study the primary and secondary structure of DNA molecules [6.567] and may be applied to the chemical synthesis of double-stranded DNAs with specific nucleotide sequences [6.295].

## 6.6.3. *Escherichia coli* DNA Ligase

The enzyme *E. coli* DNA ligase, also called poly(deoxyribonucleotide):poly-(deoxyribonucleotide) ligase (AMP-forming, NMN-forming) (E.C. 6.5.1.2) [*37259-52-2*], is obtained from *Escherichia coli*.

**Properties.** *Escherichia coli* DNA ligase consists of a single polypeptide chain with a molecular mass of $74 \times 10^3$ dalton [6.425], [6.572]. The bacterial enzyme catalyzes phosphodiester-bond synthesis coupled to cleavage of the pyrophosphate group in NAD.

One unit of *E. coli* DNA ligase is defined as the enzyme activity that converts 100 nmol poly[d(AT)] to an exonuclease III-resistant form in 30 min at 30 °C [6.545].

The NAD-dependent *E. coli* DNA ligase is much more specific than the ATP-dependent T4 DNA ligase. A number of ligations catalyzed by T4 DNA ligase are not catalyzed by *E. coli* DNA ligase, e.g., blunt-end ligation [6.623], [6.634]. The *E. coli* DNA ligase also does not act as an RNA ligase joining RNA molecules or DNA molecules containing RNA primer sequences.

**Uses.** The *E. coli* DNA ligase is used in full-length cDNA synthesis [6.565]. This method complements the present technique that employs nuclease S1 digestion of the hairpin loop prior to the second-strand synthesis. With this new technique, second-strand synthesis is mediated by the synchronous action of the enzymes *E. coli* DNA polymerase I, *E. coli* RNase H, and *E. coli* DNA ligase after first-strand synthesis with reverse transcriptase. The procedure uses a plasmid DNA vector which itself serves as a primer for the first- and, ultimately, the second-strand cDNA synthesis. Both steps are designed to enrich for recombinants containing full-length cDNAs over those with truncated cDNAs.

## 6.6.4. T4 Polynucleotide Kinase

The enzyme T4 polynucleotide kinase, also called ATP:5'-dephosphopoly-nucleotide-5'-phosphotransferase (E.C. 2.7.1.78) [*37211-65-7*], is obtained from phage T4-infected *Escherichia coli*.

**Properties.** The T4 polynucleotide kinase catalyzes the transfer for the terminal $\gamma$-phosphate group of ATP to the 5'-hydroxylated termini of polynucleotides like DNA or RNA. It also catalyzes the exchange of 5'-terminal phosphate groups [6.345], [6.602].

One unit of T4 polynucleotide kinase is defined as the enzyme activity required for the formation of 1 nmol of acid-precipitable $^{32}P$ in 30 min at 37 °C [6.602].

The enzyme T4 polynucleotide kinase migrates as a single species in sodium dodecyl sulfate–polyacrylamide gel electrophoresis. The single band with a molecular mass of $33 \times 10^3$ dalton represents one of the four identical subunits of the active enzyme complex with a total molecular mass of $140 \times 10^3$ dalton [6.506], [6.574]. Maximum activity is obtained at pH 7.6 at 37 °C and requires $Mg^{2+}$ ions and such reagents as dithiothreitol or 2-mercaptoethanol [6.574]. The reported stimulating effect of higher ionic strength or polyamides like spermine or spermidine [6.507], [6.574] results from stabilization of the active oligomeric tertiary structure of the enzyme [6.506]. Concentrations of ATP of at least 1 µmol/L and a ratio of 5:1 between ATP and protruding 5'-hydroxyl ends are required for optimal phosphorylation [6.508]. The T4 polynucleotide kinase is inhibited to 50 % by 7 mmol/L of sodium or potassium phosphate and to 75 % by 7 mmol/L of ammonium sulfate [6.574]. In addition to its kinase activity, T4 polynucleotide kinase also exhibits 3'-phosphatase activity [6.303], [6.304]. The pH optimum of this activity is between 5.0 and 6.0 and, thus, different from that of the kinase activity at pH 7.6.

**Uses.** The enzyme T4 polynucleotide kinase is used to label DNA and RNA at its 5'-termini with $^{32}P$ residues by using $\gamma$-$^{32}$P-ATP as substrate [6.368], [6.508],

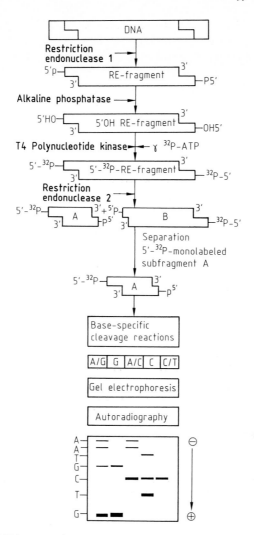

**Figure 57.** Chemical DNA sequencing

[6.602]. The 5'-terminus of DNA can be labeled with T4 polynucleotide kinase either by direct phosphorylation of 5'-hydroxyl groups or by exchange of DNA-bound, nonradioactively labeled 5'-phosphoryl groups and $^{32}$P molecules [6.282], [6.308], [6.531]. Alternative conditions for both reactions are also described for 5'-end-labeling of RNA [6.348], [6.648]. The T4 polynucleotide kinase is most commonly used in chemical sequence determination of DNA [6.530], [6.531] and RNA [6.348], [6.406], [6.648], [6.650]. Chemical DNA sequencing can determine the sequence of both the coding and the noncoding strands. Sequencing of the coding strand in the 5'−3' direction is possible after labeling the 5'-end of DNA with T4 polynucleotide kinase and $\gamma$-$^{32}$P-ATP (Fig. 57). Sequencing of the complementary noncoding strand in the 3'-5' direction is possible if a recessed 3'-end of dsDNA is elongated with Klenow enzyme and suitable $\alpha$-$^{32}$P-dNTPs as substrates. 5'-Endlabeling is also used for mapping restriction sites by partial digestion [6.520], [6.660]; DNA [6.694] or RNA fingerprinting [6.405], [6.596]; DNA footprinting via DNase protection [6.380] or methylation protection [6.443]; hybridization studies [6.685]; synthesis of substrates for DNA or RNA ligase [6.457], [6.646]; and sequence analysis of DNA [6.589], [6.595]. In addition, the 3'-phosphatase activity of T4 polynucleotide kinase may be used as a specific 3'-phosphatase [6.303].

## 6.6.5. T4 Polynucleotide Kinase, 3'-Phosphatase-Free

3'-Phosphatase-free T4 polynucleotide kinase, also called ATP:5'-dephospho-polynucleotide 5'-phosphotransferase (E.C. 2.7.1.78) [*37211-65-7*], is obtained from phage T4 *pse*T1- infected *Escherichia coli*.

**Properties.** 3'-Phosphatase-free T4 polynucleotide kinase is an altered T4 *pse* T1 gene product in which only the 3'-phosphatase activity but not the 5'-kinase activity has been affected [6.303], [6.304], [6.343], [6.665].

One unit of 3'-phosphatase-free T4 polynucleotide kinase is defined as the enzyme activity required for the formation of 1 nmol of acid-precipitable $^{32}$P in 30 min at 37 °C [6.602].

Maximum activity of the 5'-kinase reaction is obtained under the same conditions as those for wild-type T4 polynucleotide kinase. However, even under optimal conditions of the wild-type 3'-phosphatase activity, no removal of 3'-phosphatase groups of 3'-AMP is observed with the mutant T4 polynucleotide kinase at pH 5.5−6.0.

**Uses.** 3'-Phosphatase-free T4 polynucleotide kinase is of special interest in RNA analysis. The absence of the 3'-phosphatase makes this enzyme extremely useful for the preparation of unique species of RNA phosphorylated at both the 5'- and the 3'-termini. Oligoribonucleotides with both 3'- and 5'-terminal phosphates are used as donors in the T4 RNA ligase reaction [6.653], [6.697]. Although the 3'-phosphatase is not required for activity with T4 RNA ligase, it is a convenient blocking group to prevent cyclization or self-addition of the donor. Thus, a unique intermolecular product is ensured.

Another use of 3'-phosphatase-free T4 polynucleotide kinase is the labeling of CMP to give 5'-$^{32}$P-Cp, which is commonly used for 3'-endlabeling of RNA with T4 RNA ligase [6.564]. The 3'-terminally labeled RNA is very useful for fingerprinting and sequencing studies [6.348], [6.355], [6.648].

## 6.6.6. DNA Modification Methyltransferase (M · *Hpa*II)

M · *Hpa*II, also called *S*-adenosyl-L-methionine:DNA (cytosine-5)-methyltransferase (E.C. 2.1.1.37) [*9037-42-7*], is obtained from *Haemophilus parainfluenzae.*

**Properties.** M · *Hpa*II is isolated from *H. parainfluenzae* containing M · *Hpa*I as well [6.523], [6.740]. Methyltransferase *Hpa*I acts on double-stranded DNA by transferring methyl groups from *S*-adenosylmethionine to the palindromic recognition site

$$5' \overset{\text{M}}{\text{CCGG}} {}^{3'}$$
$$3' \underset{\text{M}}{\text{GGCC}} {}_{5'}$$

which results in 5-methylation of both internal cytosines [6.523], [6.726].

One unit of M · *Hpa*II is defined as the enzyme activity required to protect 1 μg of λ DNA to > 95 % against cleavage by the restriction endonuclease *Hpa*II in 1 h at 37 °C [6.523]. Single-stranded DNA is methylated by M · *Hpa*II with very low efficiency. However, M · *Hpa*II is capable of methylating hemimethylated recognition sequences [6.583]. Since divalent cations such as Mg$^{2+}$ are not essential for its activity, methylation of DNA with M · *Hpa*II may be carried out in the presence of ethylenediaminetetraacetic acid. Site-specific methylation of DNA by M · *Hpa*II protects the DNA against digestion by the restriction endonuclease *Hpa*II which recognizes and digests the identical tetrameric sequence CCGG [6.659]. However, the activity of restriction endonuclease *Msp*I, an heterohypekomer of *Hpa*II, is not affected by methylation of CCGG residues with M · *Hpa*II because *Msp*I is only sensitive to methylation of the external cytosine residues [6.442], [6.708]. In addition, the action of the restriction endonuclease *Sma*I whose recognition site CCCGGG is a subset of the *Hpa*II sites, is also inhibited by methylation of the internal cytosine residue with M · *Hpa*II. In contrast, the activity of its heterohypekomer *Xma*I (similar to *Msp*I) is not inhibited by methylation of *Hpa*II sites with M · *Hpa*II [6.741].

**Uses.** M · *Hpa*II is a useful tool for studying the in vivo effect of distinctly methylated $^{5'\text{m}}$CpG$^{3'}$ residues toward gene expression in eukaryotic cells. These 5-methylcytosine residues are located predominantly in CpG sequences [6.358], [6.409], [6.410] at a level of about 1 % of the total nucleotides [6.593]. An inverse correlation between the amount of CpG residues and gene activity was postulated [6.594], [6.683]. These results were confirmed by the gene-stimulating effect of azacytidine in vivo, because incorporation of this analogue into DNA mimics an undermethylated state of DNA [6.444]. To study the in vivo effects of distinct methylation

Table 38. Properties of important DNA- and RNA-modifying enzymes

## A. DNA polymerases

### A.1. Prokaryotic DNA polymerases

**E. coli DNA polymerases**

| Enzyme | Organism | $M_r$, $10^3$ dalton | Sub-units | References (isolation) | Specificity | Substrates | pH optimum | Cofactors | References (specificity) |
|---|---|---|---|---|---|---|---|---|---|
| DNA polymerase I (Kornberg enzyme) | Escherichia coli | 109 [Zn$^{2+}$] | α | [6.446], [6.456], [6.468] | 3'–5' polymerase<br><br>3'–5' exonuclease<br>5'–3' exonuclease pyrophosphorylase<br><br>pyrophosphate exchange | DNA matrix, 3'-OH DNA or RNA primer, dNTP<br>ssDNA<br>dsDNA, DNA/RNA hybrid<br>DNA matrix, 3'-OH DNA or RNA primer, pp$_i$<br>DNA, dNTP, pp$_i$ | 7–8 | Mg$^{2+}$ | [6.496]<br><br>[6.297], [6.497], [6.329], [6.462], [6.346]<br><br>[6.346] |
| DNA polymerase I, large fragment (Klenow enzyme) | Escherichia coli | 76 [Zn$^{2+}$] | α$_C$ | [6.296], [6.461], [6.554] | 3'–5' polymerase<br>3'–5' exonuclease pyrophosphorylase<br>pyrophosphate exchange | DNA matrix, 3'-OH DNA or RNA primer, dNTP<br>ssDNA<br>dsDNA<br>DNA matrix, 3'-OH DNA or RNA primer, pp$_i$ | 7–8 | Mg$^{2+}$<br><br>Mg$^{2+}$ | [6.554], [6.631] |
| DNA polymerase I, small fragment | Escherichia coli | 36 | α$_N$ | [6.296], [6.461], [6.632] | 5'–3' exonuclease | dsDNA | 7–8 | Mg$^{2+}$ | [6.632], [6.439] |
| DNA polymerase II | Escherichia coli | 120 | α | [6.385], [6.722] | 3'–5' polymerase<br>3'–5' exonuclease | DNA matrix, 3'-OH DNA or RNA primer, ssDNA | 7–8 | Mg$^{2+}$ | [6.469], [6.553], [6.554] |
| DNA polymerase III | Escherichia coli | [(175$_{core}$) 379$_{holo}$] | [(α,ε,θ)$_{core}$ τ,γ,δ,β$_{holo}$] | [6.470], [6.533], [6.534] | 3'–5' polymerase<br>3'–5' exonuclease<br>5'–3' exonuclease | ssDNA matrix, 3'-OH RNA primer, dNTP, ATP<br>ssDNA<br>ssDNA | 7.0 | Mg$^{2+}$ | [6.723], [6.724]<br>[6.517]<br>[6.517] |

**Phage-coded DNA polymerases**

| Enzyme | Organism | $M_r$, $10^3$ dalton | Sub-units | References (isolation) | Specificity | Substrates | pH optimum | Cofactors | References (specificity) |
|---|---|---|---|---|---|---|---|---|---|
| T4 DNA polymerase | Escherichia coli, phage T4-infected | 114 | α | [6.396], [6.551] | 3'–5' polymerase<br>3'–5' exonuclease | DNA matrix, 3'-OH DNA or RNA primer, dNTP<br>ssDNA | 8–9 | Mg$^{2+}$ | [6.606]<br>[6.396], [6.561] |
| T5 DNA polymerase | Escherichia coli, phage T5-infected | 96 | α | [6.377], [6.568] | 3'–5' polymerase<br>3'–5' exonuclease | DNA matrix, 3'-OH DNA primer, dNTP<br>ss or dsDNA | 8.5 | Mg$^{2+}$ | [6.336], [6.337]<br>[6.372] |
| T7 DNA polymerase | Escherichia coli, phage T7-infected | 96 | α,β | [6.265], [6.404] | 3'–5' polymerase<br>3'–5' exonuclease | DNA matrix, 3'-OH DNA primer, dNTP<br>ss or dsDNA | 7.6–7.8 | Mg$^{2+}$ | [6.462], [6.607]<br>[6.265], [6.434], [6.404] |

## A.2. Eukaryotic DNA polymerases

### Type α DNA polymerases

| DNA polymerase α | Source | MW (kDa) | | | Activity | Template/substrate, pH | Metal | References |
|---|---|---|---|---|---|---|---|---|
| | *Drosophila melanogaster* | 280 | 4 subunits | [6.275] | 3'–5' polymerase | DNA (RNA) matrix, 3'-OH DNA primer, dNTP, 7.5–8.5 | Mg$^{2+}$ | [6.261], [6.467] |
| | mouse myeloma cells | 190 | 2 subunits | [6.324], [6.528] | | | | |
| | rat liver | 155–250 | 5 subunits | [6.366], [6.535] | pyrophosphorylase | | | [6.324] |
| | calf thymus | 210–230 | 2 subunits | [6.411], [6.424], [6.433] | pyrophosphate exchange | | | [6.324] |
| | human HeLa cells | 320 ($\alpha_1$) | 5 subunits | [6.487] | | | | |
| | | 600 ($\alpha_2$) | 5 subunits | [6.487] | | | | |
| | | 220 ($\alpha_3$) | 2 subunits | [6.487] | | | | |
| | human KB cells | 150–160 | 2 subunits | [6.367], [6.467] | | | | |

### Type β DNA polymerases

| DNA polymerase β | Source | MW (kDa) | | | Activity | Template/substrate, pH | Metal | References |
|---|---|---|---|---|---|---|---|---|
| | calf thymus | 44 | $\alpha$ | [6.316], [6.319] | 3'–5' polymerase | DNA or RNA matrix, 3'-OH DNA primer, dNTP, 8.4–9.2 | Mg$^{2+}$ | [6.317], [6.674], [6.711] |
| | chicken embryo | 40 | $\alpha$ | [6.737] | | | | |
| | human KB cells | 43 | $\alpha$ | [6.711] | | | | |
| | Novikoff hepatoma cells | 31 | $\alpha$ | [6.674] | | | | |

### Type γ DNA polymerases

| DNA polymerase γ | Source | MW (kDa) | | | Activity | Template/substrate, pH | Metal | References |
|---|---|---|---|---|---|---|---|---|
| | chicken embryo | 180 | $\alpha_4$ | [6.736] | 3'–5' polymerase | RNA (DNA) matrix, 3'-OH DNA primer, dNTP, 8.5–9 | Mn$^{2+}$ | [6.464], [6.472], [6.527] |
| | mouse myeloma cells | 140 | $\alpha_4$ | [6.529] | | | | |
| | human HeLa cells | 110 | | [6.464], [6.667] | | | | |
| | human lymphoblasts | 120 | | [6.609] | | | | |

### Type δ DNA polymerases

| DNA polymerase δ | Source | MW (kDa) | | | Activity | Template/substrate, pH | Metal | References |
|---|---|---|---|---|---|---|---|---|
| | calf thymus | 140–200 | $\alpha, \beta$ | [6.493], [6.494] | 3'–5' polymerase; 3'–5' exonuclease | RNA matrix, 3'-OH RNA primer, dNTP | Mg$^{2+}$ | [6.300], [6.494] |
| | rabbit bone marrow | 122 | | [6.300], [6.301] | 3'–5' polymerase; 3'–5' exonuclease | RNA matrix, 3'-OH RNA primer, dNTP | Mg$^{2+}$ | [6.301], [6.493] |

Table 38. (continued)

| Enzyme | Organism | $M_r$, $10^3$ dalton | Subunits | References (isolation) | Specificity | Substrates | pH optimum | Cofactors | References (specificity) |
|---|---|---|---|---|---|---|---|---|---|
| **Virus-coded DNA polymerases** | | | | | | | | | |
| Reverse transcriptase | avian myeloblastosis virus (AMV) | 160 [$Zn^{2+}$] | $\alpha,\beta$ | [6.389], [6.703], [6.703] | 3'–5' polymerase<br><br>RNase H<br>DNA endonuclease<br>pyrophosphorylase<br>pyrophosphate exchange | RNA or DNA matrix, 3'-OH DNA or RNA primer, dNTP<br>RNA/DNA hybrid dsDNA | 8–8.5 | $Mg^{2+}$ | [6.450], [6.662], [6.703],<br><br>[6.547]<br>[6.397]<br>[6.629], [6.671]<br>[6.629], [6.671] |
| Reverse transcriptase | Moloney murine leukemia virus (Mo–MLV) | 80 [$Zn^{2+}$] | $\alpha$ | [6.388], [6.546], [6.702], [6.703] | 3'–5' polymerase<br><br>RNase H<br>DNA endonuclease<br>pyrophosphorylase<br>pyrophosphate exchange | RNA or DNA matrix, 3'-OH DNA or RNA primer, dNTP<br>RNA or DNA hybrid dsDNA | 8–8.5 | $Mn^{2+}$ | [6.38], [6.546], [6.702], [6.703] |
| HSV-1 DNA polymerase | RC-37 cells, HSV-1-infected | 144–150 | $\alpha$ | [6.289], [6.362], [6.463], [6.498] | 3'–5' polymerase<br><br>3'–5' exonuclease<br>pyrophosphate exchange | DNA matrix, 3'-OH DNA primer ssDNA | 8–8.5 | $Mg^{2+}$ | [6.289], [6.362], [6.463], [6.498] |
| Vaccina virus DNA polymerase | human HeLa cells, vaccina virus-infected | 110–115 | $\alpha$ | [6.310] | 3'–5' polymerase<br><br>3'–5' exonuclease<br>pyrophosphate exchange | DNA matrix, 3'-OH DNA primer, dNTP ssDNA | 8–9 | $Mg^{2+}$ | [6.311] |
| **Terminal transferases** | | | | | | | | | |
| Terminal transferase (Bollum enzyme) | calf thymus | 62 [$Co^{2+}$] | $\alpha$ | [6.318], [6.340] | 3'–5' polymerase | 3'-OH DNA primer, dNTP | 7.2 | $Mg^{2+}$, $Co^{2+}$ | [6.354] |

**B. RNA Polymerases**

**B.1. Prokaryotic RNA polymerases**

**Bacterial RNA polymerases**

| Enzyme | Source | MW [metal] | Subunit | References | Activity | Substrate | pH | Metal | References |
|---|---|---|---|---|---|---|---|---|---|
| RNA polymerase | Escherichia coli | 454 [Zn$^{2+}$] | * | [6.150], [6.298], [6.312], [6.569], [6.571], [6.570], [6.743], [6.476], [6.477] | 3'–5' RNA polymerase | DNA matrix, NTP | | Mg$^{2+}$ | [6.477], [6.522] |
| | Acetobacter vinelandii | 400–500 [Zn$^{2+}$] | * | [6.272], [6.642] | de novo RNA synthesis | ATP and UTP or ITP and CTP | 7–8 | Mg$^{2+}$ | [6.474] |
| | Bacillus subtilis | 400–500 [Zn$^{2+}$] | * | [6.278] | pyrophosphate exchange | poly[d(AT)], ATP, UTP, pp$_i$ | 7–8 | Mg$^{2+}$ | [6.475], [6.656] |
| | Caulobacter crescentus | 400–500 [Zn$^{2+}$] | * | [6.680] | | | | | |
| | Lactobacillus curvatus | 400–500 [Zn$^{2+}$] | * | | | | | | |
| | Micrococcus luteus | 400–500 [Zn$^{2+}$] | * | [6.420], [6.505] | | | | | |

**Phage-coded RNA polymerases**

| Enzyme | Source | MW | Subunit | References | Activity | Substrate | pH | Metal | References |
|---|---|---|---|---|---|---|---|---|---|
| T7 RNA polymerase | Escherichia coli, phage T7-infected | 98 | α | [6.314], [6.673] | 3'–5' RNA polymerase | | 7–8 | Mg$^{2+}$ | [6.314], [6.353], [6.673] |
| T3 RNA polymerase | Escherichia coli, phage T3-infected | 105 | α | [6.277], [6.309], [6.532] | | | 7–8 | Mg$^{2+}$ | [6.277], [6.532] |
| N4 RNA polymerase | Escherichia coli, phage N4-infected | 350 | α | [6.577] | 3'–5' RNA polymerase | DNA matrix, NTP | 7–8 | Mg$^{2+}$ | [6.313], [6.577] |
| SP6 RNA polymerase | Salmonella typhimurium, phage SP6-infected | 96 | α | [6.299], [6.452] | | | 7–8 | Mg$^{2+}$ | [6.299], [6.452] |
| PBS2 RNA polymerase | Bacillus subtilis, phage PBS2-infected | 260 | α,β,γ, δ,ε | [6.327] | | | 7–8 | Mg$^{2+}$ | [6.313], [6.327] |

* $[(\beta\beta'\alpha_2)_{core}\,\sigma]_{holo}$

**Table 38.** (continued)

| Enzyme | Organism | $M_r$, $10^3$ dalton | Sub-units | References (isolation) | Specificity | Substrates | pH optimum | Cofactors | References (specificity) |
|---|---|---|---|---|---|---|---|---|---|
| **Poly(A) polymerases** | | | | | | | | | |
| Poly(A) polymerase | *Escherichia coli* | 50 | α | [6.652] | terminal riboade- nylate transferase | 3'-OH RNA | 7–9 | ATP, Mg²⁺ or Mn²⁺ | [6.624], [6.652] |
| Poly(A) polymerase | calf thymus | 60 120–140 | α α₂ | [6.693] | terminal riboade- nylate transferase | 3'-OH RNA | 7.4 | ATP, Mn²⁺ or ATP, Mg²⁺ | [6.693] [6.731] |
| Poly(A) polymerase | human HeLa cells | 63–75 50–58 | α α | [6.559] | terminal riboade- nylate transferase | 3'-OH RNA | 7.9 | ATP, Mn²⁺ or ATP, Mg²⁺ | [6.559] [6.559] |
| **Polynucleotide phosphorylases** | | | | | | | | | |
| PNPase | *Escherichia coli* | 252 | α₃ | [6.580] | polyribonucleotide nucleotidyl- transferase ADP–pᵢ exchange | NDP | | Mg²⁺ | [6.684] [6.728] |
| PNPase | *Micrococcus luteus* | 237 | | [6.276] | polyribonucleotide nucleotidyl- transferase ADP–pᵢ exchange | NDP | | Mg²⁺ | [6.276] [6.684] |
| **tRNA nucleotidyltransferases** | | | | | | | | | |
| tRNA nucleotidyltrans- ferase | *Escherichia coli* B | 45–54 | α | [6.540] | tRNA nucleotidyltrans- ferase | tRNA, ATP and CTP | 9–9.4 | Mg²⁺ | [6.306] |
| tRNA nucleotidyltrans- ferase | baker's yeast | 70–71 | α | [6.677] | tRNA nucleotidyltrans- ferase | tRNA, ATP and CTP | 9.5 | Mg²⁺ | [6.678] |

## B.2. Eukaryotic RNA polymerases

### RNA polymerases

| Enzyme | Source | MW (×10³) [metal] | Subunits | References | Reaction | Requirements | pH | Metal | References |
|---|---|---|---|---|---|---|---|---|---|
| RNA polymerase I | yeast | 500–600 [Zn$^{2+}$] | 11 sub-units | [6.630], [6.700] | | | | | |
| | wheat germ | 400–500 | 7 sub-units | [6.440], [6.441] | | | | | |
| | *Acanthamoeba castellanii* | 500–600 | 11 sub-units | [6.268], [6.269] | | | | | |
| RNA polymerase II | yeast | 500–600 [Zn$^{2+}$] | 9 sub-units | [6.630], [6.700] | 3′–5′ polymerase | ss (ds) DNA matrix, NTP | 7–8 | Mn$^{2+}$ | [6.413], [6.503] |
| | wheat germ | 500–600 [Zn$^{2+}$] | 11 sub-units | [6.440], [6.441] | | | | | |
| | *Acanthamoeba castellanii* | 500–600 | 10 sub-units | [6.268], [6.269] | | | | | |
| RNA polymerase III | yeast | 500–600 [Zn$^{2+}$] | 10 sub-units | [6.630], [6.700] | | | | | |
| | wheat germ | 500–600 | 14 sub-units | [6.440], [6.441] | | | | | |
| | *Acanthamoeba castellanii* | 500–600 | 14 sub-units | [6.268], [6.269] | | | | | |

### Organelle-specific RNA polymerases

| Enzyme | Source | MW (×10³) | Subunits | References | Reaction | Requirements | pH | Metal | References |
|---|---|---|---|---|---|---|---|---|---|
| RNA polymerase | maize chloroplasts | 500 | $\alpha,\beta,\gamma,\delta$ | [6.65], [6.459] | 3′–5′ polymerase | DNA matrix, NTP | | Mg$^{2+}$ | [6.658], [6.459] |
| RNA polymerase | yeast mitochondria | 100–150 | $\alpha_{2-3}$ | [6.500] | 3′–5′ polymerase | DNA matrix, NTP | | Mg$^{2+}$ | [6.500] |

### 2′,5′-Oligoadenylate synthetases

| Enzyme | Source | MW (×10³) | References | Reaction | Requirements | pH | Metal | References |
|---|---|---|---|---|---|---|---|---|
| 2′5′-Oligoadenylate synthetase | chicken embryonic cells | 50–60 | [6.273] | 2′,5′-oligoadenylate adenyltransferase | dsRNA, ATP | 7.0–8.0 | Mg$^{2+}$ | [6.273], [6.274], [6.621] |
| 2′5′-Oligoadenylate synthetase | Ehrlich ascites tumor cells | 85 | [6.273], [6.349] | 2′,5′-oligoadenylate adenyltransferase | dsRNA, ATP | 7.0–8.0 | Mg$^{2+}$ | [6.274], [6.542], [6.621] |

## C. Nucleases

### C.1. Restriction endonucleases (see Table 34)

### C.2. Endonucleases

Table 38. (continued)

| Enzyme | Organism | $M_r$, $10^3$ dalton | Sub-units | References (isolation) | Specificity[a] | Substrates | pH optimum | Cofactors | References (specificity) |
|---|---|---|---|---|---|---|---|---|---|
| **Double-strand specific endonuclease** | | | | | | | | | |
| DNase I | bovine pancreas | 31 | α | [6.484], [6.504] | endonuclease | dsDNA | 7.0 | $Ca^{2+}$, $Mg^{2+}$ | [6.488], [6.489] [6.550] |
| **Single-strand specific endonucleases** | | | | | | | | | |
| Nuclease S1 | Aspergillus oryzae | 38 [$Zn^{2+}$] | α | [6.270], [6.705] | endonuclease | ssDNA (RNA), partially denatured dsDNA, DNA/RNA hybrid | 4.0–4.3 | $Co^{2+}$, $Zn^{2+}$ | [6.638] [6.453], [6.566] |
| P. citrinum nuclease | Penicillium citrinum | | | [6.375] | endonuclease | ssRNA (DNA) | 5.0 | $Zn^{2+}$ | [6.376] |
| U. maydis nuclease | Ustilago maydis | 42 [$Zn^{2+}$] | α | [6.432] | endonuclease | ssDNA (RNA) | 8.0 | $Mg^{2+}$, $Ca^{2+}$, $Co^{2+}$ or $Zn^{2+}$ | [6.431] |
| Mung bean nuclease | mung beans | 39 [$Zn^{2+}$] | α | [6.270] | endonuclease | ssDNA or ssRNA | 5.0 | $Zn^{2+}$ | [6.705] |
| WS nuclease | wheat germ | 43 [$Zn^{2+}$] | α | [6.478] | endonuclease | ssDNA (RNA) | 4.8–5.5 | $Zn^{2+}$ | [6.478] |
| **AP endonucleases** | | | | | | | | | |
| Endonuclease III | Escherichia coli | 27 | α | [6.384] | AP endonuclease DNA glyosylase | dsAP DNA thymine dimers | 7 | EDTA resistant | [6.341] [6.586] |
| Endonuclease IV | Escherichia coli | 33 | α | [6.518] | AP endonuclease | dsAP DNA | 8.0–8.5 | EDTA resistant | [6.374], [6.518] |
| Endonuclease V | Escherichia coli | 20 | α | [6.383] | AP endonuclease | dsAP DNA | 9.2–9.5 | $Mg^{2+}$ | [6.374], [6.383] |
| Endonuclease VII | Escherichia coli | 45 | α | [6.373] | AP endonuclease | ssAP DNA | 7.0 | $Mg^{2+}$ or $Ca^{2+}$ | [6.373], [6.374] |
| Mlu AP endoA | Micrococcus luteus | 35 | α | [6.579] | AP endonuclease | dsAP DNA | 7.5 | ($Mg^{2+}$) | [6.579] |
| Mlu AP endoB | Micrococcus luteus | 35 | α | [6.579] | AP endonuclease | dsAP DNA | 6.5–8.0 | ($Mg^{2+}$) | [6.579] |
| Hin AP endo | Haemophilus influenzae | 30 | α | [6.328] | AP endonuclease 3'-exonuclease 3'-phosphatase | dsAP DNA | 7–8 | $Mg^{2+}$ or $Mn^{2+}$ | [6.328] |
| Bsu AP endo | Bacillus subtilis | 56 | α | [6.437] | AP endonuclease | | | ($Mg^{2+}$) | [6.437] |
| CL AP endo | calf liver | 28 | α | [6.481] | AP endonuclease | | 9.5 | $Mg^{2+}$ | [6.481] |
| HP AP endo | human placenta | 27–31 | α | [6.511] | AP endonuclease | | | ($Mg^{2+}$) | [6.511] |

## C.3. Exonucleases

### E. coli exonucleases

| Enzyme | Source | MW | Subunit | Refs | Activity | Substrate | pH | Cofactor | Refs |
|---|---|---|---|---|---|---|---|---|---|
| Exonuclease I | Escherichia coli | 70–72 | α | [6.521], [6.592] | 3′–5′ exonuclease | ssDNA | 9.5 | $Mg^{2+}$ | [6.601], [6.734] |
| Exonuclease III | Escherichia coli | 28 | α | [6.446], [6.714] | 3′–5′ exonuclease / DNA 3′-phosphatase / AP endonuclease / RNase H | dsDNA / 3′-P DNA / apurinic–apyrimidinic DNA / RNA/DNA hybrid | 7.6–8.5 | $Mg^{2+}$ | [6.455], [6.538], [6.604], [6.718] |
| Exonuclease IV | Escherichia coli | | | [6.445] | 3′–5′ exonuclease | ssDNA | 8.0–9.5 | $Mg^{2+}$ | [6.445], [6.715] |
| Exonuclease V | Escherichia coli | 270 | α,β | [6.393] | 3′–5′ exonuclease / 5′–3′ exonuclease | dsDNA | 9 | ATP, $Mg^{2+}$ | [6.733] |
| Exonuclease VII | Escherichia coli | 88 | α | [6.321] | 3′–5′ exonuclease / 5′–3′ exonuclease | ssDNA | 7.9 | EDTA resistant | [6.322] |
| Exonuclease VIII | Escherichia coli | 140 | α | [6.392] | 5′–3′ exonuclease | dsDNA | 8.0–9.0 | $Mg^{2+}$ | [6.486] |

### Exonuclease V-analogous enzymes

| Enzyme | Source | MW | Subunit | Refs | Activity | Substrate | pH | Cofactor | Refs |
|---|---|---|---|---|---|---|---|---|---|
| Hind exonuclease V | Haemophilus influenzae | 290 | α,β,γ | [6.325], [6.566], [6.727] | 3′–5′ exonuclease / 5′–3′ exonuclease | dsDNA | 9 | ATP, $Mg^{2+}$ | [6.325], [6.727] |
| Bsu exonuclease V | Bacillus subtilis | 300 | α,β | [6.556], [6.636] | 3′–5′ exonuclease / 5′–3′ exonuclease | dsDNA | 9 | ATP, $Mg^{2+}$ | [6.636] |
| Pae exonuclease V | Pseudomonas aeruginosa | 300 | | [6.539], [6.556] | 3′–5′ exonuclease / 5′–3′ exonuclease | dsDNA | 9 | ATP, $Mg^{2+}$ | [6.539] |
| Sin exonuclease V | sea urchin intermedius | 450 | | [6.381], [6.556] | 3′–5′ exonuclease / 5′–3′ exonuclease | dsDNA | 9 | ATP, $Mg^{2+}$ | [6.381] |

### Other exonucleases

| Enzyme | Source | MW | Subunit | Refs | Activity | Substrate | pH | Cofactor | Refs |
|---|---|---|---|---|---|---|---|---|---|
| Nuclease Bal 31 | Alteromonas espejiani | 73 | α | [6.398], [6.400] | 3′–5′ exonuclease / 5′–3′ exonuclease | ss or dsDNA | 8.0 | $Mg^{2+}$, $Ca^{2+}$ | [6.398], [6.399], [6.400], [6.491] |
| N. crassa nuclease | Neurospora crassa | 55 [$Zn^{2+}$] | α | [6.509], [6.510] | exonuclease / endonuclease | linear ss or dsDNA / circular ssDNA or ssRNA | 7.5–8.5 | $Mg^{2+}$ | [6.370], [6.371] |

### Phage-coded exonucleases

| Enzyme | Source | MW | Subunit | Refs | Activity | Substrate | pH | Cofactor | Refs |
|---|---|---|---|---|---|---|---|---|---|
| λ exonuclease | Escherichia coli, phage λ-infected | 52 | α | [6.395], [6.585] | 5′–3′ exonuclease | dsDNA, 5′-P terminus | 8–9 | $Mg^{2+}$ | [6.512], [6.670] |

Table 38. (continued)

| Enzyme | Organism | $M_r$, $10^3$ dalton | Sub-units | References (isolation) | Specificity | Substrates | pH optimum | Cofactors | References (specificity) |
|---|---|---|---|---|---|---|---|---|---|
| **D. RNA nucleases** | | | | | | | | | |
| **D.1. E. coli ribonucleases** | | | | | | | | | |
| RNase I | Escherichia coli | low | α | [6.669] | endonuclease | ssRNA, (A/U)p/N | 8.1 | | [6.669] |
| RNase II | Escherichia coli | 68–85 | α | [6.417] | 3'–5' exonuclease | ssRNA, 3'-OH terminus | 7–8 | $Mg^{2+}$, $K^+$ | [6.562] |
| RNase III | Escherichia coli | 50 | $\alpha_2$ | [6.335], [6.351] | endonuclease | ss or dsRNA secondary structures | 7–8 | $Mg^{2+}$ or $Mn^{2+}$; $NH_4^+$, $Na^+$, or $K^+$ | [6.352], [6.611] |
| RNase IV | Escherichia coli | | | [6.665] | endonuclease | ssRNA secondary structures | | | [6.637] |
| RNase H | Escherichia coli | 40 | α | [6.332], [6.427], [6.541] | endonuclease | DNA/RNA hybrid | 7.5–9.1 | $Mg^{2+}$ or $Mn^{2+}$ | [6.333], [6.610] |
| **D.2. Other ribonucleases** | | | | | | | | | |
| RNase A | bovine pancreas | 13.7 | α | [6.485] | endonuclease | ssRNA, (C/U)/pN | 7.0–7.5 | | [6.287], [6.479] |
| RNase CL3 | chicken liver | 16.85 | α | [6.288], [6.502] | endonuclease | ssRNA, C(A/G)p/N | 8.0 | | [6.479] |
| RNase T₁ | Aspergillus oryzae | 11.1 | α | [6.357], [6.625] | endonuclease | ssRNA, Gp/N | 3.5 | | [6.357], [6.479] |
| RNase U₂ | Ustilago spherogena | 10.0 | α | [6.687] | endonuclease | ssRNA, Ap/N | 3.5 | | [6.348],[6.357], [6.479] |
| Nuclease S7 | Staphylococcus aureus | 16.8 | α | [6.576], [6.690] | endonuclease | ssRNA, Np/(A/U) | 7.5 | $Ca^{2+}$ | [6.479] |
| **D.3. tRNA-processing enzymes** | | | | | | | | | |
| RNase P | Escherichia coli | 17.5 + 120 | α, M1 RNA | [6.465] | endonuclease | tRNA or rRNA precursor | 8 | $Mg^{2+}$; $NH_4^+$, $Na^+$, or $K^+$ | [6.465], [6.466], [6.675] |
| RNase D | Escherichia coli | 40 | | [6.334] | exonuclease | tRNA precursor, 3'-terminus | 9–10 | $Mg^{2+}$ | [6.390], [6.466] |
| RNase P₂ | Escherichia coli | | | [6.62] | endonuclease | tRNA precursor | | | [6.626] |
| RNase O | Escherichia coli | | | [6.640] | endonuclease | tRNA precursor | 7.5–10 | $Mg^{2+}$ or $Mn^{2+}$ | [6.640] |

# E. DNA-modifying enzymes

## E. 1. DNA modification methyltransferases (MTases) (see Table 34)

## E.2. Phosphatases

| Enzyme | Source | MW | Subunit | Ref. | Reaction | Substrate | pH | Ion | Ref. |
|---|---|---|---|---|---|---|---|---|---|
| Alkaline phosphatase | calf intestine | 140 [$Zn^{2+}$] | $\alpha_2$ | [6.355] | phosphatase | 5'-P or 3'-P DNA or RNA | 7.5–9.5 | $Zn^{2+}$ | [6.396], [6.548] |
| Alkaline phosphatase | Escherichia coli | 80 [$Zn^{2+}$] | $\alpha_2$ | [6.691] | phosphatase | 5'-P or 3'-P DNA or RNA | 7.5–9.5 | $Zn^{2+}$ | [6.597] |
| Phosphodiesterase | calf spleen | | | [6.284] | phosphodiesterase | 5'-OH DNA or RNA | 5–7 | | [6.284] |
| Phosphodiesterase | Crotalus durissus | | | [6.599] | phosphodiesterase | 3'-OH DNA or RNA | 5–7 | | [6.490] |

## E.3. DNA ligases
### Bacterial DNA ligases

| Enzyme | Source | MW | Subunit | Ref. | Reaction | Substrate | pH | Ion | Ref. |
|---|---|---|---|---|---|---|---|---|---|
| DNA ligase | Escherichia coli | 77 | $\alpha$ | [6.544], [6.572] | polydeoxyribonucleotide synthase | 5'-P DNA or RNA, 3'-OH DNA or RNA (overlapping single-stranded ends of dsDNA or dsRNA) | 7.5–8 | $Mg^{2+}$, NAD | [6.557], [6.633], [6.681] |

### Phage-coded DNA ligases

| Enzyme | Source | MW | Subunit | Ref. | Reaction | Substrate | pH | Ion | Ref. |
|---|---|---|---|---|---|---|---|---|---|
| T4 DNA ligase | Escherichia coli, phage T4-infected | 68 | $\alpha$ | [6.555], [6.574] | polydeoxyribonucleotide synthase | 5'-P DNA or RNA, 3'-OH DNA or RNA (overlapping or blunt, single-stranded or double-stranded ends of ds DNA or dsRNA) | 7.2–7.8 | $Mg^{2+}$, ATP | [6.72], [6.363], [6.557], [6.633], [6.681] |
| | | | | | pyrophosphate exchange | ATP, pp$_i$ | | $Mg^{2+}$ | [6.717] |

## E.4. Kinases

| Enzyme | Source | MW | Subunit | Ref. | Reaction | Substrate | pH | Ion | Ref. |
|---|---|---|---|---|---|---|---|---|---|
| T4 polynucleotide kinase | Escherichia coli, phage T4-infected | 140 | $\alpha_4$ | [6.506], [6.574] | 5'-DNA kinase | 5'-OH DNA or RNA, NTP or dATP | 7.4–8.0 | $Mg^{2+}$ | [6.563], [6.600], [6.603] |
| | | | | | 3'-phosphatase | 3'-P DNA or RNA | 5.5–6.0 | $Mg^{2+}$ | [6.304] |
| T4 polynucleotide kinase, 3'-phosphatase-free | Escherichia coli, phage T4pseT1-infected | 140 | $\alpha_4$ | [6.302], [6.653] | 5'-DNA kinase | 5'-OH DNA or RNA, NTP or dATP | 7.4–8.0 | $Mg^{2+}$ | [6.302], [6.603], [6.653] |
| DNA kinase | rat liver | 80 | | [6.501] | 5'-DNA kinase | 5'-OH DNA, ATP | 5.5 | $Mg^{2+}$ | [6.501], [6.578], [6.744] |
| DNA kinase | calf thymus | 70 | | [6.271] | 5'-DNA kinase | 5'-OH DNA, ATP | 5.5 | $Mg^{2+}$ | [6.271], [6.744] |
| DNA kinase | human HeLa cells | | | [6.643] | 5'-RNA kinase | 5'-OH RNA, ATP | 5.5 | $Mg^{2+}$ | [6.643], [6.744] |

**Table 38.** (continued)

| Enzyme | Organism | $M_r$, $10^3$ dalton | Sub-units | References (isolation) | Specificity | Substrates | pH optimum | Cofactors | References (specificity) |
|---|---|---|---|---|---|---|---|---|---|
| **E.5. DNA topoisomerases** | | | | | | | | | |
| **Type I DNA topoisomerases** | | | | | | | | | |
| ω protein | Escherichia coli | 100–120 | α | [6.344] | transient ssDNA breakage: | | | | |
| M. luteus topoisomerase I | Micrococcus luteus | 100–120 | α | [6.482] | ccDNA relaxation[b] | dsDNA | 7–8 | $Mg^{2+}$ | [6.710] |
| HeLa topoisomerase I | human HeLa cells | 100 | α | [6.514] | DNA–catenane knot formation | dsDNA | | | |
| **Type II DNA topoisomerases** | | | | | | | | | |
| E. coli DNA gyrase | Escherichia coli | | α,β | [6.543] | transient dsDNA breakage: | | | | |
| M. luteus DNA gyrase | Micrococcus luteus | 420–430 | (αβ)$_2$ | [6.515] | cccDNA formation[b] DNA–dependent ATPase | dsDNA, ATP dsDNA, ATP | 7–8 | $Mg^{2+}$ | [6.386] |
| T4 topo-isomerase I | Escherichia coli, phage T4-infected | | α,β | [6.516], [6.679] | cccDNA relaxation[b] DNA–catenane knot formation | dsDNA dsDNA | | | |
| **E.6. DNA helix unwinding enzymes** | | | | | | | | | |
| rep protein | Escherichia coli, phage ΦX174-infected | 65 | α | [6.628] | DNA unwinding DNA–dependent ATPase | dsDNA, E. coli ssb protein ssDNA, ATP | | $Mg^{2+}$ | [6.628], [6.739] |
| Helicase III | Escherichia coli | 40 | α$_2$ | [6.350] | 5'–3' DNA unwinding DNA–dependent ATPase | dsDNA, E. coli ssb protein ssDNA, ATP | | $Mg^{2+}$ | [6.738] |

**E.7. Single-strand DNA-binding proteins**

| Protein | Source | MW | Structure | Ref. | Function | Substrate | pH (IP) | Cofactor | Ref. |
|---|---|---|---|---|---|---|---|---|---|
| E. coli ssb protein | Escherichia coli | | $\alpha_4$ | [6.323] | ssDNA binding | ssDNA, 8 nucleotides per monomer | 6.0 (IP)[c] | | [6.471] |
| T4 gene 32 protein | Escherichia coli, phage T4-infected | | $\alpha_{1-2/n}$ | [6.286] | ssDNA binding | ssDNA, 7 nucleotides per monomer | 5.5 (IP)[c] | | [6.471] |
| T7 DNA-binding protein | Escherichia coli, phage T7-infected | | $\alpha$ | [6.598] | ssDNA binding | ssDNA | 7 (IP)[c] | | [6.471] |
| fd gene 5 protein | Escherichia coli, phage fd-infected | | $\alpha_{1-2}$ | [6.267] | ssDNA binding | ssDNA, 4 nucleotides per monomer | 8.0 (IP)[c] | | [6.471] |
| CT HDP-I protein | calf thymus | | $\alpha$ | [6.423] | ssDNA binding | ssDNA, 7 nucleotides per monomer | 7.8 (IP)[c] | | [6.471] |

**F. RNA-modifying enzymes**

**F.1. RNA ligases**

| Enzyme | Source | MW | Structure | Ref. | Function | Substrate | pH | Cofactor | Ref. |
|---|---|---|---|---|---|---|---|---|---|
| T4 RNA ligase | Escherichia coli, phage T4-infected | 43 | $\alpha$ | [6.331], [6.646] | polyribonucleotide synthase | 5'-P RNA (DNA), 3'-OH RNA (DNA) | 7.2–8.4 | $Mg^{2+}$ | [6.552], [6.681], [6.682], [6.698], [6.709] |
| | | | | | pyrophosphate exchange | ATP, $pp_i$ | | $Mg^{2+}$ | [6.331] |

**F.2. CAP-forming enzymes**[d]

| Enzyme | Source | MW | Structure | Ref. | Function | Substrate | pH | Cofactor | Ref. |
|---|---|---|---|---|---|---|---|---|---|
| Capping enzyme | smallpox virus | 127 | $\alpha,\beta$ | [6.526], [6.645] | RNA triphosphatase<br>RNA guanylyltransferase<br>RNA (guanine-7) methyltransferase<br>GTP–$pp_i$ exchange<br>nucleoside triphosphate phosphohydrolase | pppRNA<br>GTP + ppRNA<br>GpppRNA + SAM<br>GTP + $pp_i$<br>NTP | 7.8 | $Mg^{2+}$ | [6.525], [6.644] |

**F.3. CAP-splitting enzymes**[d]

| Enzyme | Source | MW | Structure | Ref. | Function | Substrate | pH | Cofactor | Ref. |
|---|---|---|---|---|---|---|---|---|---|
| Decapping enzyme (TAP)[e] | Nicotiana tabacum var. Wisconsin 38 | 280 | $\alpha_4$ | [6.641] | acid pyrophosphatase | 7-methyl-GpppN | 6.0 | | [6.519], [6.641] |

[a] AP = apurinic. [b] cc = closed circular; ccc = covalent closed circular. [c] IP = isoelectric point. [d] CAP = 7-methyl-GpppN. [e] TAP = tobacco acid pyrophosphatase.

patterns, M · *Hpa*II is used to introduce in vitro methylated CpG residues at the specific CCGG sites of isolated DNA [6.701]. The DNA, which was previously methylated with M · *Hpa*II to different degrees, is microinjected into frog oocytes [6.419], or eukaryotic cells are transformed with the modified DNA by DNA-mediated gene transfer [6.558]. The in vivo effects of the distinct *Hpa*II methylation patterns on gene activity can be studied by analysis of gene-specific transcripts in Northern blots. To correlate the level of transcriptional activity with changes in the degree of methylation, the level of *Hpa*II-specific sites of methyl transfer within the transformed eukaryotic cells can be demonstrated directly by comparison of the restriction activities of heterohypekomeric restriction endonucleases *Hpa*II and *Msp*I on extracted DNA in Southern blots [6.307], [6.419], [6.676].

# 7. Economic Aspects

The commercial exploitation of enzymes ranges from very high-volume but low-cost enzymes containing products such as malt through traditional industrial enzymes such as detergent proteases to highly purified enzymes for scientific or medical use. Their prices range from a few cents per kilogram to several thousand dollars per gram. The market is usually divided into industrial enzymes, enzymes for medical use, and enzymes for scientific and analytical use. The following survey of the enzyme market and costs of enzymes is based on data compiled by Novo Industri market research; for further reference, see [7.1], [7.2].

**Industrial Enzymes.** The world market for enzymes in 1989 is about $\$ 500 \times 10^6$. Before 1965, the size of the enzyme market was only a few million dollars. With the introduction of detergent proteases, enzyme sales grew to about a hundred million dollars within a few years. Then virtually overnight, this enzyme market ceased because of the fear of unfavorable side effects of enzymes. After these unfounded fears subsided, enzymes were reintroduced into detergents and the consumption of detergent enzymes has grown ever since.

In 1972–1974, enzymes for making high-fructose corn syrup were introduced. This application became almost as large as the use of enzymes in detergents. At approximately the same time, very efficient types of microbial rennet were introduced. Within a short period of time, these enzymes accounted for a substantial part of the traditional calf rennet market.

From 1974 to 1986 no major new enzyme applications emerged. Still, the enzyme market has grown at an average rate of 10–15% despite the general decline of enzyme prices.

In the last few years, the free enzyme market has been reduced because of in-house production of some enzymes, mostly amylase and glucamylase, by a few large consumers. The industrial enzyme market consists of very few enzymes. The most important ones are listed in Table 39. For most of the enzymes listed, several different types are offered. For example, at least four different proteases are available for detergents. These enzymes are produced by different microorganisms and have different properties.

The most important industrial consumers of enzymes are listed in Table 40. The detergent and starch industries consume by far the largest share, whereas traditional enzyme-consuming industries such as the dairy, textile, and leather industries have not increased their consumption significantly in many years. Therefore, they play a relatively minor role.

**Table 39.** Breakdown of enzyme sales according to type

| Enzyme type | Fraction of sales, % |
|---|---|
| *Bacillus* protease | 30–35 |
| Glucamylase | 8–10 |
| *Bacillus* amylase | 10–12 |
| Glucose isomerase | 5–7 |
| Calf rennet | 10–12 |
| Microbial rennet | 2–4 |
| Pectinase | 4–5 |
| Pancreatin, trypsin | 2–4 |
| Papain, bromelain | 4–6 |
| Lipase | 2–3 |
| Others (e.g., *β*-amylase, *β*-glucanase, cellulase, dextranase, fungal amylase, fungal protease, glucose oxidase, hemicellulase, invertase, lactase, lysozyme, penicillin acylase, pullulanase) | 5–10 |

**Table 40.** Breakdown of enzyme sales according to industry

| Industry | Sales fraction, % |
|---|---|
| Detergent | 40–45 |
| Starch processing | 20–25 |
| Dairy | 12–15 |
| Brewing | 2–4 |
| Fruit and wine | 3–5 |
| Milling and baking | 1–2 |
| Textile and paper | 4–6 |
| Leather | 1–2 |
| Feed | 0–1 |
| Other | 5–10 |

Enzymes are usually sold on an activity basis, and the cost of the enzyme generally represents only a small percentage of the product value (Table 41). The cost of rennet may even be as low as 0.1 % of the product value.

If the complexity and efficiency of enzymes are considered, their cost, based on active enzyme protein, is very low (Table 42). A calculation of this type is, however, of value only to the enzyme manufacturer because it is related to the cost of enzyme production. For the user of the enzyme, only the activity counts.

Table 43 shows the apparent turnover number under application conditions for the enzymes used in high-fructose corn syrup production. If this is used to calculate the relative price of enzyme per mole of product in the enzyme reaction, then based on activity, isomerase is an expensive enzyme, but the use of the immobilized form still allows its economical application.

**Table 41.** Economy of enzyme use

| Product | Enzyme protein used, ppm of product | Enzyme cost, % of product value |
|---|---|---|
| Detergents | 150–200 | 1–4 |
| Dextrose | 150–200 | 1 |
| High-fructose syrup | 150–200 | 2–3 |
| Ethanol | 300–400 | 2–3 |
| Cheese | 3–6 | 0.1–0.3 |

**Table 42.** Cost of selected commercial enzyme preparations

| Enzyme | Approximate price of pure enzyme protein, $/kg | Enzyme fraction in commercial preparations, % | Physical form |
|---|---|---|---|
| Glucamylase | 30 | 5–10 | liquid |
| α-Amylase | 250 | 1–2 | liquid |
| Glucose isomerase | 250 | 5–10 | immobilized granules |
| Detergent protease | 100 | 3–6 | granules/liquid |
| Calf rennet (chymosin) | 5000 | 0.1–0.2 | liquid/powder |
| Microbial rennet | 500 | 0.2–0.4 | liquid/solid |

**Table 43.** Efficiency of enzymes used in high-fructose corn syrup manufacture

| | α-Amylase | Glucamylase | Glucose isomerase |
|---|---|---|---|
| Apparent turnover number under conditions of application, $\dfrac{\text{mol}_{Product}}{\text{mol}_{Enzyme} \times \text{min}}$ | $3 \times 10^4$ | $10^4$ | $2 \times 10^2$ |
| Relative price of enzyme, based on activity | 0.5 | 0.2 | 100 |
| Amount of enzyme used, ppm of product | 12 | 150 | 10–20 |
| Relative cost of enzyme | 80 | 133 | 100 |
| Product produced per kg enzyme, ton | 80 | 7 | 50–100 |

**Enzymes for Medical Use.** The medical market is a large one, with $ 300 \times 10^6$ to $ 500 \times 10^6$ retail sales. The most important products are fibrinolytic agents, anti-inflammatory preparations (proteases, lysozyme), and digestive aids. Enzymes are often combined with other ingredients, which makes identification of a specific enzyme market difficult.

**Enzymes for Scientific and Analytical Use.** The enzyme market for scientific and analytical uses is also large, probably several hundred million dollars, based on retail prices. The market is characterized by several thousand different products and by high marketing and distribution costs. Analytical enzymes are often part of a special kit in which the cost of the enzyme is relatively low.

# 8. Safety and Environmental Aspects

Workers handling enzymes should use protective clothing and eye protection. Water should be easily accessible in case spillage or splashing occurs. Adequate protective breathing apparatus should be used if dust or aerosols are encountered. Skin exposed to enzymes should be washed immediately, and eyewash and shower stations must be convenient and fully functional. If enzyme is inadvertently swallowed, adequate rinsing and water should be given, and medical advice should be sought immediately. Plant and laboratory practices should emphasize safety and good housekeeping. Ventilation systems must adequately filter and not recirculate particulate matter. Companies must fully comply with OSHA standards. Enzymes can be allergenic, depending on the susceptibility of individuals. Adequate precautions should be taken to protect individuals, and medical advice should be sought if allergic reactions are suspected.

Enzymes used in food processing have been shown to be safe through many years of manufacturing practice. Enzyme manufacturers and their customers cooperate in maintaining such safety. The Enzyme Technical Association regularly reviews this subject and the status of enzymes being considered for formal regulatory approval. Full cooperation is maintained with governmental agencies [8.1]. Approval of new enzymes requires consideration of a series of toxicological and multigenerational animal studies, depending on the degree of risk involved [8.1].

# 9. References

References for Chapter 1

[1.1] M. Dixon, E. C. Webb: *Enzymes*, Longmans, London 1979.

[1.2] A. R. Fersht: *Enzyme Structure and Mechanism*, Freeman, New York 1985.

[1.3] C. Walsh: *Enzymatic Reaction Mechanisms*, Freeman, San Francisco 1979.

[1.4] N. C. Price, L. Stevens: *Fundamentals of Enzymology*, Oxford University Press, Oxford 1982.

[1.5] P. D. Boyer (ed.): *The Enzymes*, 3rd ed., Academic Press, New York 1983.

[1.6] J. S. Fruton: *Molecules and Life*, John Wiley, New York 1972.

[1.7] H. Gutfreund (ed.): *Enzymes: 100 years. FEBS Letters*, vol. 62, Suppl., North Holland, Amsterdam 1976.

[1.8] D. H. Spackman, W. H. Stein, S. Moore, *Anal. Chem.* **30** (1958) 1190–1206.

[1.9] M. Elzinga (ed.): *Methods in Protein Sequence Analysis*, Humana Press, Clifton, New Jersey, 1982.

[1.10] F. Sanger, *Science (Washington, D.C.)* **129** (1959) 1340–1344.

[1.11] P. Edman, G. Begg, *Eur. J. Biochem.* **1** (1967) 80–91.

[1.12] D. G. Smyth, W. H. Stein, S. Moore, *J. Biol. Chem.* **238** (1963) 227–234.

[1.13] R. E. Canfield, *J. Biol. Chem.* **238** (1963) 2698–2707.

[1.14] B. E. Davidson, M. Sajgo, H. F. Noller, J. I. Harris, *Nature (London)* **216** (1967) 1181–1185.

[1.15] A. V. Fowler, I. Zabin, *Proc. Natl. Acad. Sci. U.S.A.* **74** (1977) 1507–1510.

[1.16] I. Harris, B. P. Meriwether, J. H. Park, *Nature (London)* **198** (1963) 154–157.

[1.17] E. Shaw, *Physiol. Rev.* **50** (1970) 244–296.

[1.18] M. F. Perutz, M. G. Rossmann, A. F. Cullis, H. Muirhead et al., *Nature (London)* **185** (1960) 416–422.

[1.19] D. C. Phillips, *Proc. Natl. Acad. Sci. U.S.A.* **57** (1967) 484–495.

[1.20] Enzyme Nomenclature, Academic Press, New York 1984.

[1.21] *Eur. J. Biochem.* **157** (1986) 1–26.

[1.22] C. H. W. Hirs, S. N. Timasheff (eds.): *Enzyme Structure Part I, Methods in Enzymology*, vol. 91, Academic Press, New York 1983.

[1.23] R. W. Old, S. B. Primrose: *Principles of Gene Manipulation*, Blackwell Scientific Publications, Oxford 1985.

[1.24] T. Maniatis, E. F. Frisch, J. Sambrook: *Molecular Cloning. A Laboratory Manual*, Cold Spring Harbor Laboratory, Cold Spring Harbor, New York, 1982.

[1.25] L. M. Smith, J. Z. Sanders, R. J. Kaiser, P. Hughes et al., *Nature (London)* **321** (1986) 674–679.

[1.26] A. Wada, *Nature (London)* **325** (1987) 771–772.

[1.27] G. von Heinje: *Sequence Analysis in Molecular Biology*, Academic Press, San Diego 1987.

[1.28] T. L. Blundell, L. N. Johnson: *Protein Crystallography*, Academic Press, New York 1976.

[1.29] G. E. Schulz, R. H. Schirmer: *Principles of Protein Structure*, Springer Verlag, New York 1979.

[1.30] T. E. Creighton: *Proteins*, Freeman, New York 1983.

[1.31] M. Levitt, C. Chothia, *Nature (London)* **261** (1976) 552–558.

[1.32] O. Jardetzky, G. C. K. Roberts: *NMR in Molecular Biology*, Academic Press, New York 1981.

[1.33] K. Wüthrich: *NMR of Proteins and Nucleic Acids*, Wiley, New York 1986.

[1.34] R. Porter, M. O'Connor, J. Whelan (eds): *CIBA Foundation Symposium 93, Mobility and Function in Proteins and Nucleic Acids*, Pitman, London 1983.

[1.35] R. N. Perham, *Philos. Trans. R. Soc. London B* **272** (1975) 123–136.

[1.36] R. Jaenicke, in E.-L. Winnacker, R. Huber (eds): *Protein Structure and Protein Engineering*, Mosbach Colloquium 39, Springer Berlin 1988, pp. 16–36.

[1.37] C. Guerrier-Tokada, S. Altman, *Science (Washington, D.C.)* **223** (1984) 285–286.

[1.38] A. J. Zaug, T. R. Cech, *Science (Washington, D.C.)* **231** (1986) 470–475.

[1.39] J. Darnell, H. Lodish, D. Baltimore: *Molecular Cell Biology*, Scientific American Books, New York 1986.

[1.40] D. Freifelder: *Molecular Biology*, Jones and Bartlett, Boston 1987.

[1.41] A. Kornberg: *DNA Replication*, Freeman, San Francisco 1980 (and 1982 supplement).

[1.42] D. M. Blow, A. R. Fersht, G. Winter (eds), "Design, Construction and Properties of Novel Protein Molecules," *Philos. Trans. R. Soc. London A* **317** (1986) 291–451.

[1.43] W. V. Shaw, *Biochem. J.* **246** (1987) 1–17.

[1.44] G. Biesecker, J. I. Harris, J. C. Thierry, J. E. Walker, A. Wonacott, *Nature (London)* **266** (1977) 328–333.

[1.45] D. L. Oxender, C. F. Fox (eds): *Protein Engineering*, Alan R. Liss, New York 1987.


References for Chapter 2

[2.1] H. U. Bergmeyer, J. Bergmeyer, M. Graßl (eds.): *Methods of Enzymatic Analysis*, 3rd ed., vols. I–XII, Verlag Chemie, Weinheim 1983–1986.

[2.2] P. D. Boyer (ed.): *The Enzymes*, 3rd ed., Academic Press, New York 1970.

[2.3] D. Glick (ed.): *Methods of Biochemical Analysis*, vols. 1–30 will be continued, Interscience Publ., New York 1954–1984.

[2.3a] K. Musil, O. Novakova, K. Kunz: *Biochemistry in Schematic Perspective*, Avicenum Czechoslovak Medical Press, Prague 1977.

[2.4] L. Michaelis, M. L. Menten: "Die Kinetik der Invertinwirkung," *Biochem. Z.* **49** (1913) 333–369.

[2.5] [2.1, vol. I, pp. 69–85].

[2.6] [2.1, vol. I, pp. 109–114].

[2.7] Boehringer Mannheim: *Biochemicals for the Diagnostics Industry* (1984).

[2.8] J. L. Webb: *Enzyme and Metabolic Inhibitors*, vol. I, Academic Press, New York-London 1963.

[2.9] [2.2, vol. 1, pp. 341–396].

[2.10] [2.1, vol. V, pp. 86–98].

[2.11] [2.1, vol. I, pp. 7–14].

[2.12] Boehringer Mannheim: *Biochemicals for Molecular Biology*, p. 83 (1985).

[2.13] P. Modrich, R. J. Roberts in St. M. Linn, R. J. Roberts (eds.): *Nucleases*, Cold Spring Harbor Laboratory, Cold Spring Harbor, New York, 1982, p. 109.

[2.14] [2.1, vol. I, pp. 280–305].

[2.15] [2.1, vol. III, pp. 118–126].

[2.16] [2.1, vol. III, pp. 126–133].

[2.17] [2.1, vol. III, pp. 496–501].

[2.18] [2.1, vol. IV, pp. 75–86].

[2.19] [2.1, vol. I, pp. 210–221].

[2.20] P. Trinder: "Determination of Glucose in Blood Using Glucose Oxidase with an Alternative Oxygen Acceptor," *Ann. Clin. Biochem.* **12** (1969) 24–27.

[2.21] [2.1, vol. I, pp. 326–340].

[2.22] [2.3, vol. 17, pp. 189–285].

[2.23] [2.1, vol. I, pp. 340–368].

[2.24] [2.1, vol. I, pp. 368–397].

[2.25] Th. Bücher, H. Hofner, J.-F. Ronayrenc: "Methods with Glass Electrodes," in H. U. Bergmeyer (ed.): *Methods of Enzymatic Analysis,* 2nd ed., vol. I, Verlag Chemie, Weinheim, and Academic Press, New York 1974, pp. 254–261.

[2.26] [2.1, vol. IV, pp. 15–25].

[2.27] H. Bakala, J. Wallach, M. Hauss: "Determination of Elastolytic Activity Using a Conductometric Method," *Biochimie* **60** (1978) 1205–1207.

[2.28] A. J. Lawrence, G. R. Moores: "Conductimetry in Enzyme Studies," *Eur. J. Biochem.* **24** (1972) 538–546.

[2.29] J. K. Grime: "Biochemical and Clinical Analysis by Enthalpimetric Measurements – A Realistic Alternative Approach?" *Anal. Chim. Acta* **118** (1980) 191–225.

[2.30] J. M. Bailey, P. H. Fishman, P. G. Pentchev: "Studies on Mutarotases," *J. Biol. Chem.* **242** (1967) 4263–4269.

[2.31] K. Kusai, J. Sekuzu, B. Hagihara, K. Okunuki et al., "Crystallization of Glucose Oxidase from Penicillium amagasakiense," *Biochim. Biophys. Acta* **40** (1960) 555–557.

[2.32] S. J. Bach, J. D. Killip: "Purification and Crystallization of Arginase," *Biochim. Biophys. Acta* **29** (1958) 273–280.

[2.33] [2.1, vol. IV, pp. 180–183].

[2.34] [2.1, vol. I, pp. 142–163].

[2.35] Boehringer Mannheim: *Assay Instruction Penicillin-G Amidase* (*carrier fixed*) (1980).

[2.36] [2.1, vol. II, pp. 102–115].

[2.37] [2.1, vol. II, pp. 86–99].

[2.38] O. Warburg, W. Christian: "Isolierung und Kristallisation des Gärungsferments Enolase," *Biochim. Z.* **310** (1941) 384–421.

[2.39] J. R. Whitaker, P. E. Granum: "An Absolute Method for Protein Determination Based on Difference of Absorbance at 235 and 280 nm," *Anal. Biochem.* **109** (1980) 156–159.

[2.40] G. Beisenherz, H. J. Boltze, Th. Bücher, R. Czok et al., "Diphosphofructose-Aldolase, Phosphoglyceraldehyd-Dehydrogenase, Milchsäure-Dehydrogenase, Glycerophosphat-Dehydrogenase und Pyruvat-Kinase aus Kaninchenmuskulatur in einem Arbeitsgang," *Z. Naturforsch. B Anorg. Chem. Org. Chem.* **8 B** (1953) 555–577.

[2.41] O. H. Lowry, N. J. Rosebrough, A. L. Farr, R. J. Randall: "Protein Measurement with the Folin Phenol Reagent," *J. Biol. Chem.* **193** (1951) 265–275.

[2.42] M. M. Bradford: "A Rapid and Sensitive Method for the Quantitation of Microgram Quantities of Protein Utilizing the Principle of Protein-Dye Binding," *Anal. Biochem.* **72** (1976) 248–254.

[2.43] H. R. Maurer: *Disk-Elektrophorese,* 2nd ed., De Gruyter, Berlin 1971.

[2.44] U. K. Laemmli: "Cleavage of Structural Proteins during the Assembly of the Head of Bacteriophage T 4," *Nature* (*London*) **227** (1970) 680–685.

[2.45] M.-E. Mirault, K. Scherrer, L. Hansen: "Isolation of Preribosomes from Hela Cells and Their Characterization by Electrophoresis on Uniform and Exponential-Gradient-Polyacrylamide Gels," *Eur. J. Biochem.* **23** (1971) 372–386.

[2.46] C. R. Merril, D. Goldmann, M. L. van Keuren: "Silver Staining Methods for Polyacrylamide Gel Electrophoresis," in *Methods in Enzymology,* vol. 96, Academic Press, New York 1983, p. 17.

References for Chapter 3

[3.1] S. J. Pirt: *Principles of Microbe and Cell Cultivation*, Blackwell Scientific Publ., Oxford 1975.

[3.2] J. R. Norris, D. W. Ribbons: *Methods in Microbiology*, vols. 1–8, Academic Press, London 1969–1973.

[3.3] H. J. Rehm: *Industrielle Mikrobiologie*, 2nd. ed., Springer Verlag, Berlin 1980.

[3.4] W. Crueger, A. Crueger: *Biotechnology: A Textbook of Industrial Microbiology*, Science Tech., Madison 1984.

[3.5] M. Moo-Young (ed.): *Comprehensive Biotechnology*, part 1–4, Pergamon Press, Oxford 1985.

[3.6] H. J. Rehm, G. Reed (eds.): *Biotechnology, A Comprehensive Treatise in Eight Volumes*, VCH Verlagsgesellschaft, Weinheim 1985.

[3.7] B. Alberts, D. Bray, J. Lewis , M. Raff et al.: *Molecular Biology of the Cell*, Garland Publ., New York 1983.

[3.8] R. Walker: *The Molecular Biology of Enzyme Synthesis*, J. Wiley & Sons, New York 1983.

[3.9] L. L. Randall, S. J. S. Hardy, *Microbiol. Rev.* **48** (1984) 290–298.

[3.10] B. P. Wasserman, *Food Technol.* **1984**, 78–98.

[3.11] S. Gottesman, *Annu. Rev. Genet.* **18** (1984) 415–441.

[3.12] G. Holt, G. Saunders in M. Moo-Young (ed.): *Comprehensive Biotechnology*, part 1, Pergamon Press, Oxford 1985, pp. 51–76.

[3.13] K. Aunstrup, *Abh. Akad. Wiss. DDR Abt. Math. Naturwiss. Tech.* **1981**, 445–457 (3 N, Mikrob. Enzymprod.).

[3.14] D. E. Eveleigh, B. S. Montenecourt, *Adv. Appl. Microbiol.* **25** (1979) 57–74.

[3.15] T. J. White, J. H. Meade, S. P. Shoemaker, K. E. Koths, M. A. Innis, *Food Technol.* **1984**, 90–98.

[3.16] A. L. Demain, *Adv. Appl. Microbiol.* **16** (1973) 177–202.

[3.17] W. Harder, J. G. Kuenen, A. Matin, *J. Appl. Bacteriol.* **43** (1977) 1–24.

[3.18] T. J. R. Harris, J. S. Emtage, *Microbiol. Sci.* **3** (1986) 28–31.

[3.19] T. J. Silhavy, S. A. Benson, S. D. Emr, *Microbiol. Rev.* **47** (1983) 313–344.

[3.20] Gist-brocades, EP 0 096 430, 1983 (C. P. Hollenberg, A. de Leeuw, S. Pas, J. A. van den Berg).

[3.21] J. Monod in P. S. S. Dawson (ed.): *Microbiol. Growth*, Halsted Press, Dowden, Hutchinson and Ross, Inc., Strandsburg, Penn., 1974, pp. 88–110.

[3.22] D. K. Button, *Microbiol. Rev.* **49** (1985) 270–297.

[3.23] A. G. Williams, S. E. Withers, *Appl. Microbiol. Biotechnol.* **22** (1985) 318–324.

[3.24] J. Greenman, K. T. Holland, *J. Gen. Microbiol.* **131** (1985) 1619–1624.

[3.25] E. I. Emanuilova, K. Toda, *Appl. Microbiol. Biotechnol.* **19** (1984) 301–305.

[3.26] J. Frankena, G. M. Koningstein, H. W. van Verseveld, A. H. Stouthamer, *Appl. Microbiol. Biotechnol.* **24** (1986) 106–112.

[3.27] S. Aiba, A. E. Humphrey, N. F. Millis: *Biochemical Engineering*, 2nd ed., Academic Press, New York 1973.

[3.28] A. E. Humphrey in M. Moo-Young (ed.): *Comprehensive Biotechnology*, vol. 2, Pergamon Press, Oxford 1985, pp. 3–4.

[3.29] J. C. van Suijdam: *Mycelial Pellet Suspensions*, Thesis, Delft University Press 1980.

[3.30] G. Skøt: "Proc. Conf. on Advances in Fermentation," *Chelsea Coll.*, 21–23 Sept. 1983, University of London, Wheatland Journ. Ltd., Penn House 1983, pp. 154–159.

[3.31] T. Yamanè, S. Shimizu, *Adv. Biochem. Eng. Biotechnol.* **30** (1984) 147–194.

[3.32] D. Herbert in P. S. S. Dawson (ed.): *Microbiol. Growth*, Halsted Press, Dowden, Hutchinson and Ross, Inc., Strandsburg, Penn., 1974, pp. 230–262.

[3.33] S. Aiba, *Biotechnol. Bioeng. Symp.* **9** (1979) 269–281.

[3.34] A. R. Moreira, G. van Dedem, M. Moo-Young, *Biotechnol. Bioeng. Symp.* **9** (1979) 179–203.

[3.35] J. A. Roels: *Energetics and Kinetics in Biotechnology,* Elsevier Biomedical Press, Amsterdam 1983.

[3.36] M. J. Rolf, H. C. Lim in M. Moo-Young (ed.): *Comprehensive Biotechnology,* Pergamon Press, Oxford 1985, pp. 165–174.

[3.37] D. J. Clarke, M. R. Calder, R. J. G. Carr, B. C. Blake-Coleman, S. C. Moody, T. A. Collinge, *Biosensors* **1** (1985) 213–320.

[3.38] C. L. Cooney, *Biotechnol. Bioeng. Symp.* **9** (1979) 1–11.

General References for Section 3.2

[3.39] S. Aiba, A. E. Humphrey, N. F. Mills: *Biochemical Engineering,* Chap. 13, Academic Press, New York-London 1973, pp. 346–392.

[3.40] M. Cheryan: *Ultrafiltration Handbook,* Technomic Publishing Company, Inc., Lancaster, Pennsylvania, 1986.

[3.41] C. L. Cooney (ed.): *Comprehensive Biotechnology* vol. 2, Sect. 2, Pergamon Press, Oxford-New York 1985, pp. 231–590.

[3.42] H. Determann: *Gelchromatographie,* Springer Verlag, Berlin-Heidelberg-New York 1969.

[3.43] T. J. C. Gribnau, J. Visser, R. J. F. Nivard: "Affinity Chromatography and Related Techniques," *Anal. Chem. Symp. Ser.* **9,** 1982.

[3.44] J.-C. Janson, P. Hedman: "Large-Scale Chromatography of Proteins," *Adv. Biochem. Eng.* **25** (1982) 43–99.

[3.45] T. Kremmer, L. Boross: *Gel Chromatography,* "Theory, Methodology, Applications," J. Wiley and Sons, New York 1979.

[3.46] G. Kopperschläger, H.-J. Böhme, E. Hofmann: "Cibachrom Blue F3G-A and Related Dyes as Ligands in Affinity Chromatography," *Adv. Biochem. Eng.* **25** (1982) 101–138.

[3.47] R. M. Lafferty (ed.): *Enzyme Technology,* Springer Verlag, Berlin-Heidelberg-New York 1983.

[3.48] C. J. O. R. Morris, P. Morris: *Separation Methods in Biochemistry,* Pitman Publishing, 1976.

[3.49] R. Scopes: *Protein Purification,* "Principles and Practice," Springer Verlag, New York-Heidelberg-Berlin 1982.

[3.50] L. Svarovsky (ed.): *Solid-liquid separation,* Butterworths, London 1981.

[3.51] J. Turkova: "Affinity Chromatography," *J. Chromatogr. Libr.* **12** (1978).

[3.52] I. C. D. Wang, C. L. Cooney, A. L. Demain, P. Dunnill, E. A. Humphrey, M. D. Lilly: *Fermentation and Enzyme Technology,* Chap. 12, J. Wiley and Sons, New York 1979, pp. 238–310.

[3.53] A. Wisemann (ed.): *Handbook of Enzyme Biotechnology,* J. Wiley and Sons, New York 1985.

[3.54] C. Yang, G. T. Tsao: "Affinity Chromatography," *Adv. Biochem. Eng.* **25** (1982) 19–42.

Specific References for Section 3.2

[3.55] H. Schutt, DE-OS 2 509 482, 1976.

[3.56] Brewer et al., *Trends in Biotechnol.* **3** (1985) 119–122.

[3.57] J. A. Currie, P. Dunnill, M. D. Lilly, *Biotechnol. Bioeng.* **14** (1972) 725–736.

[3.58] J. R. Woodrow, A. V. Quirk, *Enzyme Microb. Technol.* **4** (1982) 385–389.

[3.59] H. Schütte, K. H. Kroner, H. Hustedt, M. R. Kula in R. M. Lafferty (ed.): *Enzyme Technology,* Springer Verlag, Berlin 1983, pp. 115–124.

[3.60] H. Schütte, M.-R. Kula, *Biotech-Forum* **2** (1986) 68–79.

[3.61] H. Schütte R. Kraume-Flügel, M.-R. Kula, *Chem. Ing. Tech.* **57** (1985) 626–627.

[3.62] A. Wiseman, *Process Biochem.* **4** (1969) 63–65.

[3.63] J. A. Asenjo, B. A. Andrews, J. B. Hunter, S. LeCorre, *Process Biochem.* **20** (1985) 158–164.

[3.64] K. S. Lam, J. W. D. GrootWassink, *Enzyme Microb. Technol.* **7** (1985) 239–242.

[3.65]  G. B. Tanny, D. Mirelman, T. Pistole, *Appl. Environ. Microbiol.* **40** (1980) 269–273.

[3.66]  K. H. Kroner, H. Schütte, H. Hustedt, M.-R. Kula, *Process Biochem.* **19** (1984) 67–74.

[3.67]  K. H. Kroner, *Biotech-Forum* **3** (1986) 20–30.

[3.68]  M. S. Lee, T. Atkinson, *Process Biochem.* **20** (1985) 26–31.

[3.69]  R. A. Erickson, *Chem. Eng. Prog.* **91** (1984) Dec, 51–54.

[3.70]  H. Hemfort, W. Kohlstette, *Chem. Ind. London* **37** (1985) 412–416.

[3.71]  M.-R. Kula, *Appl. Biochem. Bioeng.* **2** (1979) 71–95.

[3.72]  H. Hustedt, K. H. Kroner, U. Menge, M.-R. Kula, *Enzyme Eng.* **5** (1980) 45–47.

[3.73]  M.-R. Kula, K. H. Kroner, H. Hustedt, H. Schütte: "Biochemical Engineering II," *Ann. N.Y. Acad. Sci.* **369** (1981) 341–354.

[3.74]  A. Veide, A.-L. Smeds, S.-O. Enfors, *Biotechnol. Bioeng.* **25** (1983) 1789–1800.

[3.75]  H. Hustedt, K. H. Kroner, H. Schütte, M.-R. Kula, *Biotech-Forum* **1** (1985) 10–17.

[3.76]  P. A. Albertson: *Partition of Cell Particles and Macromolecules,* Wiley-Interscience, New York 1971.

[3.77]  S. D. Flanagan, S. H. Barondes, *J. Biol. Chem.* **250** (1975) 1484–1489.

[3.78]  G. Kopperschläger, G. Johansson, *Anal. Biochem.* **124** (1982) 117.

[3.79]  A.-K. Frej, J. G. Gustafsson, P. Hedman, *Biotechnol. Bioeng.* **28** (1986) 133–137.

[3.80]  B. Atkinson, I. S. Daoud, *Adv. Biochem. Eng.* **4** (1976) 41–124.

[3.81]  K. Esser, U. Kües, *Process Biochem.* **18** (1983) 21–23.

[3.82]  J. L. Van Haecht, M. Bolipombo, P. G. Rouxhet, *Biotechnol. Bioeng.* **27** (1985) 217–224.

[3.83]  M. Mayer, R. Woernle, *Chem. Ing. Tech.* **57** (1985) 152–153.

[3.84]  D. J. Bell, M. Hoare, P. Dunnill, *Adv. Biochem. Eng.* **26** (1982) 1–72.

[3.85]  M. Hoare, P. Dunnill, D. J. Bell, *Ann. N.Y. Acad. Sci.* **413** (1983) 254–269.

[3.86]  P. Dunnill, M. D. Lilly, *Biotechnol. Bioeng. Symp.* **3** (1972) 97.

[3.87]  H. Strathmann, *Chem. Tech.* **7** (1978) 333–346.

[3.88]  *Ullmann,* 4th ed., **16,** 515–535.

[3.89]  H.-D. Saier, H. Strathmann, *Angew. Chem. Int. Ed. Engl.* **14** (1975) 452–459.

[3.90]  P. Mattock, G. F. Aitchison, EP 82/3 015 706, 1982.

[3.91]  D. E. G. Austen, T. Cartwright, C. H. Dickerson, *Vox Sang.* **44** (1983) 151–155.

[3.92]  P. Andrews in E. Glick (ed.): *Methods of Biochem. Anal.,* vol. 18, Interscience, New York-London 1970, pp. 1–53.

[3.93]  S. Shaltiel, Z. Er-el, *Proc. Nat. Acad. Sci. U.S.A.* **70** (1973) 778–781.

[3.94]  J. Porath, J. Carlson, I. Olsson, G. Belfrage, *Nature (London)* **258** (1975) 598–599.

[3.95]  J. Carlsson, I. Olsson, R. Axen, H. Drevin, *Acta Chem. Scand. Ser. B* **30** (1976) 180–182.

[3.96]  J.-C. Janson, P. Dunnill, *Proc. FEBS Meet.* **30** (1974) 81–105.

[3.97]  J.-C. Janson, *J. Agr. Food Chem.* **19** (1971) 581–588.

[3.98]  I. Chibata (ed.): *Immobilized Enzymes, Research and Development,* Kodansha, Tokyo 1978.

[3.99]  M. D. Lilly, *Enzyme Microb. Technol.* **8** (1986) 315.

[3.100] J. M. Nelson, E. G. Griffin, *J. Am. Chem. Soc.* **38** (1916) 1109.

[3.101] N. Grubhofer, L. Schleith, *Naturwissenschaften* **40** (1953) 508.

[3.102] M. A. Mitz, *Science* **123** (1956) 1076.

[3.103] P. Bernfeld, J. Wan, *Science* **142** (1963) 678.

[3.104] F. A. Quiocho, F. M. Richards, *Proc. Natl. Acad. Sci.* **52** (1964) 833.

[3.105] T. M. S. Chang, *Science* **146** (1964) 524.

[3.106] G. Gregoriadis, P. D. Leathwood, B. E. Ryman, *FEBS Lett.* **14** (1971) 95.

[3.107] E. Katchalski, I. H. Silman, R. Goldman, *Adv. Enzymol.* **34** (1971) 445.

[3.108] I. Chibata, T. Tosa, T. Sato, T. Mori, T. Matuo, *Proc. IVth Int. Ferment. Symp.,* "Fermentation Technology Today," Soc. Ferment. Technol., Japan 1972, p. 383.

[3.109] K. Mosbach (ed.): *Methods in Enzymology,* vol. 44, Academic Press, New York 1976.
        K. Mosbach (ed.): *Methods in Enzymology,* vol. 135, Academic Press, New York 1987.

[3.110] R. Axén, J. Porath, S. Ernback, *Nature (London)* **214** (1967) 1302
P. Axén, S. Ernback, *Eur. J. Biochem.* **18** (1971) 351.

[3.111] M. A. Mitz, L. J. Summaria, *Nature (London)* **189** (1961) 576.
W. Brümmer, N. Hennrich, M. Klockow, H. Lang, H. D. Orth, *Eur. J. Biochem.* **25** (1972) 129.

[3.112] N. Weliky, F. S. Brown, E. C. Dale, *Arch. Biochem. Biophys.* **131** (1969) 1.

[3.113] P. Liu-Osheroff, R. J. Guillory, *Biochem. J.* **127** (1972) 419.

[3.114] H. Filippusson, W. E. Hornby, *Biochem. J.* **120** (1970) 215.

[3.115] T. Sato, T. Mori, T. Tosa, I. Chibata, *Arch. Biochem. Biophys.* **147** (1971) 788.

[3.116] G. Kay, E. M. Crook, *Nature (London)* **216** (1967) 514.

[3.117] R. D. Falb, J. Lynn, J. Shapira, *Experientia* **28** (1973) 958.

[3.118] M. Kierstan, C. Bucke, *Biotechnol. Bioeng.* **19** (1977) 387.

[3.119] I. Takata, T. Tosa, I. Chibata, *J. Solid-Phase Biochem.* **2** (1977) 225.

[3.120] T. Tosa, T. Sato, T. Mori, K. Yamamoto, I. Takata, Y. Nishida, I. Chibata, *Biotechnol. Bioeng.* **21** (1979) 1967.

[3.121] T. Sato, Y. Nishida, T. Tosa, I. Chibata, *Biochim. Biophys. Acta* **570** (1979) 179.

[3.122] I. Takata, K. Yamamoto, T. Tosa, I. Chibata, *Enzyme Microb. Technol.* **2** (1980) 30.

[3.123] S. Fukui, A. Tanaka, *Adv. Biochem. Eng./Biotechnol.* **29** (1984) 1.

[3.124] S. Fukui, A. Tanaka, T. Iida, E. Hasegawa, *FEBS Lett.* **66** (1976) 179.

[3.125] S. Fukushima, T. Nagai, K. Fujita, A. Tanaka, S. Fukui, *Biotechnol. Bioeng.* **20** (1978) 1465.

[3.126] S. Fukui, A. Tanaka, *Endeavour* **9** (1985) 10.

[3.127] The Working Party on Immobilized Biocatalysts, *Enzyme Microb. Technol.* **5** (1983) 304.

[3.128] I. Chibata, T. Tosa, T. Sato, *Appl. Microbiol.* **27** (1974) 878.

[3.129] K. Yamamoto, T. Tosa, I. Chibata, *Biotechnol. Bioeng.* **22** (1980) 2045.

[3.130] K. Yamamoto, T. Tosa. K. Yamashita, I. Chibata, *Eur. J. Appl. Microbiol.* **3** (1976) 169.

[3.131] Y. Asano, T. Yasuda, Y. Tani, H. Yamada, *Agric. Biol. Chem.* **46** (1982) 1183.

[3.132] S. Fukui, A. Tanaka, *Annu. Rev. Microbiol.* **36** (1982) 145.

References for Chapter 4

[4.1] G. Reed: *Enzymes in Food Processing,* 2nd ed., Academic Press, New York 1975.

[4.2] *Ullmann,* 4th ed., **10,** 475–551.

[4.3] T. Godfrey, J. Reichelt: *Industrial Enzymology,* The Nature Press, New York-London 1983.

[4.4] K. Kulp in [4.1] pp. 54–117.
R. L. Wistler, J. N. Bemiller, E. F. Paschal: *Starch: Chemistry and Technology,* 2nd ed., Academic Press, New York 1965, pp. 110–135.

[4.5] R. M. Sandstedt, E. Kneen, M. J. Blish, *Cereal Chem.* **16** (1939) 712.

[4.5a] C. A. Ayre, J. A. Anderson, *Can. J. Res.* **17**C (1939) 239.

[4.6] E. T. Reese (ed.): *Enzymic Hydrolysis of Cellulose and Related Materials,* Pergamon Press, Oxford 1963.

[4.7] W. Pilnik, F. M. Rombouts: "Pectic enzymes" in *Enzymes and Food Processing,* Appl. Science Publ., London 1980.

[4.8] G. E. Perlmann, L. Lorand (eds.): *Proteolytic Enzymes, Methods in Enzymology,* vol. 19, Academic Press, New York 1970.
M. Bahn, R. D. Schmid: *Waschmittelenzyme,* Labor 2000 1966, pp. 16–23.

[4.9] S. Schwimmer, *Source Book of Food, Enzymology,* The Avi. Publ. Comp., Westport, Conn., 1981, pp. 594–610.

[4.10] K. Morihara, "Comparative Specificities of Microbial Proteinases," in *Advances in Enzymology and Related Molecular Biology,* vol. 41, Interscience Publ., New York 1974, pp. 179–243.

[4.11]  M. L. Anson, *J. Gen. Physiol.* **22** (1939) 79.

[4.12]  H. Brockerhoff, R. G. Jensen: *Lipolytic Enzymes,* Academic Press, New York 1974.

[4.13]  S. H. Hemmingsen, *Appl. Biochem. Bioeng.* **2** (1979) 157–183.

[4.14]  T. P. Shukla, *CRC Crit. Rev. Food Technol.* **1975**, no. 5, 325.

[4.15]  B. SpRößler, H. Plainer: "Immobilized Lactase for Processing Whey", *Food Technol.* **1983** (Oct.), 93–95.

[4.16]  D. Scott in [4.1].

[4.17]  R. L. Whistler in R. L. Whistler, J. N. Bemiller and E. F. Paschall (eds.): *Starch Chemistry and Technology,* 2nd ed., Academic Press, Orlando, Florida, 1984, 1–9.

[4.18]  G. G. Taylor, US 3 348 972, 1967; *Chem. Abstr.* **62** (1965) 10656 a.

[4.19]  K. K. Kawakami: *Jokichi Takamine, A Record of his American Achievements,* William Edwin Rudge, New York 1928.

[4.20]  Grain Processing Corp., US 3 560 343, 1971 (F. C. Armbruster, C. F. Harjes).

[4.21]  J. K. Dale, D. P. Langlois, US 2 201 609, 1940.

[4.22]  P. L. Farris, N. E. Lloyd, N. J. Nelson in [4.17] 11–13, 612.

[4.23]  W. E. Goldstein in C. D. Scott (ed.): *Eighth Symposion on Biotechnology for Fuels and Chemicals,* Wiley-Interscience, New York 1986.
*Proceedings of Biomass for Production of Fuels and Chemicals,* J. Wiley & Sons, 1987 (in press).

[4.24]  W. E. Goldstein in M. Schuler, W. Wiegand (eds.): *Biochemical Engineering V,* The New York Academy of Sciences, New York 1987 (in press).

[4.25]  W. D. Cowan, *BNF Nutr. Bull.* **8** (1983) no. 3, 170–179.

[4.26]  G. Durand, J. M. Navarro, *Process Biochem.* **13** (1978) no. 9, 14–17, 20–23.

[4.27]  J. F. Kennedy, J. M. S. Cabral, *Appl. Biochem. Bioeng.* **4** (1983) 189–280.

[4.28]  A. I. Laskin, G. T. Tsao, L. B. Wingard (eds.): *Enzyme Engineering* 7, The New York Academy of Sciences, New York 1984, vol. 434.

[4.29]  D. P. Atkins, J. F. Kennedy, *Starch/Stärke* **37** (1982) no. 12, 421–427.

[4.30]  C. A. Knutson, U. Khou, J. E. Cluskey, G. E. Inglett, *Cereal Chem.* **59** (1982) no. 6, 512–515.

[4.31]  J. Holm, I. Bjoerck, S. Ostrowska, A. C. Eliasson, N. G. Asp, K. Lorsson, et al., *Starch/Stärke* **35** (1983) no. 9, 294–297.

[4.32]  J. R. Reichelt in T. Godfrey, J. Reichelt (eds.): *Industrial Enzymology,* "The Application of Enzymes in Industry," 1st ed., The Nature Press, New York-London 1983, pp. 66, 148, 381.

[4.33]  B. F. Jensen, B. E. Norman, *Process Biochem.* **19** (1984) no. 4, 129–134.

[4.34]  B. E. Norman in *Maize, Recent Progress in Chemistry,* Academic Press, New York 1982, 157–179.

[4.35]  J. J. M. Labout, *Starch/Stärke* **37** (1985) no. 5, 157–161.

[4.36]  Critical Data Tables, Corn Refiners Association, Washington, D.C., 1975.

[4.37]  P. B. Poulsen, *Biotechnol. Genet. Eng. Rev.* **1** (1984) 121–140.

[4.38]  J. F. Kennedy, J. M. S. Cabral, B. Kalogerakis, *Enzyme Microb. Technol.* **7** (1985) no. 1, 22–28.

[4.39]  O. J. Lantero, in *Abstracts and Proceedings of 188th American Chemical Society Meeting,* Philadelphia, August 1984.

[4.40]  U.O.P., Inc., US 4 582 803, 1986 (H. P. G. Knapik, W. H. Mueller).

[4.41]  Miles Laboratories, Inc., US 4 438 196, 1984 (O. J. Lantero).

[4.42]  S. A. Watson in [4.17] 462.

[4.43]  J. F. Kennedy, J. M. S. Cabral, B. Kalogerakis, *Enzyme Microb. Technol.* **7** (1985) no. 1, 22–28.

[4.44]  G. Tegge, G. Richter, *Starch/Stärke* **38** (1986) no. 2, 61–67.

[4.45] R. D. Yang, D. A. Grow, W. E. Goldstein in *Proceedings of 182nd ACS National Meeting,* New York 1981.

[4.46] G. G. Stewart, in *Proceedings of Bio Expo '85,* Boston 1985.

[4.47] A. M. Sills, G. G. Stewart, *J. Inst. Brew.* **88** (1982), no. 5, 313–316.

[4.48] E. ter Haseborg, *Process Biochem.* **16** (1981) no. 5, 16–20.

[4.49] J. R. Reichelt in [4.32] 210–220.

[4.50] J. F. Robyt in [4.17] 87–152.

[4.51] W. H. B. Denner in [4.32] 136.

[4.52] H. C. S. DeWhalley in *ICUMSA Methods of Sugar Analysis,* Chap. 7, Elsevier Publ. Co., New York 1964.

[4.53] C. G. Johansson, M. Siljestroem, *Z. Lebensm. Unters. Forsch.* **179** (1984) no. 1, 24–28.

[4.54] M. Siljestroem, N. G. Asp, *Z. Lebensm. Unters. Forsch.* **181** (1985) no. 1, 4–8.

[4.55] K. Maninckr, O. B. Jorgensen, *Starch/Stärke* **35** (1983) no. 12, 419–426.

[4.56] R. Renneberg, G. Kaiser, F. Scheller, Y. Tsujisaka, *Biotechnol. Lett.* **7** (1985) no. 11, 809–812.

[4.57] K. Khaleeluddin, L. Bradford, *J. Assoc. Off. Anal. Chem.* **69** (1986) no. 1, 162–166.

[4.58] J. E. Kruger, B. A. Marchylo, *Cereal Chem.* **62** (1985) no. 1, 11–18.

[4.59] M. J. W. Povey, A. J. Rosenthal, *J. Food Technol.* **19** (1984) no. 1, 115–119.

[4.60] M. R. Dhawale, J. J. Wilson, G. G. Khachatourians, W. M. Ingledew, *Appl. Environ. Microbiol.* **44** (1982) no. 3, 747–750.
       W. H. B. Denner, R. D. Farrow, J. R. Reichelt in [4.32], 136, 157–169.

[4.61] N. B. Havewala, W. H. Pitcher in E. Kendall Pye, L. B. Wingard, Jr. (eds.): *Enzyme Engineering,* vol. 2, J. Wiley, New York 1974, pp. 315–328.

[4.62] J. A. Roels, R. van Tilburg, *Starch/Stärke* **31** (1979) 17–24.

[4.63] N. E. Lloyd, W. J. Nelson in R. L. Whistler, J. N. Bemiller, E. F. Paschall (eds.): *Starch, Chemistry and Technology,* 2nd ed., Academic Press, New York 1984, pp. 611–660.

[4.64] Y. B. Tewari, R. N. Goldberg, *J. Solution Chem.* **13** (1984) 523–547.

[4.65] R. L. Antrim, W. Colilla, B. J. Schnyder in L. B. Wingard, E. Katchalski-Katzir, L. Goldstein (eds.): *Applied Biochemistry and Bioengineering,* vol. 2, Academic Press, New York 1979, pp. 97–155.

[4.66] R. O. Marshall, US 2950288, 1959.

[4.67] Y. Takasaki, Y. Kosogi, A. Kanbayashi in D. Perlman (ed.): *Fermentation Advances,* Academic Press, New York 1969, pp. 561–590.

[4.68] Y. Takasaki, GB 1103394, 1965.

[4.69] S. H. Hemmingsen in L. B. Wingard, E. Katchalski-Katzir, L. Goldstein (eds.): *Applied Biochimistry and Bioengineering,* vol. 2, Academic Press, New York 1979, pp. 157–183.

[4.70] N. E. Lloyd, W. J. Nelson in R. L. Whistler, J. N. Bemiller, E. F. Paschall (eds.): *Starch, Chemistry and Technology,* 2nd ed., Academic Press, New York 1984, pp. 633–635.

[4.71] R. V. McAllister in W. H. Pitcher Jr. (ed.): *Immobilized Enzymes for Food Processing,* CRC Press, Boca Raton, Florida, 1980, pp. 81–111.

[4.72] L. Zittan, P. B. Poulsen, S. H. Hemingsen, *Stärke* **27** (1975) 236–241.

[4.73] J. V. Hupkes, R. van Tilburg, *Stärke* **28** (1976) 356–360.

[4.74] G. Weidenbach, D. Bonse, G. Richter, *Stärke* **36** (1984) 412–416.

[4.75] R. L. Antrim, A. Auterinen, *Starch/Stärke* **38** (1986) 132–137.

[4.76] Denki Kagaku Kogyo, US 3915797, 1973 (Y. Ishimatsu).

[4.77] P. H. Blanchard, E. O. Geiger, *Sugar Technol. Rev.* **11** (1984) 1–94.

[4.78] J. Oestergaard, S. L. Knudsen, *Stärke* **28** (1976) 350–356.

[4.79] J. Straatsma, K. Vellenga, H. G. J. de Witt, G. E. Joosten, *Ind. Eng. Chem. Process Des. Dev.* **22** (1983) 349–356.

[4.80]  S. Vuilleumier, *Sugar Azucar* **80** (1985) no. 10, 13–20.

[4.81]  B. H. Landis, K. E. Beery in H. W. Houghton (ed.): *Developments in Soft Drink Technology*, vol. 3, Elsevier Applied Science Publishers, London 1984, pp. 85–120.

[4.82]  S. Crocco, *Food Eng.* **48** (1976) no. 12, 30–31; *Chilton's Food Eng.* **49** (1977) no. 2, 41–45.

[4.83]  J. C. Fruin, B. L. Scallet, *Food Technol.* (*Chicago*) **29** (1975) no. 11, 40–45.

[4.83 a]  J. S. White, W. Parke, *Cereal Foods World* **34** (1989) no. 5, 392–398.

[4.84]  *Eur. Chem. News* **46** (1986) no. 1219, 11–12.

[4.85]  T. Godfrey, J. R. Reichelt in T. Godfrey, J. R. Reichelt (eds.): *Industrial Enzymology*, The Nature Press, New York-London 1983, pp. 1–7.

[4.86]  F. Eberstadt & Co., Inc. (ed.): *Industrial Enzyme Products and Sales 1981. Industry Sources*, New York 1981.

[4.87]  O. Röhm, DE 283 923, 1913.

[4.88]  C. Dambmann, P. Holm, V. Jensen, M. H. Nielsen, *Dev. Ind. Microbiol.* **12** (1971) 11–23.

[4.89]  R. van Tilburg, *Prog. Microbiol.* **20** (1984) 31–52.

[4.90]  Novo Industry, US 4 435 307, 1984 (P. O. Barbesgaard, G. W. Jensen, P. Holm).

[4.91]  Gist-brocades, EP 85 201 302.8, 1985 (F. Farin, J. J. M. Labout, G. J. Verschoor).

[4.92]  P. L. Layman, *Chem. Eng. News* **62** (1984) no. 4, 17–49.

[4.93]  C. A. Starace, *Soap Cosmet. Chem. Spec.* **5** (1983) 48–50 F.

[4.94]  S. E. Godtfredsen, K. Ingvorsen, B. Yde, O. Andresen in J. Tramper, H. C. van der Plas, P. Linko (eds.): *Biocatalysts in Organic Synthesis*, Elsevier, Amsterdam 1985, p. 3.

[4.95]  Y. Lin, G. E. Means, R. E. Feeney, *J. Biol. Chem.* **244** (1969) 789.

[4.96]  Atlas Electric Devices Co., Chicago, Ill., U.S.A., product information: Launder-Ometer.

[4.97]  United States Testing Co., Hoboken, N.J., U.S.A., product information: Tergotometer.

[4.98]  Eidgenössische Materialprüfungs- und Versuchsanstalt, St. Gallen, Switzerland, product information: EMPA 116.

[4.99]  G. Richter, G. Konieczny-Janda, *Seifen, Öle, Fette, Wachse* **111** (1985) 455–460.

[4.100]  J. Donahue, *Soap Cosmet. Chem. Spec.* **60** (1984) no. 1, 27–29.

[4.101]  M. M. Crutchfield, *J. Am. Oil Chem. Soc.* **55** (1978) 58–65.

[4.102]  G. J. Hignett, *Tenside Deterg.* **23** (1986) 69–72.

[4.103]  Unilever, EP 0 173 397, 1986 (J. R. Nooi, M. W. Parslow).

[4.104]  H. C. Barfoed in T. Godfrey, J. R. Reichelt (eds.): *Industrial Enzymology*, The Nature Press, New York-London 1983, pp. 284–293.

[4.105]  Unilever, EP 0 072 098, 1983 (R. B. Cox, D. C. Steer, J. R. Woodward).

[4.106]  T. Godfrey, J. R. Reichelt in T. Godfrey, J. R. Reichelt (eds.): *Industrial Enzymology*, The Nature Press, New York-London 1983, p. 321.

[4.107]  A. L. Lehninger in A. L. Lehninger (ed.): *Biochemistry*, Worth Publishers Inc., New York 1979, pp. 125–155.

[4.108]  N. Muthukumaran, S. C. Dhar, *Leather Sci.* **29** (1982) 417–424.

[4.109]  Röhm, US 3 939 040, 1976 (R. Monsheimer, E. Pfleiderer).

[4.110]  M. Wolkstein, *Food Eng. Int.* **12** (1983) 33–36.

[4.111]  E. M. Meijer, W. H. J. Boesten, H. E. Schoemaker, J. A. M. van Balken in J. Tramper, H. C. van der Plas, P. Linko (eds.): *Biocatalysts in Organic Synthesis*, Elsevier, Amsterdam 1985, p. 135.

[4.112]  Y. Isowa, M. Ohmori, T. Ichikawa, K. Mori, Y. Nonaka, K. Kihara, K. Oyama, H. Satoh, S. Nishimura, *Tetrahedron Lett.* **28** (1979) 2611–2612.

[4.113]  K. Nakanishi, T. Kamikubo, R. Matsuno, *Biotechnology* **3** (1985) 459–464.

[4.114]  K. Morihara, T. Oka, H. Tsuzuki, *Nature* (*London*) **280** (1979) 412–413.

[4.115]  K. Breddam, J. T. Johansen, *Carlsberg Res. Commun.* **49** (1984) 463–472.

[4.116]  K. Breddam, *Carlsberg Res. Commun.* **51** (1986) 83–128.

[4.117] S. Schwimmer: *Source Book of Food Enzymology,* The Avi Publ. Comp., Westport, Conn., 1981, pp. 461–510.

[4.118] J. C. Allen, W. L. Wrieden, *J. Dairy Res.* **49** (1982) no. 2, 249–263.

[4.119] C. D. Azzara, P. S. Dimick, *J. Dairy Sci.* **68** (1985) no. 7, 1804–1812.

[4.120] B. J. Kitchen, G. C. Taylor, I. C. White, *J. Dairy Res.* **37** (1970) no. 2, 279–288.

[4.121] S. Patton, E. G. Trams, State Univ. Penn. 16802, doc. 1395, **14** (1971) no. 4, 230–232.

[4.122] E. H. Reimerdes, *J. Dairy Sci.* **66** (1983) no. 8, 1591–1600.

[4.123] K. M. Shahani, W. J. Harper, R. G. Jensen, R. M. Jr. Parry. C. A. Zittle, *J. Dairy Sci.* **56** (1973) no. 5, 531–543.

[4.124] M. Tkadlecova, J. Hanus, *Nahrung* **17** (1973) no. 5, 565–577.

[4.125] M. Anderson, A. T. Andrews, *J. Dairy Res.* **44** (1977) no. 2, 223–235.

[4.126] S. E. Birkeland, L. Stepaniak, T. Sorhaug, *Appl. Environ. Microbiol.* **49** (1985) no. 2, 382–387.

[4.127] B. A. Law, *J. Dairy Res.* **46** (1979) no. 3, 573–588.

[4.128] T. R. Patel, F. M. Bartlett, J. Hamid, *J. Food Prot.* **46** (1983) no. 2, 90–94.

[4.129] E. H. Reimerdes, F. Petersen, G. Kielwein, *Milchwissenschaft* **34** (1979) no. 9, 548–551.

[4.130] D. G. Dalgleish: *Development of Dairy Chemistry. I. Proteins,* Elsevier Appl. Sci. Publ., London 1982, pp. 157–287.

[4.131] P. F. Fox, *J. Dairy Res.* **36** (1969) no. 3, 427–433.

[4.132] S. Gordin, I. Rosenthal, *J. Food Prot.* **41** (1978) no. 9, 684–688.

[4.133] P. Dupuy: *Use of Enzymes in Food Technology,* Technique et Documentation Lavoisier, Paris 1982.

[4.134] P. F. Fox: *Development in Dairy Chemistry III, Lactose and Minor Constituents,* Elsevier, London 1985.

[4.135] A. C. Olson, R. A. Korus, *ACS Symp. Ser.* **47** (1977) 100–131.

[4.136] W. H. Pitcher, Jr.: *Immobilized Enzymes for Food Processing,* CRC Press, Boca Raton, Florida, 1980.

[4.137] E. H. Reimerdes, E. Herlitz, *Food Engineering Symposium,* 5th IUFOST Congress, Abstract no. 4.1.8. (1979).

[4.138] H. H. Weetall, *Food Prod. Dev.* **7** (1973) 94, 96, 98, 100.

[4.139] H. Ruttloff, J. Huber, F. Zickler, K.-H. Mangold: *Industrial Enzymes,* VEB Fachbuchverlag, Leipzig 1978.

[4.140] Z. Puhan, *Lebensm. Wiss. Technol.* **6** (1973) no. 6, 195–201.

[4.141] S. L. Neidleman, *Biotechnol. Genet. Eng. Rev.* **1** (1984) 1–38.

[4.142] N. A. M. Eskin, H. M. Henderson, F. J. Townsend: *Biochemistry of Foods,* Academic Press, New York 1971.

[4.143] W. Hartmeier, *Gordian* **77** (1977) 202, 204, 206, 208–210, 232, 234, 236–237.

[4.144] S. El-Shibiny, I. D. Rifaat, A. H. Fahmi, M. H. Abd El-Salam, *Egypt. J. Food Sci.* **1** (1973) no. 2, 235–240.

[4.145] J. Calvy, FR 2091958, 1971.

[4.146] P. A. O'Leary, P. F. Fox, *J. Dairy Res.* **42** (1975) no. 3, 445–451.

[4.147] M. L. Green, *J. Dairy Res.* **44** (1977) no. 1, 159–188.

[4.148] M. Gutfeld, P. P. Rosenfeld, *Dairy Ind.* **40** (1975) no. 2, 52, 55.

[4.149] Nestle, CH 582196, 1976 (M. Horisberger, T. Sozzi, R. Pusaz).

[4.150] P. J. de Koning, *Dairy Ind. Int.* **43** (1978) no. 7, 7–12.

[4.151] P. J. de Koning, *Bull. Int. Dairy Fed.* **126** (1980) 11–15.

[4.152] J. H. Nelson, *Am. Dairy Rev.* **34** (1972) no. 10, 37–40.

[4.153] R. Scott, *Top. Enzyme Ferment. Biotechnol.* **3** (1979) 103–169.

[4.154] J. Shoversw, R. Kornowski, G. Fossum, *J. Dairy Sci.* **56** (1973) no. 8, 994–997.

[4.155]  M. Sternberg, *Adv. Appl. Microbiol.* **20** (1976) 135–157.

[4.156]  C. R. Southward, P. D. Elston, *N. Z. J. Dairy Sci. Technol.* **11** (1976) no. 2, 144–146.

[4.157]  N. J. Tofte Jespersen, V. Dinesen, *J. Soc. Dairy Technol.* **32** (1979) no. 4, 194–197.

[4.158]  F. Addeo, J.-P. Pelissier, L. Chianese, *J. Dairy Res.* **47** (1980) no. 3, 421–426.

[4.159]  C. Alais, *Chimia* **28** (1974) no. 10, 597–604.

[4.160]  C. Alais, D. Paquet, A. Lagrange, *19th Int. Diary Congr.* **1974,** 347–348.

[4.161]  S. Bachmann, B. Klimczak, Z. Gasyna, *Acta Aliment. Pol.* **6** (1980) no. 3, 135–143.

[4.162]  G. Bakker, W. A. Scheffers, T. O. Wiken, *Neth. Milk Dairy J.* **22** (1968) no. 1/2, 16–21.

[4.163]  A. Carlson, *J. Biotechnol. Progr.* **1** (1985) no. 1, 46–52.

[4.164]  A. W. Kowalchyk, N. F. Olson, *J. Dairy Sci.* **62** (1979) no. 8, 1233–1237.

[4.165]  D. J. McMahon, R. J. Brown, *J. Dairy Sci.* **65** (1982) no. 8, 1639–1642.

[4.166]  D. J. McMahon, R. J. Brown, *J. Dairy Sci.* **67** (1984) no. 5, 919–929.

[4.167]  N. F. Olson, V. Bottazzi, *J. Food Sci.* **42** (1977) no. 3, 669–673.

[4.168]  C. Raharintsoa, M. L. Gaulard, C. Alais, *Lait* **57** (1977) no. 569/570, 631–645.

[4.169]  P. K. Sabarwal, N. C. Ganguli, *J. Dairy Sci. Technol.* **12** (1977) no. 4, 205–212.

[4.170]  M. Sponcet, P. Sifflet, A. Michallet, *Dairy Ind. Int.* **50** (1985) no. 10, 27–30.

[4.171]  R. Vanderpoorten, M. Weckx, *Neth. Milk Dairy J.* **26** (1972) no. 2, 47–59.

[4.172]  F. M. W. Visser, *Neth. Milk Dairy J.* **31** (1977) no. 4, 265–276.

[4.173]  J. S. Yadav, S. S. Sannabhadti, R. A. Srinivasan, *J. Food Sci. Technol.* **22** (1985) no. 1, 43–46.

[4.174]  S.-E. Yun, K. Ohmiya, T. Kobayashi, S. Shimizu, *J. Food Sci.* **46** (1981) no. 3, 705–707.

[4.175]  H. D. Benedet, Y. K. Park, *J. Dairy Sci.* **65** (1982) no. 6, 899–901.

[4.176]  C. Alais, A. Lagrange, *Lait* **52** (1972) no. 517, 407–427.

[4.177]  C. Alais, P. Ducroo, R. Delecourt, *Lait* **54** (1974) no. 538, 517–527.

[4.178]  Meito Sangyo Co., DE-AS 1 442 118, 1969.

[4.179]  Novo Terapeutisk Laboratoriums AS, DE-AS 1 517 775, 1970.

[4.180]  Miles Lab. Inc., US 3 549 390, 1970.

[4.181]  F. Etoh, H. Shoun, T. Beppu, K. Arima, *Agric. Biol. Chem.* **43** (1979) no. 2, 209–215.

[4.182]  E. R. Fraile, J. O. Muse, S. E. Bernardinelli, *Eur. J. Appl. Microbiol. Biotechnol.* **13** (1981) no. 3, 191–193.

[4.183]  J. P. Ramet, F. Weber, *Lait* **61** (1981) no. 607, 381–392.

[4.184]  J. P. Ramet, F. Weber, *Lait* **61** (1981) no. 608, 458–464.

[4.185]  T. Cserhati, J. Hollo, *Gordian* **72** (1972) no. 6, 226–229.

[4.186]  T. Cserhati, J. Hollo, *Gordian* **72** (1972) no. 11, 405–407.

[4.187]  L. Goranova, M. Stefanova-Kondratenko, *Lait* **55** (1975) no. 541/542, 58–67.

[4.188]  DSO Mlechna Promishlenost, US 4 048 339, 1977.

[4.189]  L. Rao Krishna, D. K. Mathur, *Biotechnol. Bioeng.* **21** (1979) no. 4, 535–549.

[4.190]  L. Rao Krishna, D. K. Mathur, *J. Dairy Sci.* **62** (1979) no. 3, 378–383.

[4.191]  J. Labatt Ltd., US 3 542 563, 1970.

[4.192]  J. Labatt Ltd., US 3 507 750, 1970.

[4.193]  T. Cserhati, J. Hollo, *Nahrung* **16** (1972) no. 5, 431–440.

[4.194]  T. Cserhati, J. Hollo, *Gordian* **74** (1974) no. 7/8, 257–260.

[4.195]  A. Erdelyi, E. Kiss, *Acta Aliment.* **7** (1978) no. 2, 155–166.

[4.196]  J. Rymaszewski, S. Poznanski, A. Reps, J. Ichilczyk, *Milchwissenschaft* **28** (1973) no. 12, 779–784.

[4.197]  C. Pfizer & Co. Inc., DE-AS 1 442 140, 1968.

[4.198]  A. M. Hamdy, M. A. Cheded, L. A. El-Koussy, E. A. Foda, *Agric. Res. Rev.* **54** (1976) no. 7, 135–143.

[4.199]  M. Barbosa, E. Valles, L. Vassal, G. Mocquot, *Lait* **56** (1976) no. 551/552, 1–17.

[4.200]  R. Scott: *Cheesemaking Practice,* Elsevier Appl. Sci. Publ., London 1986.

[4.201]  J. T. Barach, *Food Technol.* (*Chicago*) **39** (1985) no. 10, 73–74.

[4.202]  A. Löffler, *Food Technol.* **40** (1986) 63–70.

[4.203]  E. Refstrup, F. K. Vogensen, *Maelkeritidende* **93** (1980) no. 22, 612–615.

[4.204]  United States of America, Office of Technology Assessment 1981.

[4.205]  H. Kobayashi, I. Kusakabe, K. Murakami, *Agric. Biol. Chem.* **49** (1985) no. 6, 1611–1619.

[4.206]  A. Carlson, C. G. Hill Jr., N. F. Olson, *J. Dairy Sci.* **68** (1985) no. 2, 290–299.

[4.207]  P. Garnot, J. L. Thapon, C. M. Methieu, J. L. Maubois, *J. Dairy Sci.* **55** (1972) no. 12, 1641–1650.

[4.208]  S. C. Martiny, M. K. Harboe, *J. World Galaxy* **7** (1979) 33, 35.

[4.209]  K. Pozsar-Hajnal, *Fresenius Z. Anal. Chem.* **279** (1976) no. 2, 118–119.

[4.210]  M. J. Prager, *J. Assoc. Off. Anal. Chem.* **60** (1977) no. 6, 1372–1374.

[4.211]  P. A. O'Leary, P. F. Fox, *J. Dairy Res.* **41** (1974) no. 3, 381–387.

[4.212]  M. Rampilli, D. Morgante, P. Resmini, *Sci. Tec. Latt. Casearia* **33** (1982) no. 2, 106–120.

[4.213]  E. H. Reimerdes, M.-M. Geuer, *Kiel. Milchwirtsch. Forschungsber.* **32** (1980) no. 1, 15–36.

[4.214]  P. Resmini, P. Cantagalli, S. E. Piazzi, S. Sordi, A. Bracciali, *Sci. Tec. Latt. Casearia* **30** (1979) no. 1, 17–31.

[4.215]  G. H. Richardson, *J. Dairy Sci.* **53** (1970) no. 10, 1373–1376.

[4.216]  P. G. Righetti, B. M. Molinari, G. Molinari, *J. Dairy Res.* **44** (1977) no. 1, 69–72.

[4.217]  G. A. L. Rothe, M. K. Harboe, S. C. Martiny, *J. Dairy Res.* **44** (1977) no. 1, 73–77.

[4.218]  S. G. Severinsen, *Sci. Tec. Latt. Casearia* **30** (1979) no. 2, 109–116.

[4.219]  M. J. Taylor, N. F. Olson, T. Richardson, *Process Biochem.* **14** (1979) no. 2, 10, 12–14, 16

[4.220]  M. J. Taylor, T. Richardson, N. F. Olson, *J. Milk Food Technol.* **39** (1976) no. 12, 864–871.

[4.221]  I. A. Angelo, K. M. Shahani, *J. Dairy Sci.* **62** (1979) no. 1, 64.

[4.222]  A. Carlson, *Enzyme Microb. Technol.* **6** (1984) no. 1, 46–47.

[4.223]  K. Shindo, K. Sakurada, R. Niki, S. Arima, *Milchwissenschaft* **35** (1980) no. 9, 527–530.

[4.224]  I. D. Stal'naya, L. A. Nakhapetyan, *Appl. Biochem. Microbiol.* **17** (1981) no. 1, 121–125.

[4.225]  L. Vamos-Vigyazo, M. El-Hawary, E. Kiss, *Acta Aliment.* **9** (1980) no. 4, 383–390.

[4.226]  L. P. Voutsinas, S. Nakai, *J. Dairy Sci.* **66** (1983) no. 4, 694–703.

[4.227]  D. G. Holmes, *Diss. Abstr. Int. B.* **36** (1975) no. 3, 1124.

[4.228]  D. G. Holmes, J. W. Duersch, C. A. Ernstrom, *J. Dairy Sci.* **60** (1977) no. 6, 862–869.

[4.229]  J. Dziuba, W. Chojnowski, *Milchwissenschaft* **37** (1982) no. 3, 148–150.

[4.230]  F. di Gregorio, R. Sisto, F. Morisi, *J. Dairy Res.* **46** (1979) no. 4, 673–680.

[4.231]  W. J. Harper, C. R. Lee, *J. Food Sci.* **40** (1975) no. 2, 282–284.

[4.232]  H. Kobayashi, I. Kusakabe, K. Murakami, *Anal. Biochem.* **122** (1982) no. 2, 308–312.

[4.233]  K. Murakami, H. Kobayashi, *Int. Congr. Food Sci. Technol. Abstracts,* Kyoto 1978, 230.

[4.234]  L. E. Wierzbicke, F. V. Kosikowski, *J. Dairy Sci.* **56** (1973) no. 1, 26–32.

[4.235]  D. M. Paige, T. M. Bayless: *Lactose Digestion, Clinical and Nutritional Implications,* John Hopkins Univ. Press, Baltimore 1980.

[4.236]  L. C. Blankenship, P. A. Wells, *J. Milk Food Technol.* **37** (1974) no. 4, 199–202.

[4.237]  M. W. Griffiths, D. D. Nuir, *J. Sci. Food Agric.* **29** (1978) no. 9, 753–761.

[4.238]  T. Okamoto, T. Morichi, *Agric. Biol. Chem.* **43** (1979) no. 11, 2389–2390.

[4.239]  F. C. Knopf, M. R. Okos, D. A. Fouts, A. Syverson, *J. Food Sci.* **44** (1979) no. 3, 896–900.

[4.240]  A. Olano, M. Ramos, I. Martinez-Castro, *Food Chem.* **10** (1983) no. 1, 57–67.

[4.241]  T. P. Shukla, *Crit. Rev. Food Technol.* **5** (1975) no. 3, 325–456.

[4.242]  K. Vandamme, R. Delbeke, *Lait* **61** (1981) no. 605/606, 282–293.

[4.243]  J. L. Baret, L. A. Dohan, US 4 409 247, 1983.

[4.244]  L. A. Dohan, J. L. Baret, S. Pain, P. Delalande, *Enzyme Eng.* **5** (1980) 279–293.

[4.245]  T. Finocchiaro, N. F. Olson, T. Richardson, *Adv. Biochem. Eng.* **15** (1980) 71–88.

[4.246]  T. Finocchiaro, T. Richardson, N. F. Olson, *J. Dairy Sci.* **63** (1980) no. 2, 215–222.

[4.247]  N. A. Greenberg, R. R. Mahoney, *Process Biochem.* **16** (1981) no. 2, 2, 4, 6–8.

[4.248]  M. W. Griffiths, D. Muir, *J. Sci. Food Agric.* **31** (1980) no. 4, 397–404.

[4.249]  V. K. Joshi, B. A. Friend, F. W. Wagner, K. M. Shahani, *J. Dairy Sci.* **62** (1979) 63.

[4.250]  T. P. Maculan, J. A. Hourigan, A. G. Rand, *J. Dairy Sci.* **61** (1978) 114.

[4.251]  W. Marconi, F. Bartoli, F. Morisi, A. Marani, *Enzyme Eng.* **5** (1980) 269–278.

[4.252]  R. A. A. Muzzarelli, *Enzyme Microb. Technol.* **2** (1980) no. 3, 177–184.

[4.253]  K. Nakanishi, R. Matsuno, K. Torii, K. Yamamoto, T. Kamikubo, *Enzyme Microb. Technol.* **5** (1983) no. 2, 115–120.

[4.254]  W. H. Pitcher, Jr., *Am. Dairy Rev.* **37** (1975) no. 9, 34D–34E, 34B.

[4.255]  D. Portetelle, P. Thonart, *Lebensm. Wiss. Technol.* **8** (1975) no. 6, 274–277.

[4.256]  M. de Rosa, A. Gambacorta, B. Nicolaus, V. Buonocore, E. Poerio, *Biotechnol. Lett.* **2** (1980) no. 1, 29–34.

[4.257]  E. H. Reimerdes, E. Herlitz, *Int. Congr. Food Sci. Technol. – Abstracts*, Kyoto 1978, p. 224.

[4.258]  B. Sproessler, H. Plainer, *Food Technol.* (*Chicago*) **37** (1983) no. 10, 93–95.

[4.259]  Sumitomo Chemical Co., GB 1 568 328, 1980.

[4.260]  K. Watanabe, Y. Yokote, K. Kimura, H. Samejima, *Agric. Biol. Chem.* **41** (1977) no. 3, 553–558.

[4.261]  A. Giec, L. Gruchala, E. Wasowicz, *Acta Aliment. Pol.* **7** (1981) no. 3, 4, 199–208.

[4.262]  Z. Mozaffar, K. Nakanishi, R. Matsuno, *J. Food Sci.* **50** (1985) no. 6, 1602–1606.

[4.263]  P. H. Evers, H. H. Nijpels, *Dtsch. Molk. Ztg.* **100** (1979) no. 34, 1200–1201.

[4.264]  D. H. Kleyn, *J. Dairy Sci.* **68** (1985) no. 10, 2791–2798.

[4.265]  D. P. Kotler, A. R. Tierney, N. S. Rosenszweig, *Anal. Biochem.* **110** (1981) no. 2, 393–396.

[4.266]  H. Lundback, B. Olsson, *Anal. Lett.* **18** (1985) no. B7, 871–889.

[4.267]  S.-L. Y. Chen, J. F. Frank, M. Loewenstein, *J. Assoc. Off. Anal. Chem.* **64** (1981) no. 6, 1414–1419.

[4.268]  I. J. Jeon, R. Bassette, *J. Food Prot.* **45** (1982) no. 1, 14–15.

[4.269]  H. U. Geyer, *Stärke* **26** (1974) no. 7, 225–232.

[4.270]  K. Poutanen, Y.-Y. Linko, P. Linko, *Milchwissenschaft* **33** (1978) no. 7, 435–438.

[4.271]  K. Poutanen, Y.-Y. Linko, P. Linko, *Nordeuropaeisk Mejeri Tidsskrift* **44** (1978) no. 4, 90–95.

[4.272]  M. B. Sliwkowski, H. E. Swaisgood, *J. Dairy Sci.* **63** (1980) 60.

[4.273]  H. E. Swaisgood, US 4 053 644, 1977.

[4.274]  M. Bakri, F. H. Wolfe, *Can. J. Biochem.* **49** (1971) no. 8, 882–884.

[4.275]  B. A. Friend, R. R. Eitenmiller, K. M. Shahani, *J. Food Sci.* **40** (1975) no. 4, 833–836.

[4.276]  J. J. Pahud, F. Widmer, R. Jost, *Adv. Exp. Med. Biol.* **137** (1981) 796–797.

[4.277]  V. M. Podboronov, V. M. Bondarenko, V. V. Evdokimov, *Parazitologiya* **16** (1982) 238–241.

[4.278]  V. M. Podboronov, *Antibiotiki* (*Moscow*) **27** (1982) no. 10, 770–774.

[4.279]  V. M. Podboronov, *Izv. Akad. Nauk Turkm. SSR Biol. Nauk* **2** (1983) 40–47.

[4.280]  F. Wasserfall, M. Teuber, *Appl. Environ. Microbiol.* **38** (1979) 197.

[4.281]  K. Hayashi, T. Kasumi, N. Kubo, N. Tsumura, *Agric. Biol. Chem.* **45** (1981) no. 10, 2289–2300.

[4.282]  J. J. Pahud, D. Schellenberg, J. C. Monti, J. C. Scherz, *Ann. Rech. Vet.* **14** (1983) no. 4, 493–501.

[4.283]  B. K. Dwivedi, *CRC Crit. Rev. Food Technol.* **3** (1973) no. 4, 457–478.

[4.284]  R. Grappin, T. C. Rank, N. F. Olson, *J. Dairy Sci.* **68** (1985) no. 3, 531–540.

[4.285]  J. E. Kinsella, D. H. Hwang, *CRC Crit. Rev. Food Sci. Nutr.* **8** (1976) no. 2, 191–228.

[4.286]  F. V. Kosikowski, US 3 975 544, 1976.

[4.287]  S. D. Braun, N. F. Olson, R. C. Lindsay, *J. Food Biochem.* **7** (1983) no. 1, 23–41.

[4.288]  J. H. Nelson, *J. Am. Oil Chem. Soc.* **49** (1972) no. 10, 559–562.

[4.289]  E. H. Reimerdes, *J. Dairy Res.* **46** (1979) no. 2, 223–226.

[4.290]  B. A. Law, *Prog. Ind. Microbiol.* **19** (1984) 245–283.

[4.291]  H. P. Gregor, US 4 033 822, 1977.

[4.292]  W. J. Harper, *Ital. Cheese J.* **4** (1974) no. 1, 1–5.

[4.293]  M. F. Parkin, *Food Technol. N. Z.* **10** (1975) no. 10, 12–13.
[4.294]  S. S. Wang, B. Davidson, C. Gillespie, L. R. Harris, D. S. Lent, *J. Food Sci.* **45** (1980) no. 3, 700–702.
[4.295]  E. H. Reimerdes, J. Roggenbuck, *Milchwissenschaft* **35** (1980) 195–201.
[4.296]  J. Balcom, P. Foulkes, N. F. Olson, T. Richardson, *Process Biochem.* **6** (1971) no. 8, 42–44.
[4.297]  L. K. Ferrier, T. Richardson, N. F. Olson, *Enzymologia* **42** (1972) no. 4, 273–283.
[4.298]  J. R. Whitaker, "Pectic substances, pectic enzymes and haze formation in fruit juices," *Enzyme Microb. Technol.* **6** (1984) 341–349.
[4.299]  Röhm GmbH, Darmstadt, *Pectinases in Fruit Technology* (comp. info leaflets 1985).
[4.300]  T. Nielsen, "Industr. application possibilities for lipases," *Fette, Seifen, Anstrichm.* **87** (1985) no. 1, 15–19.
[4.301]  K. Soda, H. Tanaka, N. Esaki in H. Dellwig (ed.): *Biotechnology,* vol. 3, Verlag Chemie, Weinheim-Deerfield Beach-Basel 1983, p. 479.
[4.302]  K. Aida, J. Chibata, K. Nakayama, K. Takinami, H. Yamada: *Biotechnology of amino acid production,* Kodansha Ltd., Tokyo 1986.
[4.303]  J. Chibata: *Immobilized Enzymes,* Kodansha-Halsted Press, Tokyo-New York 1978.
[4.304]  W. Leuchtenberger, M. Karrenbauer, U. Plöcker, *Enzyme Eng.* **7,** *Ann. N.Y. Acad. Sci.* **434** (1984) 78.
[4.305]  H. Schutt, G. Schmidt-Kastner, A. Arens, M. Preiss, *Biotechnol. Bioeng.* **27** (1985) 420.
[4.306]  Hoechst, EP-A 178 553, 1985 (R. Keller, M. Schlingman).
[4.307]  Novo Industri, US 4 080 259, 1978 (W. H. J. Boesten, L. R. M. Meyer-Hoffmann).
[4.308]  Stamicarbon B.V., EP-A 181 675, 1985 (P. L. Kerkhoffs, W. H. J. Boesten).
         Ube Industries, US 4 497 957, 1985 (M. Nakai et al.).
[4.309]  Y. Nishida, K. Nabe, S. Yamada, J. Chibata, *Enzyme Microb. Technol.* **6** (1984) 85.
[4.310]  Toray Ind., US 3 770 585, 1973; US 3 796 632, 1974 (T. Fukumura).
[4.311]  T. Fukumura: *Agric. Biol. Chem.* **41** (1977) 1327.
[4.312]  Ajinomoto, US 4 006 057, 1977 (K. Sano et al.).
[4.313]  K. Sano, K. Matsuda, N. Yasuda, K. Mitsugi, *Hakko to Taisha* **35** (1977) 112.
[4.314]  Snamprogetti, DE 2 621 076, 1976 (F. Cecere et al.).
         Kanegafuchi Kagaku Kogyo, GB 2 042 531, 1980 (H. Takahashi et al.).
         BASF, US 4 418 146, 1983 (R. Lungerhausen et al.).
[4.315]  Ajinomoto, US 4 211 840, 1980 (S. Nakamori et al.). AEC Société de chimie organique et biologique, BE 883 322, 1980.
[4.316]  Ajinomoto, US 4 016 037, 1977 (K. Mitsugi et al.).
[4.317]  Sclavo, EP-A 152 977, 1985 (R. Olivieri et al.).
[4.318]  Denki Kagaku Kogyo Kabushiki Kaisha, EP-A 159 866, 1985 (M. Teruzo et al.).
[4.319]  C. Syldatk, D. Cotoras, A. Möller, F. Wagner, *Biotech. Forum* **3** (1986) 10.
[4.320]  Degussa AG/Kernforschungsanlage Jülich/Gesellschaft für Biotechnologische Forschung, DE 3 307 094, 1984 (W. Leuchtenberger, C. Wandrey, M.-R. Kula).
[4.321]  W. Leuchtenberger in E. Magnien (ed.).: *Biomolecular Engineering in the European Community (BEP),* Martinus Nijhoff Publishers, Brussels-Luxembourg 1986, p. 227.
[4.322]  Degussa AG/Gesellschaft für Biotechnologische Forschung, DE 3 234 022, 1984; DE 3 320 495, 1985 (W. Leuchtenberger, M.-R. Kula, W. Hummel, M. Schütte).
[4.323]  Degussa AG/Gesellschaft für Biotechnologische Forschung, DE 3 307 095, 1984; DE 3 446 304, 1986 (W. Leuchtenberger, M.-R. Kula, W. Hummel, M. Schütte).
[4.324]  A. F. Bückmann, M.-R. Kula, R. Wichmann, C. Wandrey, *J. Appl. Biochem.* **3** (1981) 301.
[4.325]  J. Chibata, T. Tosa, T. Sato: *Appl. Microbiol.* **27** (1974) 878.
         Tanabe Seiyaku, DE 2 252 815, 1972 (J. Chibata et al.).
         J. Umemura, S. Takamatsu, T. Sato, T. Tosa, J. Chibata, *Eur. J. Appl. Microbiol. Biotechnol.* **20** (1984) 291.

[4.326] S. Yamada, K. Nabe, N. Jzuo, K. Nakamichi, J. Chibata, *Appl. Environ. Microbiol.* **42** (1981) 773;
Ajinomoto, EP-A 152 235, 1985 (K. Yokozeki et al.).

[4.327] Genex Corporation, EP-A 140 713, 1984 (J. J. Schruber, P. J. Vollmer, J. P. Montgomery); EP-A 165 757, 1985 (W. E. Swann); US 4 584 269, 1986 (P. J. Vollmer et al.); US 4 584 273, 1986 (M. A. J. Finkelman, H.-H. Yang); US 4 598 047, 1986 (J. C. McGuire).

[4.328] Degussa AG/Gesellschaft für Biotechnologische Forschung, US 4 304 858, 1981 (C. Wandrey, R. Wichmann, W. Leuchtenberger, M.-R. Kula).

[4.329] W. R. Grace & Co, DE-OS 3 427 495, 1985 (M. C. Fusee).

[4.330] Degussa AG/Kernforschungsanlage Jülich, DE-OS 3 419 585, 1985 (U. Groeger, H. Sahm, W. Leuchtenberger).

[4.331] Genetics Institute, US 4 518 692, 1985 (J. D. Rozzell).

[4.332] H. Zier, W. Hummel, H. Reichenbach, M.-R. Kula in 3rd European Congress on Biotechnology, vol. 1, Verlag Chemie, Weinheim 1984, p. 345.

[4.333] W. R. Grace & Co., DE-OS 3 500 054, 1985 (J. F. Walter); GB-A 2 161 159, 1986 (J. F. Walter, M. B. Sherwin).

[4.334] G. D. Searle & Co., EP-A 152 275, 1985 (D. J. Lewis, S. G. Farrand).

[4.335] Hoechst, DE-OS 3 423 936, 1986 (H. Voelskow et al.).
G. J. Calton et al., *Biotechnology* **4** (1986) 317.

[4.336] Purification Engineering, EP-A 132 999, 1984 (L. L. Wood, G. J. Calton).

[4.337] K. Nakamichi, K. Nabe, S. Yamada, T. Tosa, J. Chibata, *Eur. J. Appl. Microbiol. Biotechnol.* **19** (1984) 100.

[4.338] Degussa AG/Gesellschaft für Biotechnologische Forschung, DE-OS 3 621 839, 1987 (M.-R. Kula et al.).

[4.339] J. Chibata, T. Kakimoto, J. Kato, *Appl. Microbiol.* **13** (1965) 638.
A. S. Jandel, H. Hustedt, C. Wandrey, *Eur. J. Appl. Microbiol. Biotechnol.* **15** (1982) 59.
M. C. Fusee, J. E. Weber, *Appl. Environ. Microbiol.* **48** (1984) 694.

[4.340] H. Yamada, H. Kumagai, *Pure Appl. Chem.* **50** (1978) 1117.

[4.341] H. Yamada et al., *Biochem. Biophys. Res. Commun.* **100** (1981) 1104.

[4.342] H.-Y. Hsiao, T. Wei, K. Campbell, *Biotechnol. Bioeng.* **28** (1986) 857.
H.-Y. Hsiao, T. Wei, *Biotechnol. Bioeng.* **28** (1986) 1510.
Genex Corporation, GB-A 2 130 216, 1984 (D. M. Anderson).

[4.343] H. Yamada, S. S. Miyazaki, Y. Izumi, *Agric. Biol. Chem.* **50** (1986) 17.

[4.344] H. Yamada, S. S. Miyazaki, H. Shirae, Y. Izumi, *J. Ferment. Technol.* **63** (1985) 507.

[4.345] Asahi Kasei Kogyo, US 4 349 627, 1982; US 4 360 594, 1982 (A. Mimura et al.).

[4.346] F. Wagner, J. Klein, DE 2 841 642, 1978.

[4.347] AB Bofors, US 3 963 572, 1976 (S. V. Gatenbeck, P. O. Hedman).

[4.348] Mitsui Toatsu Chem., US 4 335 209, 1982 (Y. Asai, M. Shimada, K. Soda).

[4.349] H. Yamada, H. Kumagai, H. Kashima, H. Tori, *Biochem. Biophys. Res. Commun.* **46** (1972) 370.

[4.350] H. Yamada, H. Kumagai, *Adv. Appl. Microbiol.* **19** (1975) 249.

[4.351] Hüls AG, EP-A 175 007, 1984 (F. Schindler)

[4.352] Degussa AG/Kernforschungsanlage Jülich/Gesellschaft für Biotechnologische Forschung, US 4 326 031, 1982 (R. Wichmann, C. Wandrey, W. Leuchtenberger, M.-R. Kula, A. F. Bückmann).

[4.353] H. Schütte, W. Hummel, M.-R. Kula, *Eur. J. Appl. Microbiol. Biotechnol.* **19** (1984) 167.

[4.354] W. Hummel, H. Schütte, M.-R. Kula, *Eur. J. Appl. Microbiol. Biotechnol.* **18** (1983) 75.

[4.355] W. Hummel, H. Schütte, M.-R. Kula, *Eur. J. Appl. Microbiol. Biotechnol.* **21** (1985) 7.

[4.356] International Union of Biochemistry: *Nomenclature Committee, Enzyme Nomenclature*, Academic Press, Orlando 1984.

[4.357] H. U. Bergmeyer (ed.): *Methods of Enzymatic Analysis*, VCH Verlagsgesellschaft, Weinheim-New York 1985.
S. P. Colowick, N. O. Colowick (eds.): *Method. Enzymol.*, Academic Press, Orlando.
P. D. Boyer (ed.): *The Enzymes*, 3rd ed., Academic Press, New York 1970.

[4.358] R. Wu, L. Grossman, K. Moldave, *Methods Enzymol.* **65** (1978); **68** (1980); **100** (1983); **101** (1983).

[4.359] A. Akiyama, M. D. Bednarski, M.-J. Kim, E. S. Simon, H. Waldmann, G. M. Whitesides, *Chemtech* **18** (1988) 627–634.
M. A. Findeis, G. M. Whitesides, *Annu. Rep. Med. Chem.* **19** (1984) 263–272.

[4.360] J. B. Jones in J. D. Morrison (ed.): *Asymmetric Synthesis*, vol. 5, Academic Press, Orlando 1985, pp. 309–344.
J. B. Jones, *Tetrahedron* **42** (1986) 3351–3403.

[4.361] A. I. Laskin (ed.): *Enzymes and Immobilized Cells in Biotechnology*, Benjamin/Cummings, Menlo Park 1985.

[4.362] R. Porter, S. Clark (eds.): "Enzymes in Organic Synthesis," *Ciba Foundation Symposium III*, Pitman Publishing Ltd., London 1985.

[4.363] S. Butt, S. M. Roberts, *Nat. Prod. Rep.* **3** (1986) 489–503.

[4.364] G. M. Whitesides, C.-H. Wong, *Angew. Chem.* **97** (1985) 617–638; *Angew. Chem. Int. Ed. Engl.* **24** (1985) 617–638.

[4.365] I. Chibata: *Immobilized Enzymes – Research and Development*, Halsted Press, New York 1978.

[4.366] K. Mosbach (ed.): "Immobilized Enzymes," in *Methods Enzymol.*, vol. 44, Academic Press, New York 1976.

[4.367] A. M. Klibanov, *Science* **219** (1983) 722–727.

[4.368] A. Pollak, H. Blumenfeld, M. Wax, R. L. Baughn, G. M. Whitesides, *J. Am. Chem. Soc.* **102** (1980) 6324–6336.

[4.369] C. Buche, *Biochem. Soc. Trans.* **11** (1983) 13–14.

[4.370] W. E. Ladner, G. M. Whitesides, *J. Am. Chem. Soc.* **106** (1984) 7250–7251.

[4.371] M. G. Finn, K. B. Sharpless in J. D. Morrison (ed.): *Asymmetric Synthesis*, vol. 5, Academic Press, Orlando 1985, pp. 247–308.

[4.372] M. Schneider, M. Engel, H. Boesmann, *Angew. Chem.* **96** (1984) 52–54; *Angew. Chem. Int. Ed. Engl.* **23** (1984) 64–66.

[4.373] C. S. Chen, Y. Fujimoto, G. Giraukas, C. J. Sih, *J. Am. Chem. Soc.* **104** (1982) 7294–7299.

[4.374] M. Ohno, Y. Ito, M. Arita, T. Shibata, K. Adachi, H. Sawai, *Tetrahedron* **40** (1984) 145–152.

[4.375] S. Iruchijima, K. Hasegawa, G. Tsuchihashi, *Agric. Biol. Chem.* **46** (1982) 1907–1910.

[4.376] B. Borgström, H. L. Brockmann (eds): *Lipases*, Elsevier, Amsterdam 1984.

[4.377] B. Cambou, A. M. Klibanov, *J. Am. Chem. Soc.* **106** (1984) 2687–2692.

[4.378] G. Langrand, J. Baratti, G. Buono, C. Triantophylidis, *Tetrahedron Lett.* **27** (1986) 29–32.

[4.379] I. Chibata, T. Tosa, T. Sato, T. Mori, *Methods Enzymol.* **44** (1976) 746–759.
H. K. Chenault, J. Dahmer, G. M. Whitesides, *J. Am. Chem. Soc.* **111** (1989) 6354–6364.

[4.380] E. Lagerloef, L. Nathorst-Westfelt, B. Ekstroem, B. Sjoeberg, *Methods Enzymol.* **44** (1976) 759–768.

[4.381] K. Kato, K. Kawahara, T. Takahashi, S. Igarasi, *Agric. Biol. Chem.* **44** (1980) 821–825.

[4.382] W. G. Choi, S. B. Lee, D. Y. Ryu, *Biotechnol. Bioeng.* **23** (1981) 361–371.

[4.383] H. D. Jakubke, P. Kuhl, A. Koennecke, *Angew. Chem.* **97** (1985) 79–87; *Angew. Chem. Int. Ed. Engl.* **24** (1985) 85–93.

[4.384] J. Markussen, K. Schaumburg in K. Plaha, P. Malon (eds): *Peptides*, De Gruyter, Berlin 1983, p. 387.

[4.385] K. Oyama, S. Nishimura, Y. Nonaka, K. Kihara, T. Hashimoto, *J. Org. Chem.* **46** (1981) 5241–5242.

[4.386] W. Kullmann, *J. Org. Chem.* **47** (1982) 5300–5303.

[4.387] P. D. Boyer (ed.): *The Enzymes,* vol. 7, Chaps. 6–9, Academic Press, New York 1972.

[4.388] E. J. Toone, E. S. Simon, M. D. Bednarski, G. M. Whitesides, *Tetrahedron* **45** (1989) 5365–5422.

[4.389] C.-H. Wong, G. M. Whitesides, *J. Org. Chem.* **48** (1983) 3199–3205.
M. D. Bednarski, E. S. Simon, N. Bischofsberger, W.-D. Fessner, M.-J. Kim, W. Lees, T. Saito, H. Waldmann, G. M. Whitesides, *J. Am. Chem. Soc.* **111** (1989) 627–635.

[4.390] J. R. Durrwachter, H. M. Sweers, K. Nozaki, C. H. Wong, *Tetrahedron Lett.* **27** (1986) 1261–1264.

[4.391] J. J. Marshall, W. J. Whelan, *Chem. Ind. (London)* **25** (1971) 701–702.

[4.392] N. H. Mermelstein, *Food Technol.* **29** (1975) no. 6, 20.

[4.393] K. G. I. Nilson, *TIBTECH* **6** (1988) 256–264.

[4.394] G. Catelani, E. Mastorilli, *J. Chem. Soc. Perkin Trans. I* **1983**, 2717–2721.

[4.395] J. D. Rozzell: Abstr. 188, Am. Chem. Soc. Natl. Meeting, Philadelphia 1984, MBTD 31, Am. Chem. Soc., Washington, D.C., 1984.

[4.396] A. M. Klibanov, Z. Berman, B. N. Alberti, *J. Am. Chem. Soc.* **103** (1981) 6263–6264.

[4.397] J. M. Schwab, *J. Am. Chem. Soc.* **103** (1981) 1876–1878.

[4.398] H. K. Chenault, E. S. Simon, G. M. Whitesides in G. E. Russell (ed.): *Biotechnology and Genetic Engineering Reviews,* vol. 6, Intercept, Wimborne 1988, pp. 221–270.
H. K. Chenault, G. M. Whitesides, *Appl. Biochem. Biotechnol.* **14** (1987) 147–197.

[4.399] D. C. Crans, G. M. Whitesides, *J. Org. Chem.* **48** (1983) 3130–3132.

[4.400] B. L. Hirschbein, F. P. Mazenod, G. M. Whitesides, *J. Org. Chem.* **47** (1982) 3765–3766.

[4.401] K. M. Plowman, A. R. Krall, *Biochemistry* **4** (1965) 2809–2814.

[4.402] A. Pollak, R. L. Baughn, G. M. Whitesides, *J. Am. Chem. Soc.* **99** (1977) 2366–2367.
E. S. Simon, M. D. Bednarski, G. M. Whitesides, *J. Am. Chem. Soc.* **110** (1988) 7159–7163.

[4.403] V. M. Rios-Mercadillo, G. M. Whitesides, *J. Am. Chem Soc.* **101** (1979) 5828–5829.

[4.404] D. R. Walt, M. A. Findeis, V. M. Rios-Mercadillo, J. Augé, G. M. Whitesides, *J. Am. Chem. Soc.* **106** (1984) 234–239.

[4.405] C. Augé, S. David, C. Mathieu, C. Gautheron, *Tetrahedron Lett.* **25** (1984) 1467–1470.
S. Sabesan, J. C. Paulson, *J. Am. Chem. Soc.* **108** (1986) 2068–2080.

[4.406] A. Gross, O. Abril, J. M. Lewis, S. Geresh, G. M. Whitesides, *J. Am. Chem. Soc.* **105** (1983) 7428–7435

[4.407] L. G. Lee, G. M. Whitesides, *J. Am. Chem. Soc.* **107** (1985) 6999–7008.

[4.408] R. Wichmann, C. Wandrey, A. F. Buchnan, M. R. Kula, *Biotechnol. Bioeng.* **23** (1981) 2789–2802.
Z. Shaked, G. M. Whitesides, *J. Am. Chem. Soc.* **102** (1980) 7104–7105.

[4.409] W. Leuchtenberger, U. Plöcker, *Chem. Ing. Tech.* **60** (1988) 16–23.

[4.410] C.-H. Wong, D. G. Drueckhammer, H. M. Sweers, *J. Am. Chem. Soc.* **107** (1985) 4028–4031.

[4.411] C.-H. Wong, G. M. Whitesides, *J. Am. Chem. Soc.* **103** (1981) 4890–4899.

[4.412] T. Takemura, J. B. Jones, *J. Org. Chem.* **48** (1983) 791–796.

[4.413] D. R. Dodds, J. B. Jones, *J. Chem. Soc. Chem. Commun.* **1982**, 1080–1081.

[4.414] J. B. Jones, I. J. Jakovac, *Can. J. Chem.* **60** (1982) 19–28.

[4.415] B. L. Hirschbein, G. M. Whitesides, *J. Am. Chem. Soc.* **104** (1982) 4458–4460.
M. J. Kim, G. M. Whitesides, *J. Am. Chem. Soc.* **110** (1988) 2959–2964.
E. S. Simon, R. Plante, G. M. Whitesides, *Appl. Biochem. Biotechnol.* **22** (1989) 169–179.

[4.416] H. Simon, J. Bader, H. Guenther, S. Neumann, J. Thanos, *Angew. Chem.* **97** (1985) 541–554; *Angew. Chem. Int. Ed. Engl.* **24** (1985) 539–553.

[4.417] J. B. Jones, K. E. Taylor, *Can. J. Chem.* **54** (1976) 2974–2980.

[4.418] J. B. Jones, M. A. W. Finch, J. J. Jakovac, *Can. J. Chem.* **60** (1982) 2007–2011.

[4.419] A. Gross, S. Geresh, G. M. Whitesides, *Appl. Biochem. Biotechnol.* **8** (1983) 415–422.

[4.420]  U. T. Billhardt, P. Stein, G. M. Whitesides, *Bioorg. Chem.* **17** (1989) 1–12.

[4.421]  C.-H. Wong, A. Pollak, S. D. McCurry, J. M. Sue, J. R. Knowles, G. M. Whitesides, *Methods Enzymol.* **89** (1982) 108–125.

[4.422]  C.-H. Wong, S. C. Haynie, G. M. Whitesides, *J. Org. Chem.* **47** (1982) 5416–5418.

[4.423]  S. W. May, *Enzyme Microb. Technol.* **1** (1979), 15–22.

[4.424]  A. G. Katopodos, K. Wimalasena, J. Lee, S. W. May, *J. Am. Chem. Soc.* **106** (1984) 7928–7935.

[4.425]  Commercial Biotechnology: *An International Analysis,* U.S. Congress, Office of Technology Assessment, OTA-BAS-218, January 1984, U.S. Government Printing Office, Washington, D.C.

References for Chapter 5

[5.1a]  H. U. Bergmeyer, J. Bergmeyer, M. Graßl (eds.): *Methods of Enzymatic Analysis,* 3rd ed., vols. I–XII, VCH Verlagsgesellschaft, Weinheim 1983–1986.

[5.1b]  H. U. Bergmeyer (ed.): *Methods of Enzymatic Analysis,* 2nd. ed., vols. 1–4, Verlag Chemie, Weinheim, and Academic Press, New York 1974.

[5.2]  C. P. Price, *Philos. Trans. R. Soc. London Ser. B* **300** (1983) 411–422.

[5.3]  M. Dixon, E. C. Webb: *Enzymes,* 3rd ed., Academic Press, New York 1979, pp. 231–270.

[5.4]  Y. Inada, T. Yoshimoto, A. Matsushima, Y. Saito, *Trends Biotechnol.* **4** (1986) 68–73.

[5.5]  L. J. Perry, R. Wetzel, *Science* **226** (1984) 555–557.

[5.6]  J. Siedel, R. Deeg, J. Ziegenhorn in [5.1a], vol. 1, 1983, pp. 182–197.

[5.7]  H. U. Bergmeyer, D. Doering in [5.1a], vol. 1, 1983, pp.48–54.

[5.8]  P. D. Boyer, H. Lardy, K. Myrbäck (eds.): *The Enzymes,* 2nd ed., vols. 1–8, Academic Press, New York 1959–1963.

[5.9]  P. D. Boyer (ed.): *The Enzymes,* 3rd ed., vols. 1–16, Academic Press, New York 1970–1983.

[5.10]  S. P. Colowick, N. O. Kaplan (eds.): *Methods in Enzymology,* vols. 1–126, Academic Press, New York 1955–1986.

[5.11]  T. E. Barman: *Enzyme Handbook,* vols. 1, 2, 2a, Springer Verlag, Berlin 1969.

[5.12]  H. U. Bergmeyer, M. Graßl, H.-E. Walter in [5.1a], vol. 2, pp. 126–328.

[5.13]  R. Ruyssen, A. Lauwers (eds.): *Pharmaceutical Enzymes,* E. Story-Scientia Scientific Publ. Comp., Gent 1978.

[5.14]  W. P. Jencks in [5.8], vol. 6, 1962, pp. 373–385.

[5.15]  L. T. Webster, Jr., in [5.10], vol. 13, 1969, pp. 375–381.

[5.16]  S. P. Colowick in [5.10], vol. 2, 1955, pp. 598–604.

[5.17]  L. Noda in [5.9], vol. 8, 1973, pp. 279–305.

[5.18]  E. Racker in [5.10], vol. 1, 1955, pp. 500–503.

[5.19]  C.-I. Brändén, H. Jörnvall, H. Eklund, B. Furngren in [5.9], vol. 11, 1975, pp. 171–186.

[5.20]  W. B. Jacoby in [5.8], vol. 7, 1963, pp. 203–221.

[5.21]  S. L. Bradbury, J. F. Clark, C. R. Steinman, W. B. Jacoby in [5.10], vol. 41, 1975, pp. 354–360.

[5.22]  N. Tamaki, T. Hama in [5.10], vol. 89, 1982, pp. 469–473.

[5.23]  A. Meister, D. Wellner in [5.8], vol. 7, 1963, pp. 609–648.

[5.24]  K. Yagi in [5.10], vol. 17B, 1971, pp. 608–622.

[5.25]  J. Larner in [5.8], vol. 5, 1960, pp. 369–378.

[5.26]  J. H. Pazur in [5.10], vol. 28, 1972, pp. 931–934.

[5.27]  C. R. Dawson, R. J. Magee in [5.10], vol. 2, 1955, pp. 831–835.

[5.28]  G. R. Stark, C. R. Dawson in [5.8], vol. 8, 1963, pp. 297–311.

[5.29]  K. Stocker, G. Barlow in [5.10], vol. 45, 1976, pp. 214–223.

[5.30]  J. B. Sumner, A. L. Dounce in [5.10], vol. 2, 1955, pp. 775–781.

[5.31]  G. R. Schonbaum, B. Chance in [5.9], vol. 13, 1976, pp. 363–408.

[5.32]  T. Uwajima, O. Terada, *Agric. Biol. Chem.* **39** (1975) 1511–1512.

[5.33]  W. Richmond, *Clin. Chem.* **19** (1973) 1350–1356.

[5.34]  H. Tomioka, M. Kagawa, S. Nakamura, *J. Biochem. (Tokyo)* **79** (1976) 903–915.

[5.35]  J. B. Wittenberg in [5.10], vol. 5, 1962, pp. 560–562.

[5.36]  M. A. Brostrom, E. T. Browning, *J. Biol. Chem.* **248** (1973) 2364–2371.

[5.37]  S. Ikuta, S. Imamura, H. Misaki, Y. Horiuti, *J. Biochem. (Tokyo)* **82** (1977) 1741–1749.

[5.38]  S. Dagley in [5.10], vol. 13, 1969, pp. 160–163.

[5.39]  B. G. Spector in [5.9], vol. 7, 1972, pp. 378–382.

[5.40]  P. A. Srere in [5.10], vol. 13, 1969, pp. 3–11.

[5.41]  L. B. Spector in [5.9], vol. 7, 1972, pp. 358–368.

[5.42]  T. Yoshimoto, I. Oka, D. Tsuru, *Arch. Biochem. Biophys.* **177** (1976) 508–515.

[5.43]  T. Yoshimoto, I. Oka, D. Tsuru, *J. Biochem. (Tokyo)* **79** (1976) 1381–1383.

[5.44]  K. Rikitake, I. Oka, M. Ando, T. Yoshimoto et al., *J. Biochem. (Tokyo)* **86** (1979) 1109–1117.

[5.44 a]  T. Uwajiama, O. Terada, *Agr. Biol. Chem.* **44** (1980) 1787–1792.

[5.45]  V. Massey in [5.8], vol. 7, 1963, pp. 275–306.

[5.46]  V. Massey in [5.10], vol. 9, 1966, pp. 272–278.

[5.47]  K. Krisch in [5.9], vol. 5, 1971, pp. 43–69.

[5.48]  N. P. B. Dudman, B. Zerner in [5.10], vol. 35, 1975, pp. 190–208.

[5.49]  K. Fujikawa, E. W. Davie in [5.10], vol. 45, 1976, pp. 89–95.

[5.50]  J. Jesty, Y. Nemerson in [5.10], vol. 45, 1976, pp. 95–107.

[5.51]  M. Ando, T. Yoshimoto, S. Ogushi, K. Rikitake et al., *J. Biochem. (Tokyo)* **85** (1979) 1165–1172.

[5.52]  H. Schütte, J. Flossdorf, H. Sahm, M.-R. Kula, *Eur. J. Biochem.* **62** (1976) 151–161.

[5.53]  S. Hestrin, D. S. Feingold, M. Schramm in [5.10], vol. 1, 1955, pp. 251–257.

[5.54]  J. O. Lampen in [5.9], vol. 5, 1971, pp. 292–305.

[5.55]  G. Doudoroff in [5.8], vol. 5, 1962, pp. 339–341.

[5.56]  K. Wallenfels, G. Kurz in [5.10], vol. 9, 1966, pp. 112–116.

[5.57]  G. Maier, G. Kurz in [5.10], vol. 89, 1982, pp. 176–181.

[5.58]  K. Schmid, R. Schmitt, *Eur. J. Biochem.* **67** (1976) 95–104.

[5.59]  K. Wallenfels in [5.10], vol. 5, 1962, pp. 212–219.

[5.60]  K. Wallenfels, R. Weil in [5.9], vol. 7, 1972, pp. 617–663.

[5.61]  S. S. Cohen in [5.10], vol. 1, 1955, pp. 350–354.

[5.61 a]  H. E. Pauly, G. Pfleiderer, *Hoppe Seyler's Z. Physiol. Chem.* **356** (1975) 1613–1623.

[5.61 b]  K.-D. Jany, W. Ulmer, M. Fröschle, G. Pfleiderer, *FEBS Lett.* **165** (1984) 6–10.

[5.62]  J. H. Pazur in [5.10], vol. 9, 1966, pp. 82–87.

[5.63]  H. Tsuge, O. Natsuaki, K. Ohashi, *J. Biochem. (Tokyo)* **78** (1975) 835–843.

[5.64]  E. A. Noltmann, S. A. Kuby in [5.8], vol. 7, 1963, pp. 223–242.

[5.65]  C. Olive, H. R. Levy in [5.10], vol. 41, 1975, pp. 196–201.

[5.66]  E. A. Noltmann in [5.9], vol. 6, 1972, pp. 272–301.

[5.67]  H. Halvorson in [5.10], vol. 8, 1966, pp. 559–562.

[5.68]  S. Tabata, T. Ide, Y. Umemura, K. Torii, *Biochim. Biophys. Acta* **797** (1984) 231–238.

[5.69]  Y. C. Lee in [5.10], vol. 28, 1972, pp. 699–702.

[5.70]  E. L. Smith, B. M. Austen, K. M. Blumenthal, J. F. Nyc in [5.9], vol. 11, 1975, pp. 293–367.

[5.71]  H. F. Fisher in [5.10], vol. 113, 1985, pp. 16–27.

[5.72]  S. F. Velick, J. Vavra in [5.8], vol. 6, 1962, pp. 219–246.

[5.73]  A. E. Braunstein in [5.9], vol. 9, 1973, pp. 379–481.

[5.74]  W. T. Jenkins, M. Saier, Jr., in [5.10], vol. 17 A, 1970, pp. 159–163.

[5.75]  R. M. Burton in [5.10], vol. 1, 1955, pp. 397–400.

[5.76]  E. C. C. Lin, B. Magasanik, *J. Biol. Chem.* **235** (1960) 1820–1823.

[5.77]  J. W. Thorner, H. Paulus in [5.9], vol. 8, 1973, pp. 487–508.

[5.78]  M. Comer, C. J. Bruton, T. Atkinson, *J. Appl. Biochem.* **1** (1979) 259–270.

[5.79]  T. Baranowski in [5.8], vol. 7, 1963, pp. 85–96.

[5.80]  N. J. Jacobs, P. J. VanDemark, *Arch. Biochem. Biophys.* **88** (1960) 250–255.

[5.81]  S. P. Colowick in [5.9], vol. 9, 1973, pp. 1–48.

[5.82]  E. Barnard in [5.10], vol. 42, 1975, pp. 6–20.

[5.83]  J. T. Schulze, J. Gazith, R. H. Gooding in [5.10], vol. 9, 1966, pp. 376–381.

[5.84]  H. A. Krebs, K. Gawehn, D. H. Williamson, H. U. Bergmeyer in [5.10], vol. 14, 1969, pp. 222–227.

[5.85]  G. W. E. Plaut in [5.8], vol. 7, 1963, pp. 105–126.

[5.86]  W. W. Cleland, V. W. Thompson, R. E. Barden in [5.10], vol. 13, 1969, pp. 30–33.

[5.87]  F. Gasser, M. Doudoroff, R. Contopoulos, *J. Gen. Microbiol.* **62** (1970) 241–250.

[5.88]  W. J. Reeves Jr., G. M. Fimognari in [5.10], vol. 9, 1966, pp. 288–294.

[5.89]  J. J. Holbrook, A. Liljas, S. J. Steindel, M. G. Rossman in [5.9], vol. 11, 1975, pp. 191–292.

[5.90]  M. Sugiuri, T. Oikawa, K. Hirano, T. Inukai, *Biochim. Biophys. Acta* **488** (1977) 353–358.

[5.91]  J. W. Hastings, T. O. Baldwin, M. Z. Nicoli in [5.10], vol. 57, 1978, pp. 135–152.

[5.92]  W. D. McElroy in [5.8], vol. 6, 1962, pp. 433–442.

[5.93]  M. De Luca, W. D. McElroy in [5.10], vol. 57, 1978, pp. 3–15.

[5.94]  S. Englard, L. Siegel in [5.10], vol. 13, 1969, pp. 99–116.

[5.95]  L. J. Banaszak, R. A. Bradshaw, in [5.9], vol. 11, 1975, pp. 369–396.

[5.96]  E. A. Noltmann in [5.9], vol. 6, 1972, pp. 302–314.

[5.97]  M. I. Dolin, *J. Biol. Chem.* **225** (1957) 557–573.

[5.98]  K. G. Paul in [5.8], vol. 8, 1963, pp. 227–274.

[5.99]  A. Nason in [5.8], vol. 7, 1963, pp. 587–608.

[5.100] A. C. Maehly in [5.10], vol. 2, 1955, pp. 801–813.

[5.101] H. N. Fernley in [5.9], vol. 4, 1971, pp. 417–447.

[5.102] M. Fosset, D. Chappelet-Tordo, M. Lazdunski, *Biochem.* **13** (1974) 1783–1788.

[5.103] M. Rippa, M. Signorini in [5.10], vol. 41, 1975, pp. 237–240.

[5.104] A. C. Ottolenghi in [5.10], vol. 14, 1969, pp. 188–197.

[5.105] E. A. Dennis in [5.9], vol. 16, 1983, pp. 307–353.

[5.106] S. Imamura, Y. Horiuti, *J. Biochem.* (*Tokyo*) **85** (1979) 79–95.

[5.107] K. C. Robbins, L. Summaria in [5.10], vol. 45, 1976, pp. 257–273.

[5.108] K. C. Robbins, L. Summaria, R. C. Wohl in [5.10], vol. 80, 1981, pp. 379–387.

[5.109] T. Bücher, G. Pfleiderer in [5.10], vol. 1, 1955, pp. 435–440.

[5.110] F. J. Kayne in [5.9], vol. 8, 1973, pp. 353–382.

[5.111] L. P. Hager, F. Lipmann in [5.10], vol. 1, 1955, pp. 482–486.

[5.112] D. R. Sanadi in [5.8], vol. 7, 1963 pp. 307–344.

[5.113] M. Suzuki, *J. Biochem.* (*Tokyo*) **89** (1981) 599–607.

[5.114] J. B. Wolff, N. O. Kaplan in [5.10], vol. 1, 1955, pp. 348–350.

[5.115] S. Cha in [5.10], vol. 13, 1969, pp. 62–69.

[5.116] W. A. Bridger in [5.9], vol. 10, 1974 pp. 581–606.

[5.117] D. L. Kessler, K. V. Rajagopalan, *J. Biol. Chem.* **247** (1972) 6566–6573.

[5.118] S. Magnusson in [5.9], vol. 3, 1971, pp. 278–321.

[5.119] R. L. Lundblad, H. S. Kingdon, K. G. Mann in [5.10], vol. 45, 1976, pp. 156–176.

[5.120] M. Laskowski in [5.10], vol. 2, 1955, pp. 26–36.

[5.121] B. Keil in [5.9], vol. 3, 1971, pp. 250–275.

[5.122] H. R. Mahler in [5.8], vol. 8, 1963, pp. 285–296.

[5.123] J. B. Sumner in [5.10], vol. 2, 1955, pp. 378–379.

[5.124] F. J. Reithel in [5.9], vol. 4, 1971, pp. 1–21.

[5.125] B. L. Horecker, L. A. Heppel in [5.10], vol. 2, 1955, pp. 482–485.

[5.126] R. C. Bray in [5.9], vol. 12, 1975, pp. 299–419.

[5.127] H. U.Bergmeyer (ed.): *Methods of Enzymatic Analysis,* 3rd ed., vol. 1, Verlag Chemie, Weinheim 1983.

[5.128] H. U. Bergmeyer in [5.127], pp. 163–181.

[5.129] J. Siedel, R. Deeg, J. Ziegenhorn in [5.127], pp. 182–197.

[5.130] H. U. Bergmeyer in [5.127], pp. 233–260.

[5.131] M. Oellerich in [5.127], pp. 233–260.

[5.132] M. J. O'Sullivan, J. W. Bridges, V. Marks, *Ann. Clin. Biochem.* **16** (1979) 221–239.

[5.133] E. Ishikawa, M. Imagawa, S. Hasida, S. Yoshitake, Y. Hamagushi, T. Ueno, *J. Immunoassay* **4** (1983) 209–327.

[5.134] P. D. Weston, J. A. Devries, R.    Wrigglesworth, *Biochim. Biophys. Acta* **612** (1980) 40–49.

[5.135] A. L. Hurn, S. M. Chantler in S. P. Colowick, N. O. Kaplan (eds.): *Methods in Enzymology,* vol. 70, Academic Press, New York 1980, pp. 104–142.

[5.136] C. J. Stanley, A. Johannsson, C. H. Self, *J. Immunol. Methods* **83** (1985) 89–95.

[5.137] K. Wulff in [5.127], pp. 340–368.

[5.138] I. Weeks, J. Beheshti, F. McCapra, A. K. Campbell, J. S. Woodhead, *Clin. Chem.* (*Winston Salem N.C.*) **29** (1983) 1474–1479.

[5.139] I. Weeks. A. K. Campbell, J. S. Woodhead, *Clin. Chem.* (*Winston Salem N.C.*) **29** (1983) 1480–1483.

[5.140] J. Wannlund, M. DeLuca in M. A. DeLuca, W. D. McElroy (eds.): *Bioluminescence and Chemiluminescence,* Academic Press, New York 1981, pp. 693–696.

[5.141] H. Fricke, C. J. Strasburger, W. G. Wood, *J. Clin. Chem. Clin. Biochem.* **20** (1982) 91–94.

[5.142] Th. P. Whitehead, G. H. G. Thorpe, T. J. N. Carter, C. Croucutt,  J. L. Kricka, *Nature* (*London*) **305** (1983) 158–159.

[5.143] G. H. Thorpe, L. J. Kricka, E. Gillespie, S. Moseley, R. Amess, N. Baggett, Th. P. Whitehead, *Anal. Biochem.* **145** (1985) 96–100.

[5.144] G. H. G. Thorpe, L. J. Kricka, S. Moseley, Th. P. Whitehead, *Clin. Chem.* (*N.Y.*) **31** (1985) 1335–1341.

[5.145] Enzyme Technology Company, EP-A 0103 784, 1983 (R.C. Stout).

[5.146] Boehringer Mannheim GmbH: *Methods of Biochemical Analysis and Food Analysis,* Mannheim 1986.

[5.147] Boehringer Mannheim GmbH: *Die enzymatische Lebensmittelanalytik* (series).

[5.148] G. Henniger: "Enzymatische Lebensmittelanalytik – Enzymatic Food Analysis," *Z. Lebensm.-Technol. Verfahrenstechn.* **30** (1979), 137–144 and 182–185.

[5.149] H. Bünte, *Langenbecks Arch. Dtsch. Z. Chir.* **296** (1960) 129–137.

[5.150] K. Schultis, E. Wagner, *Dtsch. Med. Wochenschr.* **93** (1968) 1685–1691.

[5.151] F. Kümmerle, K. Beck, R. Tenner, *Dtsch. Med. Wochenschr.* **94** (1969) 691–694.

[5.152] R. Ruyssen, A. Lauwers (eds.): *Pharmaceutical Enzymes, Properties and Assay Methods,* E. Story-Scientia, Scientific Publ. Comp., Gent 1978, pp. 57–84.

[5.153] M. Kataria, D. Bhaskar Rao, *Br. J. Clin. Pract.* **23** (1969) 15–17.

[5.154] A. Tomasi, O. Ghidini, A. Battocchina, *Med. Welt* **23** (1972) 1779–1780.

[5.155] S. Karani, M. Kataria, A. E. Barber, *Br. J. Clin. Pract.* **25** (1971) 8.

[5.156] Bundesverband der Pharmazeutischen Industrie (ed.): *Rote Liste 1986,* Edito Cantor, Aulendorf 1986.

[5.157] H. Wokalek, R. Niedner, *Extracta Media Practica* **5,** 1st suppl. (1984) 41–48.

[5.158] P. Helali, K. J. Villiger, E. Vogt, *Schweiz. Rundschau. Med. Prax.* **69** (1980) 703–711.

[5.159] H. Fischer, F. Gilliet, M. Hornemann, F. Leyh, H. Lindemayr, W. Stegmann, S. Welke, *Extracta Medica Practica* **5,** 1st suppl. (1984) 77–81.

[5.160] M. R. Ewart, M. W. C. Hatton, J. M. Basford, K. S. Dogson, *Biochem. J.* **118** (1970) 603–609.

[5.161] K. Stocker in F. Markwardt (ed.): *Handbook of Experimental Pharmacology,* vol. 46, Springer Verlag, Berlin 1978, pp. 451–484.

[5.162] G. K. Wolf, *Eur. J. Clin. Pharmacol.* **9** (1976) 387–392.

[5.163] H. Giertz, L. Flohé in W. Forth, D. Henschler, W. Rummel (eds): *Allgemeine nd spezielle Pharmakologie und Toxikologie,* 5th ed., B.I. Wissenschaftsverlag, Mannheim 1988, pp. 176–215.

[5.164] D. K. Mc Clintock, P. H. Bell, *Biochem. Biophys. Res. Commun.* **43** (1971) 694–702.

[5.165] K. N. N. Reddy, G. Markus, *J. Biol. Chem.* **247** (1972) 1683–1691.

[5.166] R. N. Brodgen, T. M. Speight, G. S. Avery, *Drugs* **5** (1973) 357–445.

[5.167] The GISSI Study Group, *Lancet* **1986,** I 397–402.

[5.168] W. A. Günzler, G. J. Steffens, F. Ötting, G. Buse, L. Flohé, *Hoppe-Seyler's Z. Physiol. Chem.* **363** (1982) 133–141.

[5.169] G. J. Steffens, W. A. Günzler, F. Ötting, E. Fraukus, L. Flohé, *Hoppe-Seyler's Z. Physiol. Chem.* **363** (1982) 1043–1058.

[5.170] W. A. Günzler, G. J. Steffens, F. Ötting, S-M. A. Kim, E. Frankus, L. Flohé, *Hoppe-Seyler's Z. Physiol. Chem.* **363** (1982) 1155–1165.

[5.171] K. Danø, P. A. Andreasen, H. Grøndahl–Hansen, P. Kristensen, L. S. Nielsen, L. Skriver, *Adv. Cancer Res.* **44** (1985) 139–266.

[5.172] W. A. Günzler, J. Cramer, E. Frankus, E. Friderichs et al., *Arzneim. Forsch.* **35** (1985) no. 1, 652–662.

[5.173] F. Van de Werf, D. Collen, *Eur. Heart J.* **6** (1985) 902–904.

[5.174] D. Pennica, W. E. Holmes, W. J. Kohr, R. N. Harkins et al., *Nature (London)* **301** (1983) 214–221.

[5.175] The TIMI Study Group, *N. Engl. J. Med.* **312** (1985) 932–936.

[5.176] H. R. Lijnen, C. Zamarron, M. Blaber, M. E. Winkler, D. Collen, *J. Biol. Chem.* **261** (1986) 1253–1258.

[5.177] F.-W. Hanbücken, J. Schneider, W. A. Günzler, E. Friderichs, H. Giertz, L. Flohé, *Arzneim. Forsch. Drug Res.* **37** (II) (1987) 993–997.

[5.178] PRIMI Trial Study Group, *Lancet* **1** (1989) 863–868.

[5.179] L. Flohé, *Eur. Heart J.* **6** (1985) 905–908.

[5.180] V. Gurewich, R. Pannell, S. Louie, P. Kelley, R. K. Suddith, R. Greenlee, *J. Clin Invest.* **73** (1984) 1731–1739.

[5.181] W. E. Holmes, D. Pennica, M. Blaber, M. W. Rey, W. A. Günzler, G. J. Steffens, H. L. Heyneker, *Bio/Technology* **3** (1985) 923–929.

[5.182] L. Flohé, *Drugs in the Future* **11** (1986) 851–852.

[5.183] K. Lechner: *Blutgerinnungsstörungen,* Springer Verlag, Berlin 1982.

[5.184] W. I. Wood, D. J. Capon, C. C. Simonsen, D. L. Eaton et. al., *Nature (London)* **312** (1984) 330–336.

[5.185] J. J. Toole, J. L. Knopf, J. M. Wozney, L. A. Sultzman et al., *Nature (London)* **312** (1984) 342–346.

[5.186] R. L. Capizzi, Y. C. Cheng, *Med. Pediatr. Oncol. Suppl.* **1** (1982) 221–228.

[5.187] R. J. Esterhay, P. H. Wiernik, W. R. Grove, S. D. Markus, M. N. Wesley, *Blood* **59** (1982) 334–345.

[5.188] M. E. Nesbit, I. Ertel, G. D. Hammond, *Cancer Treat. Rep.* **65**, Suppl. 4 (1981) 101–107.

[5.189] H. Riehm, H. Gadner, K. Welte, *Klin. Paediat.* **189** (1977) 89–102.

[5.190] L. Sieger, G. Higgins, J. Ortega, N. Shore, K. Williams, *Med. Pediat. Oncol.* **2** (1976) 327–332.

[5.191] H. Wehinger, H. O. Fürste, *Klin. Pädiatr.* **193** (1981) 159–161.

[5.192] R. H. Sills, D. A. Nelson, J. A. Stockman, *Med. Pediatr. Oncol.* **4** (1978) 311–313.

[5.193] R. J. Spiegel, C. K. Echelberger, D. G. Poplack, *Med. Pediat. Oncol.* **8** (1980) 123–125.

[5.194] W. Roggendorf, M. Brock, H.-H. Görge, C. Curio, *J. Neurosurg.* **60** (1984) 518–522.

[5.195] R. D. Frazer, *Spine (Philadelphia)* **7** (1982) 606–612.

[5.196] L. Flohé, H. Giertz, R. Beckmann in I. L. Bonta, M. A. Bray, M. J. Parnham (eds.): *Handbook of Inflammation,* vol. 5, Elsevier, Amsterdam 1985, pp. 225–281.

[5.197] J. M. Mc Cord, I. Fridovich, *J. Biol. Chem.* **244** (1969) 6049–6055.

[5.198] K. M. Wilsmann in G. Rotilio (ed.): *Superoxide and Superoxide Dismutase in Chemistry, Biology and Medicine,* Elsevier, Amsterdam 1986, pp. 500–507.

[5.199] H. Marberger, W. Huber, K. B. Menander-Huber, G. Bartsch, *Eur. J. Rheumatol. Inflammation* **4** (1981) 244–249.

[5.200] H. Marberger, W. Huber, G. Bartsch, T. Schulte, P. Swoboda, *Int. Urol. Nephrol.* **6** (1974) 61–74.

[5.201] L. Flohé in W. Bors, M. Saran, D. Tait (eds.): *Oxygen Radicals in Chemistry and Biology,* W. de Gruyter, Berlin 1984, pp. 809–812.

[5.202] D. J. Hearse, A. S. Manning, J. M. Downey, D. M. Yellon, *Acta Physiol. Scand.* **126**, Suppl. 548 (1986) 65–78.

[5.203] G. B. Bulkley, J. B. Morris in G. Rotilio (ed.): *Superoxide and Superoxide Dismutase in Chemistry, Biology and Medicine,* Elsevier, Amsterdam 1986, pp. 565–570.

[5.204] B. R. Lucchesi, K. M. Mullane, *Annu. Rev. Pharmacol. Toxicol.* **26** (1986) 201–224.

[5.205] L. Flohé, S.-M. A. Kim, F. Ötting, D. Saunders et al. in G. Rotilio (ed.): *Superoxide and Superoxide Dismutase in Chemistry, Biology and Medicine*, Elsevier, Amsterdam 1986. pp. 266–269.

References for Chapter 6

[6.1] D. M. Glover: *Gene Cloning, The Mechanisms of DNA Manipulation*, Chapman and Hall, London 1984.

[6.2] D. M. Glover: *DNA Cloning, A Practical Approach,* vol. I, IRL Press Limited, Oxford 1985.

[6.3] D. M. Glover: *DNA Cloning, A Practical Approach*, vol. II, IRL Press Limited, Oxford 1985.

[6.4] L. Grossmann, K. Moldave: "Nucleic Acids," in S. P. Colowick, N. O. Kaplan (eds.): *Methods of Enzymology,* vol. 65, Part I, Academic Press, New York 1980.

[6.5] P. B. Hackett, J. A. Fuchs, J. W. Messing: *An Introduction to Recombinant DNA Techniques,* The Benjamin Cummings Publishing Company, Inc., Menlo Park, CA, 1984.

[6.6] B. D. Hames, S. J. Higgins: *Transcription and Translation. A Practical Approach.* IRL Press Limited, Oxford 1984.

[6.7] I. J. Higgins, D. J. Best, J. Jones: *Biotechnology, Principles and Applications*, Blackwell Scientific Publishers, Oxford 1985.

[6.8] A. D. B. Malcolm: "The use of restriction enzymes in genetic engineering," in R. Williamson (ed.): *Genetic Engineering,* vol. 2, Academic Press, New York 1981, p. 129.

[6.9] T. Maniatis, E. F. Fritsch, J. Sambrook: *Molecular Cloning. A Laboratory Manual,* Cold Spring Harbor Laboratory, Cold Spring Harbor, New York, 1982.

[6.10] J. H. Miller: *Experiments in Molecular Genetics,* Cold Spring Harbor Laboratory, Cold Spring Harbor, New York, 1972.

[6.11] S. J. O'Brien: *Genetic Maps 1984,* Cold Spring Harbor Laboratory, Cold Spring Harbor, New York, 1984.

[6.12] R. W. Old, S. B. Primrose in N. G. Carr, J. L. Ingraham, S. C. Rittenberg (eds.): *Principles of Gene Manipulation. Studies in Microbiology,* vol. 2., Blackwell Scientific Publishers, Oxford 1985.

[6.13] R. H. Pritchard, I. B. Holland: *Basic Cloning Techniques,* Blackwell Scientific Publishers, Oxford 1985.

[6.14] A. Pühler, K. N. Timmis: *Advanced Molecular Genetics,* Springer Verlag, Berlin 1984.

[6.15] R. L. Rodriguez, R. C. Tait: *Recombinant DNA Techniques. An Introduction,* Addison-Wesley, Publishing Company, London 1983.

[6.16] S. M. Weissmann: *Methods of DNA and RNA Sequencing,* Praeger Publishers, New York 1983.

[6.17] E.-L. Winnacker: *Gene und Klone. Eine Einführung in die Gentechnologie,* VCH Verlagsgesellschaft, Weinheim 1985.

[6.18] R. Wu: "Recombinant DNA," in S. P. Colowick, N. O. Kaplan (eds.): *Methods in Enzymology,* vol. 68, Academic Press, New York 1980.

[6.19] R. Wu, L. Grossmann, K. Moldave: "Recombinant DNA," in S. P. Colowick, N. O. Kaplan (eds.): *Methods in Enzymology,* vol. 100, Part B, Academic Press, New York 1983.

[6.20] R. Wu, L. Grossmann, K. Moldave: "Recombinant DNA," in S. P. Colowick, N. O. Kaplan (eds.): *Methods in Enzymology,* vol. 101, Part C, Academic Press, New York 1983.

[6.21] M. Zabeau, R. J. Roberts: "The role of restriction endonucleases in molecular genetics," in J. H. Taylor (ed.): *Molecular Genetics.* Academic Press, New York 1979, p. 1.

[6.22] J. G. Chirikjian: *Gene Amplification and Analysis,* vol. 1, Elsevier, Amsterdam 1980.

[6.23] C. Kessler, V. Manta: "Specificity of restriction endonucleases and DNA modification methyltransferases – a review (Edition 3)," *Gene* (1989) (in the press).

[6.24] S. M. Linn, R. J. Roberts: *Nucleases,* Cold Spring Harbor Laboratory, Cold Spring Harbor, New York, 1982.

[6.25] M. McClelland, M. Nelson: "The effect of site specific methylation on restriction endonuclease digestion," *Nucleic Acids Res.* **13** (1985) r201.

[6.26] A. Razin, H. Cedar, A. D. Riggs: *DNA Methylation,* Springer Verlag, New York 1984.

[6.27] R. J. Roberts: "Restriction and modification enzymes and their recognition sequences," *Nucleic Acids Res.* **13** (1985) r165.

[6.28] I. Schildkraut: "Screening for and Characterizing Restriction Endonucleases," in J. K. Setlow, A. Hollaender (eds.): *Genetic Engineering,* vol. 6, Plenum Press, New York 1984.

[6.29] T. A. Trautner: Methylation of DNA, in *Current Topics in Microbiology and Immunology,* Springer Verlag, New York 1984.

[6.30] K. Armstrong, W. R. Bauer, *Nucleic Acids Res.* **10** (1982) 993.

[6.31] J. R. Arrand, P. A. Myers, R. J. Roberts, *J. Mol. Biol.* **118** (1978) 127.

[6.32] J. R. Arrand, P. A. Myers, R. J. Roberts, cited in [6.23], [6.27].

[6.33] G. Bertani, J. J. Weigle, *J. Bacteriol.* **65** (1953) 113.

[6.34] T. A. Bickle in S. M. Linn, R. J. Roberts (eds.): *Nucleases,* Cold Spring Harbor Laboratory, Cold Spring Harbor, New York, 1982, p. 85.

[6.35] T. A. Bickle, K. Ineichen, *Gene* **9** (1980) 205.

[6.36] T. A. Bickle, V. Pirrotta, R. Imber in S. P. Colowick, N. O. Kaplan (eds.): *Methods in Enzymology,* vol. 65, Academic Press, New York 1980, p. 132.

[6.37] A. H. A. Bingham, T. Atkinson, D. Sciaky, R. J. Roberts, *Nucleic Acids Res.* **5** (1978) 3457.

[6.38] R. M. Blumenthal, S. A. Gregory, J. S. Cooperider, *J. Bacteriol.* **164** (1985) 501.

[6.39] I. G. Bogdarina, L. M. Vagabova, Ya. I. Burynaov, *FEBS Lett.* **68** (1976) 177.

[6.40] B. J. Bolton, M. J. Comer, C. Kessler, unpublished results.

[6.41] B. J. Bolton, G. Nesch, M. J. Comer, W. Wolf, C. Kessler, *FEBS Lett.* **182** (1985) 130.

[6.42] R. Borsetti, R. Grandoni, I. Schildkraut, cited in [6.23], [6.27].

[6.43] J. Botterman, M. Zabeau, *Gene* **37** (1985) 229.

[6.44] L. Bougueleret, M. Schwarzstein, A. Tsugita, M. Zabeau, *Nucleic Acids Res.* **12** (1984) 3659.

[6.45] L. Bougueleret, M. L. Tenchini, J. Botterman, M. Zabeau, *Nucleic Acids Res.* **13** (1985) 3823.

[6.46]  H. W. Boyer, L. T. Chow, A. Dugaiczyk, J. Hedgpeth, H. M. Goodman, *Nature (London)* **244** (1973) 40.

[6.47]  S. Bron, K. Murray, *Mol. Gen. Genet.* **143** (1975) 25.

[6.48]  J. E. Brooks, R. M. Blumenthal, T. R. Gingeras, *Nucleic Acids Res.* **11** (1983) 837.

[6.49]  N. L. Brown, C. A. Hutchison III, M. Smith, *J. Mol. Biol.* **140** (1980) 143.

[6.50]  N. L. Brown, M. McClelland, P. R. Whitehead, *Gene* **9** (1980) 49.

[6.51]  N. L. Brown, M. Smith, *FEBS Lett.* **65** (1976) 284.

[6.52]  N. L. Brown, M. Smith, *Proc. Natl. Acad. Sci. USA* **74** (1977) 3213.

[6.53]  A. R. Buchmann, L. Burnett, P. Berg in J. Tooze (ed.): *DNA Tumor Viruses,* Cold Spring Harbor Laboratory, Cold Spring Harbor, New York, 1980, p. 779.

[6.54]  S.-C. Cheng, R. Kim, K. King, S.-H. Kim, P. Modrich, *J. Biol. Chem.* **259** (1984) 11571.

[6.55]  D. G. Comb, P. Parker, R. Grandoni, I. Schildkraut, cited in [6.23], [6.27].

[6.56]  D. G. Comb, I. Schildkraut, cited in [6.23], 6.27].

[6.57]  D. G. Comb. I. Schildkraut, G. Wilson, L. Greenough, cited in [6.23], [6.27].

[6.58]  D. G. Comb, G. Wilson, cited in [6.23], [6.27].

[6.59]  D. L. Daniels, J. L. Schroeder, W. Szybalski, F. Sanger, F. R. Blattner in W. R. Hendrix, J. W. Roberts, F. W. Stahl, R. A. Weisberg (eds.): *Lambda II,* Cold Spring Harbor Laboratory, Cold Spring Harbor, New York, 1983, p. 469.

[6.60]  D. L. Daniels, J. L. Schroeder, W. Szybalski, F. Sanger, A. R. Coulson, G. F. Hong, D. F. Hill, G. B. Petersen, F. R. Blattner in W. R. Hendrix, J. W. Roberts, F. W. Stahl, R. A. Weisberg (eds.): *Lambda II,* Cold Spring Harbor Laboratory, Cold Spring Harbor, New York, 1983, p. 519.

[6.61]  C. M. De Wit, B. M. M. Dekker, A. C. Neele, A. De Waard, *FEBS Lett.* **180** (1985) 219.

[6.62]  D. F. Dombroski, A R. Morgan, *J. Biol. Chem.* **260** (1985) 415.

[6.63]  B. Dreiseikelmann, R. Eichenlaub, W. Wackernagel, *Biochim. Biophys. Acta* **562** (1979) 418.

[6.64]  A. Dugaiczyk, J. Hedgpeth, H. W. Boyer, H. M. Goodman, *Biochemistry* **13** (1974) 503.

[6.65]  C. H. Duncan, G. A. Wilson, F. E. Young, *J. Bacteriol.* **134** (1978) 338.

[6.66]  G. Duyk, J. Leis, M. Longiaru, A. M. Skalka, *Proc. Natl. Acad. Sci. USA* **80** (1983) 6745.

[6.67]  M. Ehrlich, R. Y.-H. Wang, *Science* **212** (1981) 1350.

[6.68]  B. Endlich, S. Linn in P. D. Boyer (ed.): *The Enzymes*, vol. 14, Academic Press, New York 1981, p. 137.

[6.69]  S. A. Endow, *J. Mol. Biol.* **114** (1977) 441.

[6.70]  S. A. Endow, R. J. Roberts, *J. Mol. Biol.* **112** (1977) 521.

[6.71]  G. F. Fitzgerald, C. Daly, L. R. Brown, T. R. Gingeras, *Nucleic Acids Res.* **10** (1982) 8171.

[6.72]  S. Forsblum, R. Rigler, M. Ehrenberg, U. Pettersson, L. Philipson, *Nucleic Acids Res.* **3** (1976) 3255.

[6.73]  R. Fuchs, R. Blakesley: "Guide to the Use of Type II Restriction endonucleases," in S. P. Colowick, N. O. Kaplan (eds.): *Methods in Enzymology,* vol. 100, Academic Press, New York 1983, p. 3.

[6.74]  L. Y. Fuchs, L. Covarrubias, L. Escalante, S. Sanchez, F. Bolivar, *Gene* **10** (1980) 39.

[6.75]  R. C. Gardner, A. J. Horwarth, J. Messing, R. J. Shepherd, *DNA* **1** (1982) 109.

[6.76]  D. E. Garfin, H. M. Goodman, *Biochem. Biophys. Res. Commun.* **59** (1974) 108.

[6.77]  R. E. Gelinas, P. A. Myers, R. J. Roberts, *J. Mol. Biol.* **114** (1977) 169.

[6.78]  R. E. Gelinas, P. A. Myers, G. H. Weiss, R. J. Roberts, K. Murray, *J. Mol. Biol,* **114** (1977) 433.

[6.79]  J. George, J. G. Chirikjian, *Proc. Natl. Acad. Sci. USA* **79** (1982) 2432.

[6.80]  T. R. Gingeras, J. E. Brooks, *Proc. Natl. Acad. Sci. USA* **80** (1983) 402.

[6.81]  T. R. Gingeras, L. Greenough, I. Schildkraut, R. J. Roberts, *Nucleic Acids Res.* **9** (1981) 4525.

[6.82]  T. R. Gingeras, J. P. Milazzo, R. J. Roberts, *Nucleic Acids Res.* **5** (1978) 4105.

[6.83] T. R. Gingeras, P. A. Myers, J. A. Olson, F. A. Hanberg, R. J. Roberts, *J. Mol. Biol.* **118** (1978) 113.

[6.84] M. Goppelt, J. Langowski, A. Pingoud, W. Haupt, C. Urbanke, H. Mayer, G. Maass, *Nucleic Acids Res.* **9** (1981) 6115.

[6.85] M. Goppelt, A. Pingoud, G. Maass, H. Mayer, H. Köster, R. Frank, *Eur. J. Biochem.* **104** (1980) 101.

[6.86] R. P. Grandoni, I. Schildkraut, cited in [6.23], [6.27].

[6.87] P. W. Gray, R. B. Hallick, *Biochemistry* **16** (1977) 1665.

[6.88] P. J. Greene, M. C. Betlach, H. W. Boyer, H. M. Goodman, *Methods Mol. Biol.* **7** (1974) 87.

[6.89] P. J. Greene, M. Gupta, H. W. Boyer, W. E. Brown, J. M. Rosenberg, *J. Biol. Chem.* **256** (1981) 2143.

[6.90] P. J. Greene, H. L. Heyneker, F. Bolivar, R. L. Rodriguez, M. C. Betlach, A. A. Covarrubias et al., *Nucleic Acids Res.* **5** (1978) 2373.

[6.91] R. Grosskopf, C. Kessler, unpublished results.

[6.92] R. Grosskopf, W. Wolf, C. Kessler, *Nucleic Acids Res.* **13** (1985) 1517.

[6.93] Y. Gruenbaum, H. Cedar, A. Razin, *Nucleic Acids Res.* **9** (1981) 2509.

[6.94] Y. Gruenbaum, R. Stein, H. Cedar, A. Razin, *FEBS Lett.* **124** (1981) 67.

[6.95] S. M. Hadi, B. Bächi, S. Iida, T. A. Bickle, *J. Mol. Biol.* **165** (1983) 19.

[6.96] T. M. Haqqi, S. Ahmad, N. S. Ahmad, M. Ahmad, A. Hasnain, M. Siddigi, S. M. Hadi, *Indian. J. Biochem. Biophys.* **22** (1985) 252.

[6.97] S. Hattman, in P. D. Boyer (eds.): "DNA Methylation," in *The Enzymes,* vol. 14, Academic Press, New York 1981, p. 517.

[6.98] S. Hattman, J. E. Brooks, M. Masurekar, *J. Mol. Biol.* **126** (1978) 367.

[6.99] S. Hattman, T. Keister, A. Gottehrer, *J. Mol. Biol.* **124** (1978) 701.

[6.100] J. Hedgpeth, H. M. Goodman, H. W. Boyer, *Proc. Natl. Acad. Sci. USA* **69** (1972) 3448.

[6.101] B. Hofer, G. Ruhe, A. Koch, H. Köster, *Nucleic Acids Res.* **10** (1982) 2763.

[6.102] M.-T. Hsu, P. Berg, *Biochemistry* **17** (1978) 131.

[6.103] A. W. Hu, A. H. Marschel, cited in [6.23], [6.27].

[6.104] L.-H. Huang, C. M. Farnet, K. C. Ehrlich, M. Ehrlich, *Nucleic Acids Res.* **10** (1982) 1579.

[6.105] S. G. Hughes, K. Murray, *Biochem. J.* **185** (1980) 65.

[6.106] S. Iida, J. Meyer, B. Bächi, M. Stählhammar-Carlemalm, S. Schrickel, T. A. Bickle, W. Arber, *J. Mol. Biol.* **165** (1983) 1.

[6.107] D. Jacobs, N. L. Brown, cited in [6.23], [6.27].

[6.108] A. Janulaitis, M. Petrusyte, V. Butkus, *FEBS Lett.* **161** (1983) 213.

[6.109] A. Janulaitis, P. Povilionis, K. Sasnauskas, *Gene* **20** (1982) 197.

[6.110] A. A. Janulaitis, P. S. Stakenas, J. B. Bitinaite, B. P. Jaskeleviciene, cited in [6.23], [6.27].

[6.111] S. Jentsch, U. Günthert, T. A. Trautner, *Nucleic Acids Res.* **9** (1981) 2753.

[6.112] D. A. Kaplan, D. P. Nierlich, *J. Biol. Chem.* **250** (1975) 2395.

[6.113] L. Kauc, A. Piekarowicz, *Eur. J. Biochem.* **92** (1978) 417.

[6.114] S. Kelly, D. R. Kaddurah, H. O. Smith, *J. Biol. Chem.* **260** (1985) 15 339.

[6.115] T. J. Kelly Jr., H. O. Smith, *J. Mol. Biol.* **51** (1970) 393.

[6.116] C. Kessler, unpublished results.

[6.117] C. Kessler, B. J. Bolton, M. J. Comer, *J. Cell. Biochem.* **10 D** (1986) 100, #058.

[6.118] C. Kessler, G. Nesch, R. Brack, *Gene* **16** (1981) 321.

[6.119] G. V. Kholmina, B. A. Rebentish, Yu. S. Skoblov, A. A. Mironov, N. K. Yankovskii, Yu. I. Kozlov, L. I. Glatman, A. F. Moroz, V. G. Debabov, *Dokl. Acad. Nauk SSSR* **253** (1980) 495.

[6.120] A. Kiss, G. Posfai, C. C. Keller, P. Venetianer, R. J. Roberts, *Nucleic Acids Res.* **13** (1985) 6403.

[6.121]  K. Kita, N. Hiraoka, F. Kimizuka, A. Obayashi, H. Kojima, H. Takahashi, H. Saito, *Nucleic Acids Res.* **13** (1985) 7015.

[6.122]  D. Kleid, Z. Humayun, A. Jeffrey, M. Ptashne, *Proc. Natl. Acad. Sci. USA* **73** (1976) 293.

[6.123]  C. Korch, P. Hagblom, S. Normark, *J. Bacteriol.* **155** (1983) 1324.

[6.124]  R. Kostriken, M. Zoller, F. Heffron in L. Leive (ed.): *Microbiology,* Cold Spring Harbor Laboratory, Cold Spring Harbor, New York 1985, p. 295.

[6.125]  V. G. Kosykh, Ya. I. Buryanov, A. A. Bayev, *Mol. Gen. Genet.* **178** (1980) 717.

[6.126]  A. S. Kraev, A. N. Kravets, B. K. Chernov, K. S. Skryabin, A. A. Baev, *Mol. Biol.* (*Moscow*) **19** (1985) 278.

[6.127]  V. M. Kramarov, V. V. Smolyaninov, *Biokhimiya* (*Moscow*) **46** (1981) 1526.

[6.128]  L. M. Kunkel, M. Silberklang, B. J. McCarthy,  *J. Mol. Biol.* **132** (1979) 133.

[6.129]  M. Kuosmanen, H. Pösö, *FEBS Lett.* **179** (1985) 17.

[6.130]  S. Lacks, B. Greenberg,  *J. Biol. Chem.* **250** (1975) 4060.

[6.131]  S. Lacks, B. Greenberg, *J. Mol. Biol.* **114** (1977) 153.

[6.132]  J. A. Langdale, P. A. Myers, R. J. Roberts, cited in [6.23], [6.27].

[6.133]  J. A. Lautenberger, M. H. Edgell, C. A. Hutchison III, *Gene* **12** (1980) 171.

[6.134]  S. H. Lee, H. M. Rho, *Korean J. Genet.* **7** (1985) 42.

[6.135]  D. W. Leung, A. C. P. Lui, H. Merilees, B. C. McBride, M. Smith, *Nucleic Acids Res.* **6** (1979) 17.

[6.136]  B.-C. Lin, M. C. Chien, S.-Y. Lou, *Nucleic Acids Res.* **8** (1980) 6189.

[6.137]  S. Linn, W. Arber, *Proc. Natl. Acad. Sci. USA* **59** (1968) 1300.

[6.138]  A. C. P. Lui, B. C. McBride, G. F. Vovis, M. Smith, *Nucleic Acids Res.* **6** (1979) 1.

[6.139]  P. A. Luke, S. E. Halford, *Gene* **37** (1985) 241.

[6.140]  S. E. Luria, M. L. Human, *J. Bacteriol.* **64** (1952) 557.

[6.141]  S. P. Lynn, L. K. Cohen, S. Kaplan, J. F. Gardner, *J. Bacteriol.* **142** (1980) 380.

[6.142]  R. A. Makula, R. B. Meagher, cited in [6.23], [6.27].

[6.143]  E. Malyguine, P. Vannier, P. Yot, *Gene* **8** (1980) 163.

[6.144]  M. B. Mann, H. O. Smith, cited in [6.23], [6.27]

[6.145]  M. B. Mann, H. O. Smith, *Nucleic Acids Res.* **4** (1977) 4211.

[6.146]  M. S. May, S. Hattman, *J. Bacteriol.* **122** (1975) 129.

[6.147]  H. Mayer, R. Grosschedl, H. Schütte, G. Hobom, *Nucleic Acids Res.* **9** (1981) 4833.

[6.148]  M. McClelland, *Nucleic Acids Res.* **9** (1981) 6795.

[6.149]  M. McClelland, L. G. Kessler, M. Bittner, *Proc. Natl. Acad. Sci. USA* **81** (1984) 983.

[6.150]  M. Meijer, E. Beck, F. G. Hansen, H. E. N. Bergmans et al., *Proc. Natl. Acad. Sci. USA* **76** (1979) 580.

[6.151]  M. Meselson, R. Yuan, *Nature*  (*London*) **217** (1968) 1110.

[6.152]  J. H. Middleton, M. H. Edgell, C. A. Hutchison III, cited in [6.23], [6.27].

[6.153]  J. H. Middleton, P. V. Stankus, M. H. Edgell, C. A. Hutchison III, cited in [6.23], [6.27].

[6.154]  P. S. Miller, D. M. Cheng, N. Dreon, K. Jayaraman, L. S. Kan, E. E. Leutzinger, S. M. Pulford, P. O. P. Ts-O,  *Biochemistry* **19** (1980) 4688.

[6.155]  K. Mise, K. Nakajima, *Gene* **33** (1985) 357.

[6.156]  P. Modrich, *Q. Rev. Biophys.* **12** (1979) 315.

[6.157]  P. Modrich, R. J. Roberts in S. M. Linn, R. J. Roberts (eds.): *Nucleases,* Cold Spring Harbor Laboratory, Cold Spring Harbor, New York 1982, p. 101.

[6.158]  P. L. Molloy, R. H. Symons, *Nucleic Acids Res.* **8** (1980) 2939.

[6.159]  R. Morgan, cited in [6.23], [6.27].

[6.160]  K. Murray, S. G. Hughes, J. S. Brown, S. A. Bruce, *Biochem. J.* **159** (1976) 317.

[6.161]  P. A. Myers, R. J. Roberts, cited in [6.23], [6.27].

[6.162]  G. Nardone, J. George, J. G. Chirikjian, J. Biol. Chem. **259** (1984) 10 357.

[6.163] K. Nath, B. A. Azzolina in J. G. Chirikjian (ed.): *Gene Amplification and Analysis,* vol. 1, Elsevier, Amsterdam 1981, p. 113.

[6.164] D. Nathans, H. O. Smith, *Ann. Rev. Biochem.* **44** (1975) 273.

[6.165] M. Nelson, C. Christ, I. Schildkraut, *Nucleic Acids Res.* **12** (1984) 5165.

[6.166] A. K. Newman, R. A. Rubin, S. H. Kim, P. Modrich, *J. Biol. Chem.* **256** (1981) 2131.

[6.167] M.-G. Nilsson, C. Skarped, G. Magnusson, *FEBS Lett.* **145** (1982) 360.

[6.168] K. Nishigaki, Y. Kaneko, H. Wakuda, Y. Husimi, T. Tanaka, *Nucleic Acids Res.* **13** (1985) 5747.

[6.169] R. Nussinov, *Nucleic Acids Res.* **8** (1980) 4545.

[6.170] C. D. O'Connor, E. Metcalf, C. J. Wrighton, T. J. R. Harris, J. A. Saunders, *Nucleic Acids Res.* **12** (1984) 6701.

[6.171] M. Österlund, H. Luthman, S. V. Nilsson, G. Magnusson, *Gene* **20** (1982) 121.

[6.172] R. Old, K. Murray, G. Roizes, *J. Mol. Biol.* **92** (1975) 331.

[6.173] J. A. Olson, P. A. Myers, R. J. Roberts, cited in [6.23], [6.27].

[6.174] N. Panayotatos, *Gene* **31** (1984) 291.

[6.175] K. W. C. Peden, *Gene* **22** (1983) 277.

[6.176] M. Petrosyte, A. Janulaitis, *Eur. J. Biochem.* **121** (1982) 377.

[6.177] A. Pingoud, *Eur. J. Biochem.* **147** (1985) 105.

[6.178] A. Pingoud, C. Urbanke, J. Alves, H.-J. Ehbrecht, M. Zabeau, C. Gualerzi, *Biochemistry* **23** (1984) 5697.

[6.179] V. Pirrotta, *Nucleic Acids Res.* **3** (1976) 1747.

[6.180] V. Pirrotta, T. A. Bickle in S. P. Colowick, N. O. Kaplan (eds.): *Methods in Enzymology,* vol. 65, Academic Press, New York 1980, p. 89.

[6.181] A. J. Podhaiska, W. Szybalski, *Gene* **40** (1985) 175.

[6.182] F. M. Pohl, R. Thomae, A. Karst, *Eur. J. Biochem.* **123** (1982) 141.

[6.183] A. Pope, S. P. Lynn, J. F. Gardner, cited in [6.23], [6.27].

[6.184] I. J. Purvis, B. E. B. Moseley, *Nucleic Acids Res.* **11** (1983) 5467.

[6.185] B.-Q. Qiang, I. Schildkraut, *Nucleic Acids Res.* **12** (1984) 4507.

[6.186] B.-Q. Qiang, I. Schildkraut, L. Visentin, cited in [6.23], [6.27].

[6.187] A. Razin, A. D. Riggs, *Science* **210** (1980) 604.

[6.188] J. Reaston, M. G. C. Duyvesteyn, A. de Waard, *Gene* **20** (1982) 103.

[6.189] B. Rexer, unpublished results.

[6.190] R. J. Roberts, *CRC Crit. Rev. Biochem.* **4** (1976) 123.

[6.191] R. J. Roberts, J. B. Breitmeyer, N. F. Tabachnik, P. A. Myers, *J. Mol. Biol.* **91** (1975) 121.

[6.192] R. J. Roberts, P. A. Myers, A. Morrison, K. Murray, *J. Mol. Biol.* **102** (1976) 157.

[6.193] R. J. Roberts, P. A. Myers, A. Morrison, K. Murray, *J. Mol. Biol.* **103** (1976) 199.

[6.194] R. J. Roberts, G. A. Wilson, F. E. Young, *Nature (London)* **265** (1977) 82.

[6.195] J. M. Rosenberg, P. Greene, *DNA* **1** (1982) 117.

[6.196] E. C. Rosenvold, cited in [6.23], [6.27].

[6.197] F. Sanger, A. R. Coulson, G. F. Hong, D. F. Hill, G. B. Petersen, *J. Mol. Biol.* **162** (1982) 729.

[6.198] S. Sato, C. A. Hutchison III, J. I. Harris, *Proc. Natl. Acad. Sci. USA* **74** (1977) 542.

[6.199] S. Sato, K. Nakazawa, T. Shinomiya, *J. Biochem. (Tokyo)* **88** (1980) 737.

[6.200] I. Schildkraut, C. D. B. Banner, C. S. Rhodes, S. Parekh, *Gene* **27** (1984) 327.

[6.201] I. Schildkraut, D. Comb, cited in [6.23], [6.27].

[6.202] I. Schildkraut, R. Wise, R. Borsetti, B.-Q. Quiang, cited in [6.23], [6.27].

[6.203] K. Schmid, M. Thomm, A. Laminet, F. G. Laue, C. Kessler, K. O. Stetter, R. Schmitt, *Nucleic Acids Res.* **12** (1984) 2619.

[6.204] B. Schoner, S. Kelly, H. O. Smith, *Gene* **24** (1983) 227.

[6.205] D. Sciaky, R. J. Roberts, cited in [6.23], [6.27].

[6.206] J. Seurinck, A. Van de Voorde, M. van Montagu, *Nucleic Acids Res.* **11** (1983) 4409.

[6.207] P. A. Sharp, B. Sudgen, J. Sambrook, *Biochemistry* **12** (1973) 3055.

[6.208] H. Shimotsu, H. Takahashi, H. Saito, *Gene* **11** (1980) 219.

[6.209] T. Shinomiya, M. Kobayashi, S. Sato, T. Uchida, *J. Biochem. (Tokyo)* **92** (1982) 1823.

[6.210] T. Shinomiya, S. Sato, *Nucleic Acids Res.* **8** (1980) 43.

[6.211] D. I Smith, F. R. Blattner, J. Davies, *Nucleic Acids Res.* **3** (1976) 343.

[6.212] H. O. Smith, *Science* **205** (1979) 455.

[6.213] H. O. Smith, S. V. Kelly: "The methylases of the type II restriction and modification systems," in H. Cedar, A. D. Riggs, A. Razin (eds.): *DNA Methylation,* Springer Verlag, New York 1984, p. 39.

[6.214] H. O. Smith, D. Nathans, *J. Mol. Biol.* **81** (1973) 419.

[6.215] H. O. Smith, K. W. Wilcox, *J. Mol. Biol.* **51** (1970) 379.

[6.216] K. Sugimoto, A. Oka, H. Sugisaki, M. Takanami, A. Nishimura, Y. Yasuda, Y. Hirota, *Proc. Natl. Acad. Sci. USA* **76** (1979) 575.

[6.217] H. Sugisaki, *Gene* **3** (1978) 17.

[6.218] H. Sugisaki, S. Kanazawa, *Gene* **16** (1981) 73.

[6.219] H. Sugisaki, Y. Maekawa, S. Kanazawa, M. Takanami, *Nucleic Acids Res.* **10** (1982) 5747.

[6.220] J. S. Sussenbach, C. H. Monfoort, R. Schiphof, E. E. Stobberingh, *Nucleic Acids Res.* **3** (1976) 3193.

[6.221] J. S. Sussenbach, P. H. Steenbergh, J. A. Rost, W. J. van Leuwen, J. D. A. van Embden, *Nucleic Acids Res.* **5** (1978) 1153.

[6.222] J. G. Sutcliffe, *Cold Spring Harbor Symp. Quant. Biol.* **43** (1979) 77.

[6.223] M. Szekeres, cited in [6.23], [6.27].

[6.224] W. Szybalski, *Gene* **40** (1985) 169.

[6.225] H. Takahashi, H. Kojima, H. Saito, *Biochem. J.* **231** (1985) 229.

[6.226] G. Theriault, P. H. Roy, K. A. Howard, J. S. Benner, J. E. Brooks, A. F. Waters, T. R. Gingeras, *Nucleic Acids Res.* **13** (1985) 8441.

[6.227] M. Thomas, R. W. Davis, *J. Mol. Biol.* **91** (1975) 315.

[6.228] T. I. Tikchonenko, E. V. Kramarov, B. A. Zavizion, B. S. Naroditsky, *Gene* **4** (1978) 195.

[6.229] J. Timko, A. H. Horwitz, J. Zelinka, G. Wilcox, *J. Bacteriol.* **145** (1981) 873.

[6.230] J. Tomassini, R. Roychoudhury, R. Wu, R. J. Roberts, *Nucleic Acids Res.* **5** (1978) 4055.

[6.231] C.-P. D. Tu, R. Roychoudhury, R. Wu, *Biochem. Biophys. Res. Commun.* **72** (1976) 355.

[6.232] S. Urieli-Shoval, Y. Gruenbaum, A. Razin, *J. Bacteriol.* **153** (1983) 274.

[6.233] H. van Heuverswyn, W. Fiers, *Gene* **9** (1980) 195.

[6.234] M. van Montagu, D. Sciaky, P. A. Myers, R. J. Roberts, cited in [6.23], [6.27].

[6.235] B. F. Vanyushin, A. N. Belozerskii, N. A. Kokurina, D. K. Kadirova, *Nature (London)* **218** (1968) 1066.

[6.236] L. P. Visentin, R. J. Watson, S. Martin, M. Zuker, cited in [6.23], [6.27].

[6.237] R. Wagner Jr., M. Meselson, *Proc. Natl. Acad. Sci. USA* **73** (1976) 4135.

[6.238] R. Y. Walder, J. L. Hartley, J. E. Donelson, J. A. Walder, *Proc. Natl. Acad. Sci. USA* **78** (1981) 1503.

[6.239] R. Y. Walder, J. A. Walder, J. E. Donelson, *J. Biol. Chem.* **259** (1984) 8015.

[6.240] R. J. Watson, I. Schildkraut, B. Q. Qiang, S. M. Martin, L. P. Visentin, *FEBS Lett.* **150** (1982) 114.

[6.241] R. Watson, M. Zuker, S. M. Martin L. P. Visentin, *FEBS Lett.* **118** (1980) 47.

[6.242] R. D. Wells, R. D. Klein, C. K. Singleton: "Type II Restriction Enzymes," in P. D. Boyer (ed.): *The Enzymes,* vol. 14, Academic Press, New York 1981, p. 158.

[6.243] P. R. Whitehead, N. L. Brown, *FEBS Lett.* **143** (1982) 296.

[6.244] P. R. Whitehead, N. L. Brown, *FEBS Lett.* **155** (1983) 97.

[6.245] M. H. Wigler, *Cell* **24** (1981) 285.

[6.246] G. A. Wilson, F. E. Young, *J. Mol. Biol.* **97** (1975) 123.

[6.247] J. L. Woodhead, N. Bhave, A. D. B. Malcolm, *Eur. J. Biochem.* **115** (1981) 293.

[6.248] Y. Yamada, M. Murakami, *Agric. Biol. Chem.* **49** (1985) 3017.

[6.249] C. Yanisch-Perron, J. Vieira, J. Messing, *Gene* **33** (1985) 103.

[6.250] O. J. Yoo, K. L. Agarwal, *J. Biol. Chem.* **255** (1980) 10 559.

[6.251] R. Yuan, *Annu. Rev. Biochem.* **50** (1981) 285.

[6.252] M. Zbeau, R. Greene, P. A. Myers, R. J. Roberts, cited in [6.23], [6.27].

[6.253] M. Zabeau, R. J. Roberts, cited in [6.23], [6.27]

[6.254] B. S. Zain, R. J. Roberts, *J. Mol. Biol.* **115** (1977) 249.

[6.255] P. D. Boyer: *The Enzymes,* "Protein Synthesis, DNA Synthesis and Repair, RNA Synthesis, Energy-Linked ATPases, Synthetases," vol. 10, Academic Press, New York 1974.

[6.256] P. D. Boyer: *The Enzymes,* "Nucleic Acids," vol. 14, Part A, Academic Press, New York 1981.

[6.257] P. D. Boyer: *The Enzymes,* "Nucleic Acids," vol. 15, Part B, Academic Press, New York 1981.

[6.258] J. G. Chirikjian, T. S. Papas: *Gene Amplification and Analysis,* vol. 2, Elsevier, Amsterdam 1981.

[6.259] S. T. Jacob in L. S. Hnilica (ed.): "Enzymes of Nucleic Acid Synthesis and Modification, DNA Enzymes," *CRC Series in the Biochemistry and Molecular Biology of the Cell Nucleus,* vol. I, CRC Press, Inc., Boca Raton, Florida, 1983.

[6.260] S. T. Jacob "Enzymes of Nucleic Acid Synthesis and Modification, RNA Enzymes," in L. S. Hnilica (ed.): *CRC Series in the Biochemistry and Molecular Biology of the Cell Nucleus,* vol. II, CRC Press, Inc., Boca Raton, Florida, 1983.

[6.261] A. Kornberg: *DNA Replication,* Freeman and Co., San Fransisco 1980.

[6.262] T. Maniatis, E. F. Fritsch, J. Sambrook: *Molecular Cloning. A Laboratory Manual,* Cold Spring Harbor Laboratory, Cold Spring Harbor, New York 1982.

[6.263] J. D. Watson: *Molecular Biology of the Gene,* 3rd ed., Benjamin, Inc., Menlo Park, CA, 1977.

[6.264] R. B. Wicker: "DNA Replication," in A. I. Laskin, J. A. Last (eds.): *Methods in Molecular Biology,* vol. 7, Marcel Dekker, New York 1974.

[6.265] S. Adler, P. Modrich, *J. Biol. Chem.* **254** (1979) 11 605.

[6.266] B. M. Alberts, L. Frey, *Nature (London)* **227** (1970) 1313.

[6.267] B. M. Alberts, L. Frey, H. Delius, *J. Mol. Biol.* **68** (1972) 139.

[6.268] J. M. D'Alessio, P. J. Perna, M. R. Paule, *J. Biol. Chem.* **254** (1979) 11 282.

[6.269] J. M. D'Alessio, S. R. Spindler, M. R. Paule, *J. Biol. Chem.* **253** (1978) 4669.

[6.270] T. Ando, *Biochim. Biophys. Acta* **114** (1966) 158.

[6.271] G. E. Austin, D. Sirakoff, B. Roop, G. H. Moyer, *Biochim. Biophys, Acta* **522** (1978) 412.

[6.272] J. Avila, J. M. Hermoso, E. Vinuela, M. Salas, *Eur. J. Biochem.* **21** (1971) 526.

[6.273] L. A. Ball, *Virology* **94** (1979) 282.

[6.274] A. L. Ball: "2′,5′-Oligoadenylate Synthetase," in P. D. Boyer (ed.): *The Enzymes,* vol. 15, Academic Press, New York 1982, p. 281.

[6.275] G. R. Banks, J. A. Boezi, I. R. Lehman, *J. Biol. Chem.* **254** (1979) 9886.

[6.276] E. K. Barbehenn, J. E. Craine, A. Chrambach, C. B. Klee, *J. Biol. Chem.* **257** (1982) 1007.

[6.277] H. Beier, R. Hausmann, *Nature (London)* **251** (1974) 538.

[6.278] I. K. Bendis, L. Shapiro, *J. Bacteriol.* **115** (1973) 848.

[6.279] A. J. Berk, P. A. Sharp, *Cell* **12** (1977) 721.

[6.280] A. J. Berk, P. A. Sharp, *Cell* **14** (1978) 695.

[6.281] A. J. Berk, P. A. Sharp, *Proc. Natl. Acad. Sci. USA* **75** (1978) 1274.

[6.282] K. L. Berkner, W. R. Folk: "Polynucleotide Kinase Exchange as an Assay for Class II Restriction Endonucleases," in L. Grosman, K. Moldave (eds.): *Methods in Enzymology,* vol. 65, Academic Press, New York 1980, p. 28.

[6.283] I. Berkower, J. Leis, J. Hurwitz, *J. Biol. Chem.* **248** (1973) 5914.

[6.284]  A. Bernardi, G. Bernardi: "Spleen Acid Exonuclease," in P. D. Boyer (ed.): *The Enzymes,* vol. 4, Academic Press, New York 1971, p. 329.

[6.285]  U. Bertazzoni, S. D. Ehrlich, G. Bernardi, *Biochim. Biophys. Acta* **312** (1973) 192.

[6.286]  M. Bittner, R. L. Burke, B: M. Alberts, *J. Biol. Chem.* **254** (1979) 9565.

[6.287]  P. Blackburn, S. Moore: "Pancreatic Ribonuclease," in P. O. Boyer (ed.): *The Enzymes,* vol. 15, Academic Press, New York 1982, p. 317.

[6.288]  M. S. Boguski, P. A. Hieter, C. C. Levy, *J. Biol. Chem.* **255** (1980) 2160.

[6.289]  A. Bolden, J. Aucker, A. Weissbach, *J. Virol.* **16** (1975) 1584.

[6.290]  F. Bolivar, K. Backman: "Plasmids of Escherichia coli as Cloning Vectors," in L. Grosman, K. Moldave (eds.): *Methods in Enzymology,* vol. 68, Academic Press, New York 1980, p. 245.

[6.291]  F. J. Bollum: "Terminal Deoxynucleotidyl Transferase," in P. D. Boyer (ed.): *The Enzymes,* vol. 10, Academic Press, New York 1974, p. 145.

[6.292]  C. Brack, M. Hirama, R. Lenhard-Schuller, S. Tonegawa, *Cell* **15** (1978) 1.

[6.293]  R. Breathnach, J. L. Mandel, P. Chambon, *Nature (London)* **270** (1977) 314.

[6.294]  D. M. Brown, A. R. Todd in E. Chargaff, J. N. Davidson (eds.): *The Nucleic Acids,* vol. 1, Academic Press, New York, 1955, p. 409.

[6.295]  E. L. Brown, R. Belagaje, M. J. Ryan, H. G. Khorana in L. Grosman, K. Moldave (eds.): *Methods in Enzymology,* vol. 59, Academic Press, New York 1979, p. 109.

[6.296]  D. Brutlag, M. R. Atkinson, P. Setlow, A. Kornberg, *Biochem. Biophys. Res. Commun.* **37** (1969) 982.

[6.297]  D. Brutlag, A. Kornberg, *J. Biol. Chem.* **247** (1972) 241.

[6.298]  Z. Burton, R. R. Burgess, J. Lin, D. Moore, S. Holder, C. A. Gross, *Nucleic Acids Res.* **9** (1981) 2889.

[6.299]  E. T. Butler, M. J. Chamberlin, *J. Biol. Chem.* **257** (1982) 5772.

[6.300]  J. J. Byrnes, V. L. Black, *Biochemistry* **17** (1978) 4226.

[6.301]  J. J. Byrnes, K. M Downey, V. L. Black, A. G. So, *Biochemistry* **15** (1976) 2817.

[6.302]  V. Cameron, D. Soltis, O. C. Uhlenbeck, *Nucleic Acids Res.* **5** (1978) 825.

[6.303]  V. Cameron, D. Soltis, O. C. Uhlenbeck, *Nucleic Acids Res.* **5** (1978) 825.

[6.304]  V. Cameron, O. C. Uhlenbeck, *Biochemistry* **16** (1977) 5120.

[6.305]  V. W. Campbell, D. A. Jackson, *J. Biol. Chem.* **255** (1980) 3726.

[6.306]  D. S. Carre, F. Chapeville, *Biochim. Biophys. Acta* **361** (1974) 176.

[6.307]  H. Cedar, A. Solage, G. Glaser, A. Razin, *Nucleic Acids Res.* **6** (1979) 2125.

[6.308]  G. Chaconas, J. H. van de Sande in L. Grosman, K. Moldave (eds.): *Methods in Enzymology,* vol. 65, Academic Press, New York 1980, p. 680.

[6.309]  P. R. Chakraborty, P. Sarkar, H. H. Huang, U. Maitra, *J. Biol. Chem.* **248** (1973) 6637

[6.310]  M. D. Challberg, P. T. Englund, *J. Biol. Chem.* **254** (1979) 7812.

[6.311]  M. D. Challberg, P. T. Englund, *J. Biol. Chem.* **254** (1979) 7820.

[6.312]  M. Chamberlin in R. Losick, M. Chamberlin (eds.): *RNA Polymerase,* Cold Spring Harbor Laboratory, Cold Spring Harbor, New York 1976, p. 17.

[6.313]  M. Chamberlin in P. D. Boyer (ed.): *The Enzymes,* vol. 15, Academic Press, New York 1982, p. 61.

[6.314]  M. Chamberlin, J. McGrath, L. Waskell, *Nature (London)* **228** (1970) 227.

[6.315]  J. J. Champoux, E. Gilboa, D. Baltimore, *J. Virol.* **49** (1984) 686.

[6.316]  L. M. S. Chang, *J. Biol. Chem.* **248** (1973) 3789.

[6.317]  L. M. S. Chang, *J. Biol. Chem.* **248** (1973) 6983.

[6.318]  L. M. S. Chang, F. J. Bollum, *J. Biol. Chem.* **246** (1971) 909.

[6.319]  L. M. S. Chang, F. J. Bollum, *J. Biol. Chem.* **256** (1981) 494.

[6.320]  S. H. Chang, C. K. Brum, M. Silberklang, U. L. Rajbhandary, L. I. Hecker, W. E. Barnett, *Cell* **9** (1976) 717.

[6.321] J. W. Chase, C. C. Richardson, *J. Biol. Chem.* **249** (1974) 4545.

[6.322] J. W. Chase, L. D. Vales in J. G. Chirikjian, T. S. Papas (eds.): *Gene Amplification and Analysis,* vol. 2, Elsevier, Amsterdam 1981, p. 147.

[6.323] J. W. Chase, R. F. Whittier, J. Auerbach, A. Sancar, W. D. Rupp, *Nucleic Acids Res.* **8** (1980) 3215.

[6.324] Y.-C. Chen, E. W. Bohn, S. R. Blanck, S. H. Wilson, *J. Biol. Chem.* **254** (1979) 11 678.

[6.325] A. V. Chestukhin, M. F. Shemyakin, N. A. Kalinina, A. A. Prozorov, *FEBS Lett.* **24** (1972) 121.

[6.326] G.–M. Church, W. Gilbert, *Proc. Natl. Acad. Sci. USA* **81** (1984) 1991.

[6.327] S. Clark, *J. Virol.* **25** (1978) 224.

[6.328] J. E. Clements, S. G. Rogers, B. Weiss, *J. Biol. Chem.* **253** (1978) 2990.

[6.329] N. R. Cozzarelli, R. B. Kelly, A. Kornberg, *J. Mol. Biol.* **45** (1969) 513.

[6.330] C. S. Craik, C. Largman, T. Fletscher, S. Roczniak, P. J. Barr, R. Fletterick, W. J. Rutter, *Science* **228** (1985) 291.

[6.331] J. W. Cranston, R. Silber, V. G. Malathi, J. Hurwitz, *J. Biol. Chem.* **249** (1974) 7447.

[6.332] R. J. Crouch, *J. Biol. Chem.* **249** (1974) 1314.

[6.333] R. J. Crouch in J. G. Chirikjian, T. S. Papas (eds.): *Gene Amplification and Analysis,* vol. 2, Elsevier, Amsterdam 1981, p. 217.

[6.334] H. Cudny, M. P. Deutscher, *Proc. Natl. Acad. Sci. USA* **77** (1980) 837.

[6.335] J. L. Darlix, *Eur. J. Biochem.* **51** (1975) 369.

[6.336] S. K. Das, R. K. Fujimura, *J. Biol. Chem.* **252** (1977) 8700.

[6.337] S. K. Das, R. K. Fujimura, *J. Biol. Chem.* **254** (1979) 1227.

[6.338] P. Davanloo, A. H. Rosenberg, J. J. Dunn, F. W. Studier, *Proc. Natl. Acad. Sci. USA* **81** (1984) 2035.

[6.339] K. C. Deen, T. A. Landers, M. Berninger, *Anal. Biochem.* **135** (1983) 456.

[6.340] M. R. Deibel Jr., M. S. Coleman, *J. Biol. Chem.* **254** (1979) 8634.

[6.341] B. Demple, S. Linn, *Nature (London)* **287** (1980) 203.

[6.342] G.–R. Deng, R. Wu in S. P. Colowick, N. O. Kaplan (eds.): *Methods in Enzymology,* vol. 100, Academic Press, New York, 1983. p. 96.

[6.343] R. E. Depew, N. R. Cozzarelli, *J. Virol.* **13** (1974) 888.

[6.344] R. E. Depew, L. F. Liu, J. C. Wang, *J. Biol. Chem.* **253** (1978) 511.

[6.345] R. E. Depew, T. J. Snopek, N. R. Cozzarelli, *Virology* **64** (1975) 144.

[6.346] M. P. Deutscher, A. Kornberg, *J. Biol. Chem.* **244** (1969) 3019.

[6.347] H. Donis-Keller, *Nucleic Acids Res.* **7** (1979) 179.

[6.348] H. Donis-Keller, A. M. Maxam, W. Gilbert, *Nucleic Acids Res.* **4** (1977) 2527.

[6.349] J. P. Dougherty, H. Samantha, P. J. Farrelland, P. Lengyel, *J. Biol. Chem.* **255** (1980) 3813.

[6.350] M. Duguet, G. Yarranton, M. Gefter, *Cold Spring Harbor Symp. Quant. Biol.* **43** (1979) 335.

[6.351] J. J. Dunn, *J. Biol. Chem.* **251** (1976) 3807.

[6.352] J. Dunn in P. P. Boyer (ed.): *The Enzymes,* vol. 15, Academic Press, New York 1982, p. 485.

[6.353] J. J. Dunn, F. W. Studier, *J. Mol. Biol.* **148** (1981) 303.

[6.354] A. F. C. Efstratiadis, L. Villa-Komaroff in J. K. Setlow, A. Hollaender (eds.): *Genetic Engineering,* vol. 1, Plenum Press, New York 1979, p. 1.

[6.355] A. Efstratiadis, J. N. Vournakis, H. Donis-Keller, G. Chaconas, D. K. Dougall, F. C. Kafatos, *Nucleic Acids Res.* **4** (1977) 4165.

[6.356] F. Egami, K. Takahashi, T. Uchida in J. N. Davidson, W. E. Cohen (eds.): *Progress in Nucleic Acid Research and Molecular Biology,* vol. 3, Academic Press, New York 1964, p. 59.

[6.357] F. Egami, T. Oshima, T. Uchida in F. Chapeville, A. L. Haenni (eds.): *Molecular Biology, Biochemistry and Biophysics,* vol. 32, Springer Verlag, Berlin 1980, p. 250.

[6.358] M. Ehrlich, R. Y. -H. Wang, *Science* **212** (1981) 1350.

[6.359] R. N. Eisenman, W. S. Mason, M. Linial, *J. Virol* **86** (1980) 62.
[6.360] P. T. Englund, M. P. Deutscher, T. Jovin, R. B. Kelly, N. R. Cozzarelli, A. Kornberg, *Cold Spring Harbor Symp. Quant. Biol.* **33** (1968) 1.
[6.361] T. E. Englund, O. C. Uhlenbeck, *Nature (London)* **275** (1978) 560.
[6.362] B. Eriksson, A. Larsson, E. Helgstrand, N. G. Johansson, B. Öberg, *Biochim. Biophys. Acta* **607** (1980) 53.
[6.363] G. C. Fareed, E. M. Wilt, C. C. Richardson, *J. Biol. Chem.* **246** (1971) 925.
[6.364] J. Favaloro, R. Treisman, R. Kamen in L. Grosman, K. Moldave (eds.): *Methods in Enzymology,* vol. 65, Academic Press, New York 1980, p. 718.
[6.365] A. P. Feinberg, B. Vogelstein, *Anal. Biochem.* **132,** (1983) 6.
[6.366] O. Fichot, M. Pascal, M. Mechali, A. M. de Recondo, *Biochim. Biophys. Acta* **561** (1979) 29.
[6.367] P. A. Fisher, D. Korn, *J. Biol. Chem.* **252** (1977) 6528.
[6.368] W. R. Folk in J.-G. Chirikjian, T. S. Papas (eds.): *Gene Amplification and Analysis,* vol. 2, Elsevier, Amsterdam 1981, p. 299.
[6.369] M. Fosset, D. Chappelet-Tordo, M. Lazdunski, *Biochemistry* **13** (1974) 1783.
[6.370] M. J. Fraser in L. Grosman, K. Moldave (eds.): *Methods in Enzymology,* vol. 65, Academic Press, New York 1980, p. 255.
[6.371] M. J. Fraser, R. Tjeerde, K. Matsumoto, *Can. J. Biochem.* **54** (1976) 971.
[6.372] G. D. Frenkel, C. C. Richardson, *J. Biol. Chem.* **246** (1971) 4839.
[6.373] E. C. Friedberg, C. T. M. Anderson, T. Bonura, R. Cone, E. H. Radany, R. J. Reynolds, *Prog. Nucleic Acid Res. Mol. Biol.* **26** (1981) 197.
[6.374] E. C. Friedberg, T. Bonura, E. H. Radany, J. D. Love in P. D. Boyer (ed.): *The Enzymes,* vol. 14, Academic Press, New York 1982, p. 251.
[6.375] M. Fujimoto, A. Kuninaka, H. Yoshino, *Agric. Biol. Chem.* **38** (1974) 777.
[6.376] M. Fujimoto, A. Kuninaka, H. Yoshino, *Agric. Biol. Chem.* **38** (1974) 785.
[6.377] R. K. Fujimara, B. C. Roop, *J. Biol. Chem.* **251** (1976) 2168.
[6.378] K. Fujinaga, J. T. Parsons, J. W. Beard, D. Beard, M. Green, *Proc. Natl.. Acad. Sci. USA* **67** (1970) 1432.
[6.379] T. Fukui, A. Ishihama, E. Ohtsuka, M. Ikehara, R. Fukuda, *J. Biochem. (Tokyo)* **91** (1982) 331.
[6.380] D. J. Galas, A. Schmitz, *Nucleic Acids Res.* **5** (1978) 3157.
[6.381] Y. M. Gafurov, L. L. Terentev, V. A. Rasskazov, *Biochemistry (Engl. Transl.)* **44** (1979) 996.
[6.382] F. Gannon, J. M. Jeltsch, F. Perrin, *Nucleic Acids Res.* **8** (1980) 4405.
[6.383] F. T. Gates III, S. Linn, *J. Biol. Chem.* **252** (1977) 1647.
[6.384] T. F. Gates III, S. Linn, *J. Biol. Chem.* **252** (1977) 2802.
[6.385] M. L. Gefter, I. J. Molineux, T. Kornberg, H. G. Khorana, *J. Biol. Chem.* **247** (1972) 3321.
[6.386] M. Gellert in P. D. Boyer (ed.): *The Enzymes,* vol. 14, Academic Press, New York 1982, p. 345.
[6.387] G. F. Gerard: "Reverse Transcriptase," in S. T. Jacob (ed.): *Enzymes of Nucleic Acid Synthesis and Modification,* vol. 1, CRC Press, West Palm Beach, Florida, 1983, p. 1.
[6.388] G. F. Gerard, D. P. Grandgenett, *J. Virol.* **15** (1975) 785.
[6.389] G. F. Gerard, D. P. Grandgenett in J. R. Stephenson (ed.): *Molecular Biology of RNA Tumor Viruses,* Academic Press, New York 1980, p. 345.
[6.390] R. K. Ghosh, M. P. Deutscher in D. Söll, J. N. Abelson, P. R. Schimmel (eds.): *Transfer RNA: Biological Aspects,* Cold Spring Habor Laboratory, Cold Spring Harbor, New York, 1980, p. 59
[6.391] P. K. Ghosh, V. B. Reddy, M. Piatak, P. Lebowitz, S. M. Weissman in S. P. Colowick, N. O. Kaplan (eds.): *Methods in Enzymology,* vol. 65, Academic Press, New York 1980, p. 580.

[6.392] J. R. Gillen, A. E. Karu, H. Nagaishi, A. J. Clark, *J. Mol. Biol.* **133** (1977) 27.
[6.393] P. J. Goldmark, S. Linn, *J. Biol. Chem.* **247** (1972) 1849.
[6.394] H. M. Goodman, R. J. McDonald in S. P. Colowick, N. O. Kaplan (eds.): *Methods in Enzymology,* vol. 68, Academic Press, New York 1980, p. 75.
[6.395] S. Gottesman, M. Gottesman, *Proc. Natl. Acad. Sci. USA* **72** (1975) 2188.
[6.396] M. Goulian, Z. J. Lucas, A. Kornberg, *J. Biol. Chem.* **243** (1968) 627.
[6.397] D. P. Grandgenett, A. C. Vora, R. D. Schiff, *Virolog.* **89** (1978) 119.
[6.398] H. B. Gray Jr., D. A. Ostrander, J. L. Hodnett, R. J. Legerski, D. L. Robberson, *Nucleic Acids Res.* **2** (1975) 1459.
[6.399] H. B. Gray Jr., W. B. Upholt, J. Vinograd, *J. Mol. Biol.* **62** (1971) 1.
[6.400] H. B. Gray Jr., T. P. Winston, J. L. Hodnett, R. J. Legerski, D. W. Nees, C.-F. Wei, D. L. Robberson in J. G. Chirikjian, T. S. Papas (eds.): *Gene Amplification and Analysis,* vol. 2, Elsevier, Amsterdam 1981, p. 169.
[6.401] M. R. Green, T. Maniatis, D. A. Melton, *Cell* **32** (1983) 681.
[6.402] D. S. Gregerson, J. Albert, T. W. Reid, *Biochemistry* **19** (1980) 301.
[6.403] J. Griffith, J. A. Huberman, A. Kornberg, *J. Mol. Biol.* **55** (1971) 209.
[6.404] P. Grippo, C. C. Richardson, *J. Biol. Chem.* **246** (1971) 6867.
[6.405] H. J. Gross, H. Domdey, C. Lossow, P. Jank, M. Raba, H. Alberty, H. L. Sänger, *Nature (London)* **273** (1978) 203.
[6.406] H. J. Gross, G. Krupp, H. Domdey, M. Raba, P. Jank, C. Lossow, H. Alberty, K. Ramm, H. L. Sänger, *Eur. J. Biochem.* **121** (1981) 249.
[6.407] L. I. Grossman, R. Watson, J. Vinograd, *Proc. Natl. Acad. Sci. USA* **70** (1973) 3339.
[6.408] R. Grosschedl, M. L. Birnstiel, *Proc. Natl. Acad. Sci. USA* **77** (1980) 1432.
[6.409] Y. Gruenbaum, T. Naveh-Many, H. Cedar, A. Razin, *Nature (London)* **292** (1981) 860.
[6.410] Y. Gruenbaum, R. Stein, H. Cedar, A. Razin, *FEBS Lett.* **124** (1981) 67.
[6.411] F. Grummt, G. Walfl, H. M. Jantzen, K. Hamprecht, U. Hübscher, C. C. Kuenzle, *Proc. Natl. Acad. Sci. USA* **76** (1979) 6081.
[6.412] U. Gubler, B. J. Hoffman, *Gene* **25** (1983) 263.
[6.413] T. J. Guiltoyle in S. T. Jacob (ed.): *Enzymes of Nucleic Acid Synthesis and Modification,* vol. 2, CRC Press, Boca Raton 1983, p. 1.
[6.414] L.-H. Guo, R. Wu, *Nucleic Acids Res.* **10** (1982) 2065.
[6.415] L.-H. Guo, R. Wu in L. Grosman, K. Moldave (eds.): *Methods in Enzymology,* vol. 100, Academic Press, New York 1983, p. 60.
[6.416] L.-H. Guo, R. C. A. Yang, R. Wu, *Nucleic Acids Res.* **11** (1983) 5521.
[6.417] R. S. Gupta, T. Kasai, D. Schlessinger, *J. Biol. Chem.* **252** (1977) 8945.
[6.418] R. A. Hallewell, S. Emtage, *Gene* **9** (1980) 27.
[6.419] R. M. Harland, *Proc. Natl. Acad. Sci. USA* **70** (1982) 2323.
[6.420] G. R. Hartmann, *Angew. Chem.* **88** (1976) 197.
[6.421] G. Hartmann, K. O. Honikel, F. Knüsel, J. Nüesch, *Biochim. Biophys. Acta* **145** (1967) 843.
[6.422] J. N. Heins, H. Taniuchi, C. B. Anfisen in J. Cantoni, D. Davies (eds.): *Procedures in Nucleic Acid Research,* vol. 1, Harper and Row, New York 1966, p. 79
[6.423] G. Herrick, B. M. Alberts, *J. Biol. Chem.* **251** (1976) 2124.
[6.424] I. P. Hesslewood, A. M. Holmes, W. F. Wakeling, I. R. Johnston, *Eur. J. Biochem.* **84** (1978) 123.
[6.425] N. P. Higgins, N. R. Cozzarelli in L. Grosman, K. Moldave (eds.): *Methods in Enzymology,* vol. 68, Academic Press, New York 1980, p. 50.
[6.426] G. Hillenbrand, G. Morelli, E. Lanka, E. Scherzinger, *Cold Spring Harbor Symp. Quant. Biol.* **43** (1978) 449.
[6.427] G. Hillenbrand, W. L. Staudenbauer, *Nucleic Acids Res.* **10** (1982) 833.
[6.428] D. C. Hinkle, C. C. Richardson, *J. Biol. Chem.* **250** (1975) 5523.

[6.429] C. H. W. Hirs, S. Moore, W. H. Stein, *J. Biol. Chem.* **219** (1956) 623.

[6.430] S. E. Hitchcock, *J. Biol. Chem.* **255** (1980) 5668.

[6.431] W. K. Holloman, *J. Biol. Chem.* **248** (1973) 8114.

[6.432] W. K. Holloman, R. Holliday, *J. Biol. Chem.* **248** (1973) 8107.

[6.433] A. M. Holmes, I. P. Hesslewood, I. R. Johnston, *Eur. J. Biochem.* **62** (1976) 229.

[6.434] K. Hori, D. F. Mark, C. C. Richardson, *J. Biol. Chem.* **254** (1979) 11 591.

[6.435] K. Horikoshi, *Biochim. Biophys. Acta* **240** (1971) 532.

[6.436] W. M. Huang, I. R. Lehman, *J. Biol. Chem.* **247** (1972) 3139.

[6.437] T. Inove, F. Kada, *J. Biol. Chem.* **253** (1978) 8559.

[6.438] D. Ish-Horowicz, J. F. Burke, *Nucleic Acids Res.* **9** (1981) 2989.

[6.439] H. Jacobsen, H. Klenow, K. Overgaard-Hansen, *Eur. J. Biochem.* **45** (1974) 623.

[6.440] J. J. Jendrisak in C. Leaver (ed.): *Genome Organization and Expression in Plants,* Plenum Press, New York 1980, p. 93.

[6.441] J. J. Jendrisak, *Plant Physiol.* **67** (1981) 438.

[6.442] S. Jentsch, U. Günthert, T. A. Trautner, *Nucleic Acids Res.* **9** (1981) 2753.

[6.443] L. Johnsrud, *Proc. Natl. Acad. Sci. USA* **75** (1978) 5314.

[6.444] P. A. Jones, S. A. Taylor, *Nucleic Acids. Res.* **9** (1981) 2933.

[6.445] S. E. Jorgensen, J. F. Koerner, *J. Biol. Chem.* **241** (1966) 3090.

[6.446] T. M. Jovin, P. T. Englund, *J. Biol. Chem.* **244** (1969) 2996.

[6.447] T. M. Jovin, P. T. Englund, L. Bertsch, *J. Biol. Chem.* **244** (1969) 6996.

[6.448] T. M. Jovin, P. T. Englund, A. Kornberg, *J. Biol. Chem.* **244** (1969) 3009.

[6.449] E. Junowicz, J. H. Spencer, *Biochim. Biophys. Acta* **312** (1973) 85.

[6.450] D. L. Kacian, K. F. Watson, A. Burny, S. Spiegelmann, *Biochim. Biophys. Acta* **246** (1971) 265.

[6.451] D. Kamp, R. Kahmann, D. Zipser, R. J. Roberts, *Mol. Gen. Genet.* **154** (1977) 231.

[6.452] G. A. Kassavetis, E. T. Butler, D. Roulland, M. J. Chamberlin, *J. Biol. Chem.* **257** (1982) 5779.

[6.453] A. C. Kato, K. Bartok, M. J. Fraser, D. T. Denhart, *Biochim. Biophys. Acta* **308** (1973) 68.

[6.454] K. Kato, J. M. Gonclaves, G. E. Houts, F. J. Bollum, *J. Biol. Chem.* **242** (1967) 2780.

[6.455] W. Keller, R. Crouch, *Proc. Natl. Acad. Sci. USA* **69** (1972) 3360.

[6.456] W. S. Kelley, K. H. Stump, *J. Biol. Chem.* **254** (1979) 3206.

[6.457] H. G. Khorana, K. L. Agarwal, H. Büchi, M. H. Caruthers, N. K. Gupta, K. Kleppe et al., *J. Mol. Biol.* **72** (1972) 209.

[6.458] A. A. Kiessling, M. Goulian, *Biochem. Biophys. Res. Commun.* **71** (1976) 1069.

[6.459] G. H. Kidd, L. Bogorad, *Proc. Natl. Acad. Sci. USA* **76** (1979) 4890.

[6.460] N. Kitamura, B. L. Semler, P. G. Rothberg, G. R. Larsen, C. J. Alder, A. J. Dorner et al., *Nature (London)* **291** (1981) 547.

[6.461] H. Klenow, I. Henningsen, *Proc. Natl. Acad. Sci. USA* **65** (1970) 168.

[6.462] R. P. Klett, A. Cerami, E. Reich, *Proc. Natl. Acad. Sci. USA* **60** (1968) 943.

[6.463] K. Knopf, *Eur. J. Biochem.* **98** (1979) 231.

[6.464] K.-W. Knopf, M. Yamada, A. Weissbach, *Biochemistry* **15** (1976) 4540.

[6.465] R. Kole, S. Altman, *Biochemistry* **20** (1981) 1902.

[6.466] R. Kole, S. Altman in P. D. Boyer (ed.): *The Enzymes,* vol. 15, Academic Press, New York 1982, p. 469.

[6.467] D. Korn, P. A. Fisher, J. Battey, T. S. F. Wang, *Cold Spring Harbor Symp. Quant. Biol.* **43** (1979) 613.

[6.468] A. Kornberg, I. R. Lehman, M. Bessman, E. S. Simms, *Biochim. Biophys. Acta* **21** (1956) 197.

[6.469] T. Kornberg, M. Gefter, *Biochem. Biophys. Res. Commun.* **40** (1970) 1348.

[6.470] T. Kornberg, M. L. Gefter, *Proc. Natl. Acad. Sci. USA* **68** (1971) 761.

[6.471] S. C. Kowalczykowski, D. G. Bear, P. H. von Hippel in P. D. Boyer (ed.): *The Enzymes,* vol. 14, Academic Press, New York 1982, p. 373.

[6.472] T. Kozu, T. Kurihara, T. Seno, *J. Biochem. (Tokyo)* **89** (1981) 551.

[6.473] A. R. Krainer, T. Maniatis, B. Ruskin, M. R. Green, *Cell* **36** (1984) 993.

[6.474] J. S. Krakow, E. Fronk, *J. Biol. Chem.* **244** (1969) 5988.

[6.475] J. Krakow, M. Karstadt, *Proc. Natl. Acad. Sci. USA* **58** (1967) 2094.

[6.476] J. S. Krakow, K. von der Helm, *Cold Spring Harbor Symp. Quant. Biol.* **35** (1971) 73.

[6.477] J. Krakow, G. Rhodes, T. M. Jovin in R. Losick, M. Chamberlin (eds.): *RNA Polymerase,* Cold Spring Harbor Laboratory, Cold Spring Harbor, New York, 1976, p. 127.

[6.478] W. D. Kroecker, J. L. Fairley, *J. Biol. Chem.* **250** (1975) 3773.

[6.479] G. Krupp, H. J. Gross, *Nucleic Acids Res.* **6** (1979) 3481.

[6.480] G. Krupp, H. J. Gross in P. E. Agris, R. A. Kopper (eds.): *The Modified Nucleosides of Transfer RNA,* vol. 2, Lisa, New York 1983, p. 11.

[6.481] J. P. Kuebler, D. A. Goldthwait, *Biochemistry* **16** (1977) 1370.

[6.482] V. T. Kung, J. C. Wang, *J. Biol. Chem.* **252** (1977) 5398.

[6.483] M. Kunitz, *J. Gen. Physiol.* **24** (1940) 15.

[6.484] M. Kunitz, *J. Gen. Physiol.* **33** (1950) 363.

[6.485] M. Kunitz, M. R. McDonald, *Biochem. Prep.* **3** (1953) 9.

[6.486] S. R. Kushner, H. Nagaishi, A. J. Clark, *Proc. Natl. Acad. Sci. USA* **71** (1974) 3593.

[6.487] P. Lamothe, B. Baril, A. Chi, L. Lee, E. Baril, *Proc. Natl. Acad. Sci. USA* **78** (1981) 4723.

[6.488] M. S. Laskowski in P. D. Boyer (ed.): *The Enzymes,* vol. 5, Academic Press, New York 1961, p. 123.

[6.489] M. S. Laskowski in P. D. Boyer (ed.): *The Enzymes,* vol. 4, Academic Press, New York 1971, p. 289.

[6.490] M. S. Laskowski in P. D. Boyer (ed.): *The Enzymes,* vol. 4, Academic Press, New York 1971, p. 313.

[6.491] P. P. Lau, H. B. Gray Jr., *Nucleic Acids Res.* **6** (1979) 331.

[6.492] E. Lazarides, U. Lindberg, *Proc. Natl. Acad. Sci. USA* **71** (1974) 4742.

[6.493] M. Y. W. T. Lee, C.-K. Tan, K. M. Downey, A. G. So, *Biochemistry* **23** (1984) 1906.

[6.494] M. Y. W. T. Lee, C.-K, Tan, A. So, K. M. Downey, *Biochemistry* **19** (1980) 2096.

[6.495] R. J. Legerski, J. L. Hodnett, H. B. Gray Jr., *Nucleic Acids Res.* **5** (1978) 1445.

[6.496] I. R. Lehman, M. J. Bessman, E. S. Simms, A. Kornberg, *J. Biol. Chem.* **233** (1958) 163.

[6.497] I. R. Lehman, C. C. Richardson, *J. Biol. Chem.* **239** (1964) 233.

[6.498] S. Leinbach, J. M. Reno, L. Lee, A. F. Isbell, J. A. Baezi, *Biochemistry* **15** (1976) 426.

[6.499] J. Leis, G. Duyk, S. Johnson, M. L. Ongiaru, A. Skalka, *J. Virol.* **45** (1983) 727.

[6.500] D. Levens, A. Lustig, M. Rabinowitz, *J. Biol. Chem.* **256** (1981) 1474.

[6.501] C. J. Levin, S. Zimmermann, *J. Biol. Chem.* **251** (1976) 1767.

[6.502] C. C. Levy, T. P. Karpetsky, *J. Biol. Chem.* **255** (1980) 2153.

[6.503] M. K. Lewis, R. R. Burgess in P. D. Boyer (ed.): *The Enzymes,* vol. 15, Academic Press, New York 1982, p. 109.

[6.504] T.-H. Liao, J. Salnikow, S. Moore, W. H. Stein, *J. Biol. Chem.* **248** (1973) 1489.

[6.505] U. I. Lill, E. M. Behrendt, G. R. Hartmann, *Eur. J. Biochem.* **52** (1975) 411.

[6.506] J. R. Lillehaug, *Eur. J. Biochem.* **73** (1977) 499.

[6.507] J. R. Lillehaug, K. Kleppe, *Biochemistry* **14** (1975) 1225.

[6.508] J. R. Lillehaug, R. K. Kleppe, K. Kleppe, *Biochemistry* **15** (1976) 1858.

[6.509] S. Linn, I. R. Lehman, *J. Biol. Chem.* **240** (1965) 1287.

[6.510] S. Linn, I. R. Lehman, *J. Biol. Chem.* **240** (1965) 1294.

[6.511] W. S. Linsley, E. E. Penhoet, S. Linn, *J. Biol. Chem.* **252** (1977) 1235.

[6.512] J. W. Little in J. Chirikjian, T. S. Papas (eds.): *Gene Amplification and Analysis,* vol. 2, Elsevier, Amsterdam 1981, p. 135.

[6.513] W. Linxweiler, W. Hörz, *Nucleic Acids Res.* **10** (1982) 4845.

[6.514] L. F. Liu: "Mechanism Studies of DNA Replication and Genetic Replication," *ICN-UCLA Symp. Mol. Cell. Biol.* **19** (1980) 53.

[6.515] L. F. Liu, J. C. Wang, *Proc. Natl. Acad. Sci. USA* **75** (1978) 2098.

[6.516] L. F. Liu, C.-C. Liu, B. M. Alberts, *Nature (London)* **281** (1979) 456.

[6.517] D. Livingston, C. C. Richardson, *J. Biol. Chem.* **250** (1975) 470.

[6.518] S. Ljungquist, *J. Biol. Chem.* **252** (1977) 2808.

[6.519] R. E. Lockard, L. Riesner, J. N. Vournakis in J. G. Chirikijan, T. S. Papas (eds.): *Gene Amplification and Analysis,* vol. 2, Elsevier, Amsterdam 1981, p. 229.

[6.520] J. Maat, A. J. H. Smith, *Nucleic Acids Res.* **5** (1978) 4537.

[6.521] V. MacKay, S. Linn, *Biochim. Biophys. Acta* **349** (1974) 131.

[6.522] A. Maitra, D. Novogrodsky, D. Baltimore, J. Hurwitz, *Biochem. Biophys. Res. Commun.* **18** (1965) 801.

[6.523] M. B. Mann, H. O. Smith, *Nucleic Acids Res.* **4** (1977) 4211.

[6.524] H. G. Mannherz, R. S. Goody, M. Konrad, E. Nowak, *Eur. J. Biochem.* **104** (1979) 367.

[6.525] S. Martin, B. Moss, *J. Biol. Chem.* **250** (1975) 9330.

[6.526] S. Martin, E. Paoletti, B. Moss, *J. Biol. Chem.* **250** (1975) 9322.

[6.527] A. Matsukage, E. W. Bohn, S. H. Wilson, *Biochemistry* **14** (1975) 1006.

[6.528] A. Matsukage, M. Sivarajan, S. H. Wilson, *Biochemistry* **15** (1976) 5305.

[6.529] A. Matsukage, T. Tanabe, M. Yamaguchi, Y. N. Taguchi, M. Nishizawa, T. Takahashi, T. Takahashi, *Biochim. Biophys. Acta* **655** (1981) 269.

[6.530] A. M. Maxam, W. Gilbert, *Proc. Natl. Acad. Sci. USA* **74** (1977) 560.

[6.531] A. M. Maxam, W. Gilbert in L. Grosman, K. Moldave (eds.): *Methods in Enzymology,* vol. 65, Academic Press, New York 1980, p. 499.

[6.532] N. J. McGraw, J. N. Bailey, G. R. Cleaves, D. R. Dembrinski, C. R. Gocke, L. K. Joliffe et al., *Nucleic Acids Res.* **13** (1985) 6753.

[6.533] C. McHenry, W. Crow, *J. Biol. Chem.* **254** (1979) 1748.

[6.534] C. McHenry, A. Kornberg, *J. Biol. Chem.* **254** (1977) 6478.

[6.535] M. Mechali, J. Abadiedebat, A.-M. de Recondo, *J. Biol. Chem.* **255** (1980) 2114.

[6.536] D. A. Melton, *Proc. Natl. Acad. Sci. USA* **82** (1985) 144.

[6.537] D. A. Melton, P. A. Krieg, M. R. Rebagliati, T. Maniatis, K. Zinn, M. R. Green, *Nucleic Acids Res.* **12** (1984) 7035.

[6.538] C. Milcarek, B. Weiss, *J. Mol. Biol.* **68** (1972) 303.

[6.539] R. V. Miller, A. J. Clark, *J. Bacteriol.* **127** (1976) 794.

[6.540] J. P. Miller, G. R. Phillip, *J. Biol. Chem.* **246** (1971) 1274.

[6.541] H. I. Miller, A. D. Riggs, G. N. Gill, *J. Biol. Chem.* **248** (1973) 2621.

[6.542] M. A. Minks, S. Benvin, P. A. Maroney, C. Baglioni, *J. Biol. Chem.* **254** (1979) 5058.

[6.543] K. Mizuuchi, D. H. O'Dea, M. Gellert, *Proc. Natl. Acad. Sci. USA* **75** (1978) 5960.

[6.544] P. Modrich, Y. Anraku, I. R. Lehman, *J. Biol. Chem.* **248** (1973) 7495.

[6.545] P. Modrich, I. R. Lehman, *J. Biol. Chem.* **245** (1970) 3626.

[6.546] K. Mölling, *J. Virol.* **18** (1976) 418.

[6.547] K. Mölling, D. P. Bolognesi, W. B. Auer, W. Büsen, H. W. Plassmann, P. Hausen, *Nature (London)* **234** (1971) 240.

[6.548] E. Mössner, *Z. Physiol. Chem.* **361** (1980) 543.

[6.549] E. Mössner, M. Boll, G. Pfleiderer, *Hoppe-Seyler's Z. Physiol. Chem.* **361** (1980) 543.

[6.550] S. Moore in P. D. Boyer (ed.): *The Enzymes,* vol. 14, Academic Press, New York 1982, p. 281.

[6.551] C. F. Morris, H. Hama-Inaba, D. Mace, N. K. Senha, B. Alberts, *J. Biol. Chem.* **254** (1979) 6787.

[6.552] M. I. Moseman-McCoy, T. H. Lubben, R. I. Gumport, *Biochim. Biophys. Acta* **562** (1979) 149.

[6.553] R. E. Moses, C. C. Richardson, *Biochem. Biophys. Res. Commun.* **41** (1970) 1557.

[6.554] R. E. Moses, C. C. Richardson, *Biochem. Biophys. Res. Commun.* **41** (1970) 1565.

[6.555] N. E. Murray, S. A. Bruce, K. Murray, *J. Mol. Biol.* **132** (1979) 493.

[6.556] K. M. T. Muskavitch, S. Linn in P. D. Boyer (ed.): *The Enzymes,* vol. 14, Academic Press, New York 1982, p. 233.

[6.557] K. Nath, J. Hurwitz, *J. Biol. Chem.* **249** (1974) 3680.

[6.558] T. Naveh-Many, H. Cedar, *Proc. Natl. Acad. Sci. USA* **78** (1981) 4246.

[6.559] J. Nevins, W. Joklik, *J. Biol. Chem.* **252** (1977) 6939.

[6.560] N. G. Nossal, *J. Biol. Chem.* **249** (1974) 5668.

[6.561] N. G. Nossal, M. S. Hershfield, *J. Biol. Chem.* **246** (1971) 5414.

[6.562] N. G. Nossal, M. F. Singer, *J. Biol. Chem.* **243** (1968) 913.

[6.563] M. Novogrodsky, A. Tal, A. Traub, J. Hurwitz, *J. Biol. Chem.* **241** (1966) 2933.

[6.564] E. Ohtsuka, S. Tanaka, T. Tanaka, T. Miyake, A. F. Markham, E. Nakagawa et al., *Proc. Natl. Acad. Sci. USA* **78** (1981) 5493.

[6.565] H. Okayama, P. Berg, *Mol. Cell. Biol.* **2** (1982) 161.

[6.566] A. E. Oleson, M. Sasakuma, *Arch. Biochem. Biophys.* **204** (1980) 361.

[6.567] B. M. Olivera, I. R. Lehman, *J. Mol. Biol.* **36** (1968) 275.

[6.568] C. W. M. Orr, S. T. Herriott, M. J. Bessman, *J. Biol. Chem.* **240** (1965) 4625.

[6.569] Yu. A. Ovchinnikov, V. M. Lipkin, N. N. Modyanov, O. Yu. Chertov, Yu. V. Smirnov, *FEBS Lett.* **76** (1977) 108.

[6.570] Yu. A. Ovchinnikov, G. S. Monasturskaya, V. V. Gubanov, S. O. Guryev, O. Yu. Chertov, N. N. Modyanov et al., *Eur. J. Biochem.* **116** (1981) 621.

[6.571] Yu. A. Ovchinnikov, G. S. Monasturskaya, V. V. Gubanov, S. O. Guryev, I. S. Salomatina, T. M. Shuvaeva et al., *Dokl. Akad. Nauk SSSR* **261** (1981) 763.

[6.572] S. M. Panasenko, R. J. Alazard, I. R. Lehman, *J. Biol. Chem.* **253** (1978) 4590.

[6.573] N. Panayotatos, K. Truong, *Nucleic Acids Res.* **9** (1981) 5679.

[6.574] A. Panet, J. H. van de Sande, P. C. Loewen, H. G. Khorana, A. J. Raae, J. R. Lillehaug, K. Kleppe, *Biochemistry* **12** (1973) 5045.

[6.575] D. A. Peattie, *Proc. Natl. Acad. Sci. USA* **76** (1979) 1760.

[6.576] H. R. B. Pelham, R. J. Jackson, *Eur. J. Biochem.* **67** (1976) 247.

[6.577] A. Pesce, C. Casoli, G. Schito, *Nature (London)* **262** (1976) 412.

[6.578] B. H. Pheiffer, S. H. Zimmermann, *Biochemistry* **18** (1979) 2960.

[6.579] J. Pierre, J. Laval, *Biochemistry* **19** (1980) 5018.

[6.580] C. Portier, *Eur. J. Biochem.* **55** (1975) 573.

[6.581] P. A. Price, *J. Biol. Chem.* **250** (1975) 1981.

[6.582] P. A. Price, W. H. Stein, S. Moore, *J. Biol. Chem.* **244** (1969) 929.

[6.583] A. Quint, H. Cedar, *Nucleic Acids Res.* **9** (1981) 633.

[6.584] A. J. Raae, P. K. Kleppe, K. Kleppe, *Eur. J. Biochem.* **60** (1975) 437.

[6.585] C. M. Radding in L. Grosman, K. Moldave (eds.): *Methods in Enzymology,* vol. 21, Academic Press, New York 1971, p. 273.

[6.586] M. Radman, *J. Biol. Chem.* **251** (1976) 1438.

[6.587] U. L. Rajbhandary, *Fed. Proc. Fed. Am. Soc. Exp. Biol.* **39** (1980) 2815.

[6.588] K. Randerath, R. C. Gupta, E. Randerath in L. Grosman, K. Moldave (eds.): *Methods in Enzymology,* vol. 65, Academic Press, New York 1980, p. 638.

[6.589] K. Randerath, M. V. Reddy, R. C. Gupta, *Proc. Natl. Acad. Sci. USA* **78** (1981) 6126.

[6.590] R. N. Rao, S. G. Rogers, *Gene* **3** (1978) 247.

[6.591] R. L. Ratcliff in P. D. Boyer (ed.): *The Enzymes,* vol. 14, Academic Press, New York 1981, p. 105.

[6.592] R. K. Ray, R. Reuben, I. Molineux, M. Gefter, *J. Biol. Chem.* **249** (1974) 5379.

[6.593] A. Razin, H. Cedar, *Proc. Natl. Acad. Sci. USA* **74** (1977) 2725.

[6.594] A. Razin, A. D. Riggs, *Science* **210** (1980) 604.
[6.595] M. V. Reddy, R. C. Gupta, K. Randerath, *Anal. Biochem.* **117** (1981) 271.
[6.596] R. Reddy, D. Henning, P. Epstein, H. Busch, *Nucleic Acids Res.* **9** (1981) 5645.
[6.597] T. W. Reid, I. B. Wilson in P. D. Boyer (ed.): *The Enzymes,* vol. 4, Academic Press, New York 1971, p. 373.
[6.598] R. C. Reuben, M. L. Gefter, *Proc. Natl. Acad. Sci. USA* **70** (1973) 1846.
[6.599] G. M. Richards, D. J. Tutas, W. J. Wechter, M. Laskowski, *Biochemistry* **6** (1967) 2908.
[6.600] C. C. Richardson, *Proc. Natl. Acad. Sci. USA* **54** (1965) 158.
[6.601] C. C. Richardson, *J. Biol. Chem.* **241** (1966) 2084.
[6.602] C. C. Richardson, *Proc. Natl. Acad. Sci. USA* **54** (1968) 158.
[6.603] C. C. Richardson in P. D. Boyer (ed.): *The Enzymes,* vol. 14, Academic Press, New York 1982, p. 299.
[6.604] C. C. Richardson, A. Kornberg, *J. Biol. Chem.* **239** (1964) 242.
[6.605] C. C. Richardson, I. R. Lehman, A. Kornberg, *J. Biol. Chem.* **239** (1964) 251.
[6.606] C. C. Richardson, A. Kornberg, *J. Biol. Chem.* **244** (1964) 2996.
[6.607] C. C. Richardson, L. J. Romano, R. Kolodner, J. E. Leclerq, F. Tamanoni, M. J. Enlges et al., *Cold Spring Harbor Symp. Quant. Biol.* **43** (1978) 449.
[6.608] P. W. J. Rigby, M. Dieckmann, C. Rhodes, P. Berg, *J. Mol. Biol.* **113** (1977) 237.
[6.609] M. Robert-Guroff, A. W. Schrecker, B. J. Brinkmann, R. C. Gallo, *Biochemistry* **16** (1977) 2866.
[6.610] H. D. Robertson, J. J. Dunn, *J. Biol. Chem.* **250** (1975) 3050.
[6.611] H. D. Robertson, R. E. Webster, N. D. Zinder, *J. Biol. Chem.* **243** (1968) 82.
[6.612] S. G. Rogers, B. Weiss in L. Grosman, K. Moldave (eds.): *Methods in Enzymology,* vol. 65, Academic Press, New York 1980, p. 201.
[6.613] S. G. Rogers, B. Weiss, *Gene* **11** (1980) 187.
[6.614] J. K. Rose, *Cell* **19** (1980) 415.
[6.615] F. Rougeon, P. Kourilsky, B. Mach, *Nucleic Acids Res.* **2** (1975) 2365.
[6.616] R. Roychoudhury, E. Jay, R. Wu, *Nucleic Acids Res.* **3** (1976) 863.
[6.617] B. Royer-Pokora, L. K. Gordon, W. A. Haseltine, *Nucleic Acids Res.* **9** (1981) 4595.
[6.618] G. W. Rushizky, V. A. Shaternikov, J. H. Mozejko, H. A. Sober, *Biochemistry* **14** (1975) 4221.
[6.619] J. Salnikow, S. Moore, W. H. Stein, *J. Biol. Chem.* **245** (1970) 5685.
[6.620] W. Salser in A. M. Chakrabarty (ed.): *Genetic Engineering,* vol. 1, CRC Press, West Palm Beach, Florida, 1979, p. 53.
[6.621] H. Samanta, J. P. Dougherty, P. Lengyel, *J. Biol. Chem.* **255** (1980) 9807.
[6.622] F. Sanger, S. Nicklen, A. R. Coulson, *Proc. Natl. Acad. Sci. USA* **74** (1977) 5463.
[6.623] H. Sano, G. Feix, *Biochemistry* **13** (1974) 5110.
[6.624] H. Sano, G. Feix, *Eur. J. Biochem.* **71** (1976) 577.
[6.625] K. Sato, F. Egami, *J. Biochem.* (*Tokyo*) **44** (1957) 753.
[6.626] P. Schedl, J. Roberts, P. Primakoff, *Cell* **8** (1976) 581.
[6.627] R. D. Schiff, O. P. Grandgenett, *J. Virol.* **36** (1980) 889.
[6.628] J. F. Scott, A. Kornberg, *J. Biol. Chem.* **253** (1978) 3292.
[6.629] G. Seal, L. A. Loeb, *J. Biol. Chem.* **251** (1976) 975.
[6.630] A. Sentenac, J. M. Buhler, A. Ruet, J. Huet, F. Iborra, P. Fromageot in B. F. C. Clark, H. Klenow, J. Zeuthen (eds.): *Gene Expression. Protein Synthesis and Control. RNA Synthesis and Control. Chromatin Structure and Function,* Pergamon Press, New York 1977, p. 187.
[6.631] P. Setlow, D. Brutlag, A. Kornberg, *J. Biol. Chem.* **247** (1972) 224.
[6.632] P. Setlow, A. Kornberg, *J. Biol. Chem.* **247** (1972) 232.
[6.633] V. Sgaramella, *Proc. Natl. Acad. Sci. USA* **69** (1972) 3389.
[6.634] V. Sgaramella, H. G. Khorana, *Mol. Biol.* **72** (1972) 493.

[6.635] V. Sgaramella, J. H. van de Sande, H. G. Khorana, *Proc. Natl. Acad. Sci. USA* **67** (1970) 1468.

[6.636] M. F. Shemyakin, A. A. Grepachevsky, A. V. Chestukhin, *Eur. J. Biochem.* **98** (1979) 417.

[6.637] V. Shen, D. Schlessinger in P. D. Boyer (ed.): *The Enzymes,* vol. 15, Academic Press, New York 1982, p. 501.

[6.638] T. E. Shenk, C. Rhodes, P. W. J. Rigby, P. Berg, *Proc. Natl. Acad. Sci USA* **72** (1975) 989.

[6.639] H. M. Shepard, B. Polisky in L. Grosman, K. Moldave (eds.): *Methods in Enzymology,* vol. 68, Academic Press, New York 1980, p. 503.

[6.640] Y. Shimura, H. Sakano, F. Nagawa, *Eur. J. Biochem.* **86** (1978) 267.

[6.641] J. Shinshi, M. Miwa, K. Kato, M. Noguchi, T. Matsushima, T. Sugimura, *Biochemistry* **15** (1976) 2185.

[6.642] R. G. Shorenstein, R. Losick, *J. Biol. Chem.* **248** (1973) 6163.

[6.643] S. Shuman, J. Hurwitz, *J. Biol. Chem.* **254** (1979) 10 396.

[6.644] S. Shuman, J. Hurwitz in P. D. Boyer (ed.): *The Enzymes,* vol. 15, Academic Press, New York 1982, p. 245.

[6.645] S. Shuman, M. Surks, H. Furneaux, J. Hurwitz, *J. Biol. Chem.* **255** (1980) 11 588.

[6.646] R. Silber, V. G. Malathi, J. Hurwitz, *Proc. Natl. Acad. Sci. USA* **69** (1972) 3009.

[6.647] M. Silberklang, A. M. Gillum, U. L. Rajbhandary, *Nucleic Acids Res.* **4** (1977) 4091.

[6.648] M. Silberklang, A. M. Gillum, U. L. Rajbhandary in L. Grosman, K. Moldave (eds).: *Methods in Enzymology,* vol. 59, Academic Press, New York 1979, p. 58.

[6.649] M. Silberklang, A. Prochiantz, A.-L. Haenni, U. L. Rajbhandary, *Eur. J. Biochem.* **72** (1977) 465.

[6.650] A. Simoncsits, G. G. Brownlee, R. S. Brown, J. R. Rubin, H. Guilley, *Nature (London)* **269** (1977) 833.

[6.651] M. Simsek, J. Ziegenmeyer, J. Heckman, U. L. Rajbhandary, *Proc. Natl. Acad. Sci. USA* **70** (1973) 1041.

[6.652] A. Sippel, *Eur. J. Biochem.* **37** (1973) 31.

[6.653] K. Sirotkin, W. Cooley, J. Runnels, L. R. Snyder, *J. Mol. Biol.* **123** (1978) 221.

[6.654] J. P. Slater, A. S. Mildvan, L. A. Loeb, *Biochem. Biophys. Res. Commun.* **44** (1971) 37.

[6.655] A. J. H. Smith, *Nucleic Acids Res.* **6** (1979) 831.

[6.656] D. Smith, R. Ratcliff, D. Williams, A. Martinez, *J. Biol. Chem.* **242** (1967) 590.

[6.657] G. E. Smith, M. D. Summers, *Anal. Biochem.* **109** (1980) 123.

[6.658] H. J. Smith, L. Bogorad, *Proc. Natl. Acad. Sci. USA* **71** (1974) 4839.

[6.659] H. O. Smith, *Science* **205** (1979) 455.

[6.660] H. O. Smith, M. L. Birnstiel, *Nucleic Acids Res.* **3** (1976) 2387.

[6.661] M. Smith, S. Gillam in J. K. Setlow, A. Hollaender (eds.): *Genetic Engineering. Principle and Methods,* vol. 3, Plenum Press, New York 1981, p. 1.

[6.662] D. Smoler, I. Molineux, D. Baltimore, *J. Biol. Chem.* **246** (1971) 7697.

[6.663] S. Soderhall, T. Lindahl, *FEBS Lett.* **67** (1976) 1.

[6.664] D. A. Soltis, O. C. Uhlenbeck, *J. Biol. Chem.* **257** (1982) 11 332.

[6.665] D. A. Soltis, O. C. Uhlenbeck, *J. Biol. Chem.* **257** (1982) 11 340.

[6.666] E. M. Southern, *J. Mol. Biol.* **98** (1975) 503.

[6.667] S. Spadari, A. Weissbach, *J. Biol. Chem.* **249** (1974) 5809.

[6.668] P. F. Spahr, R. F. Gasteland, *Proc. Natl. Acad. Sci. USA* **59** (1968) 876.

[6.669] P. F. Spahr, B. R. Hollingworth, *J. Biol. Chem.* **236** (1961) 823.

[6.670] K. S. Sriprakash, N. Lundh, M. M. Huh, C. M. Radding, *J. Biol. Chem.* **250** (1975) 5438.

[6.671] A. Srivastava, M. J. Modak, *J. Biol. Chem.* **255** (1980) 2000.

[6.672] E. Stackebrandt, W. Ludwig, K. H. Schleifer, H. J. Gross, *J. Mol. Evol.* **17** (1981) 227.

[6.673] S. Stahl, K. Zinn, *J. Mol. Biol.* **148** (1981) 481.

[6.674]  D. M. Stalker, D. W. Mosbaugh, R. R. Meyer, *Biochemistry* **15** (1976) 3114.

[6.675]  B. C. Stark, R. Kole, E. J. Bowman, S. Altman, *Proc. Natl. Acad. Sci. USA* **75** (1978) 3711.

[6.676]  R. Stein, Y. Gruenbaum, Y. Pollack, A. Razin, H. Cedar, *Proc. Natl. Acad. Sci. USA* **78** (1982) 4246.

[6.677]  H. Sternbach, F. von der Haar, E. Schlimme, E. Gaertner, F. Cramer, *Eur. J. Biochem.* **22** (1971) 166.

[6.678]  H. Sternbach, M. Srpinzl, J. B. Hobbs, F. Cramer, *Eur. J. Biochem.* **67** (1976) 215.

[6.679]  G. L. Stetler, G. J. King, W. M. Huang, *Proc. Natl. Acad. Sci. USA* **76** (1979) 3737.

[6.680]  K. O. Stetter, W. Zillig, *Eur. J. Biochem.* **48** (1974) 527.

[6.681]  A. Sugino, H. M. Goodman, H. L. Heyneker, J. Shine, H. W. Boyer, N. R. Cozarelli, *J. Biol. Chem.* **249** (1977) 3680.

[6.682]  A. Sugino, T. J. Snopek, N. Cozarelli, *J. Biol. Chem.* **252** (1977) 1732.

[6.683]  D. Sutter, W. Doerfler, *Proc. Natl. Acad. Sci. USA* **77** (1980) 253.

[6.684]  K. S. Szeto, D. Soell, *Nucleic Acids Res.* **1** (1974) 171.

[6.685]  J. W. Szostak, J. I. Stiles, B.-K. Tye, P. Chiu, F. Sherman, R. Wu in L. Grosman, K. Moldave (eds.): *Methods in Enzymology,* vol. 68, Academic Press, New York 1980, p. 419.

[6.686]  S. Tabor, C. C. Richardson, *Proc. Natl. Acad. Sci. USA* **82** (1985) 1074.

[6.687]  K. Takahashi, *J. Biochem.* **49** (1961) 1.

[6.688]  J. L. Telford, A. Kressmann, R. A. Koski, R. Grosschedl, F. Müller, S. G. Clarkson, M. L. Birnstiel, *Proc. Natl. Acad. Sci. USA* **76** (1979) 2590.

[6.689]  S. M. Tilghman, P. J. Curtis, D. C. Tiemeier, P. Leder, C. Weissmann, *Proc. Natl. Acad. Sci. USA* **75** (1978) 1309.

[6.690]  H. Tanuchi, C. B. Anfinsen, A. Sodja, *J. Biol. Chem.* **242** (1967) 4752.

[6.691]  A. Torriani in S. P. Colowick, N. O. Kaplan (eds.): *Methods in Enzymology,* vol. 12, Academic Press, New York 1968, p. 212.

[6.692]  H. C. Towle, J. F. Jolly, J. A. Boezi, *J. Biol. Chem.* **250** (1975) 1723.

[6.693]  C. Tsiapalis, J. Dorson, F. Bollum, *Eur. J. Biochem.* **250** (1975) 4486.

[6.694]  C.-P. D. Tu, S. N. Cohen, *Gene* **10** (1980) 177.

[6.695]  C.-P. D. Tu, R. Wu in L. Grosman, K. Moldave (eds.): *Methods in Enzymology,* vol. 65, Academic Press, New York 1980, p. 620.

[6.696]  A. Ueno, A. Ishihama, *J. Biochem. (Tokyo)* **91** (1982) 323.

[6.697]  O. C. Uhlenbeck, V. Cameron, *Nucleic Acids Res.* **4** (1977) 85.

[6.698]  O. C. Uhlenbeck, R. I. Gumpori in P. D. Boyer (ed.): *The Enzymes,* vol. 15, Academic Press, New York 1982, p. 31.

[6.699]  A. Ullrich, J. Shine, J. Chirgwin, R. Pictet, E. Tischer, W. J. Rutter, H. M. Goodman, *Science* **196** (1977) 1313.

[6.700]  P. Valenzuela, G. I. Bell, F. Weinberg, W. Rutter in G. Stein, J. Stein, L. J. Kleinsmith (eds.): *Methods in Cell Biology,* vol. 19, Academic Press, New York 1978, p. 1.

[6.701]  L. Vardimon, I. Kuhlmann, W. Doerfler, H. Cedar, *Eur. J. Cell Biol.* **25** (1981) 13.

[6.702]  I. M. Verma, *J. Virol.* **15** (1975) 843.

[6.703]  I. M. Verma, *Biochim. Biophys. Acta* **473** (1977) 1.

[6.704]  C. Vincent, P. Tchen, M. Cohen-Solal, P. Kourilsky, *Nucleic Acids Res.* **10** (1982) 6787.

[6.705]  V. M. Vogt, *Eur. J. Biochem.* **33** (1973) 192.

[6.706]  V. M. Vogt in L. Grosman, K. Moldave (eds.): *Methods in Enzymology,* vol. 65, Academic Press, New York 1980, p. 248.

[6.707]  J. N. Vournakis, A. Efstratiadis, F. C. Kafatos, *Proc. Natl. Acad. Sci. USA* **72** (1975) 2959.

[6.708]  C. Waalwijk, R. A. Flavell, *Nucleic Acids Res.* **5** (1978) 3231.

[6.709]  G. C. Walker, O. C. Uhlenbeck, E. Bedows, R. I. Gumport, *Proc. Natl. Acad. Sci. USA* **72** (1975) 122.

[6.710] J. C. Wang, *J. Mol. Biol.* **55** (1971) 523.

[6.711] T. S.-F. Wang, W. D. Sedwick, D. Korn, *J. Biol. Chem.* **249** (1974) 841.

[6.712] R. F. Weaver, C. Weissmann, *Nucleic Acids Res.* **7** (1979) 1175.

[6.713] U. Weidle, C. Weissmann, *Nature (London)* **303** (1983) 442.

[6.714] B. Weiss, *J. Biol. Chem.* **251** (1976) 1896.

[6.715] B. Weiss in P. D. Boyer (ed.): *The Enzymes,* vol. 14, Academic Press, New York 1982, p. 203.

[6.716] B. Weiss, *J. Biol. Chem.* **251** (1976) 1968.

[6.717] B. Weiss, A. Jacquemin-Sablon, T. R. Live, C. G. Fareed, C. C. Richardson, *J. Biol. Chem.* **243** (1968) 4543.

[6.718] B. Weiss, S. G. Rogers, A. F. Taylor in P. C. Hanawalt, E. C. Friedberg, C. F. Fox (eds.): *DNA Repair Mechanisms,* Academic Press, New York 1978, p. 191.

[6.719] B. Weiss, A. Thompson, C. C. Richardson, *J. Biol. Chem.* **243** (1968) 4556.

[6.720] G. Wengler, G. Wengler, *Nature (London)* **282** (1979) 754.

[6.721] O. Westergaard, D. Brutlag, A. Kornberg, *J. Biol. Chem.* **248** (1973) 1361.

[6.722] R. B. Wickner, B. Ginsberg, I. Berkower, J. Hurwitz, *J. Biol. Chem.* **247** (1972) 489.

[6.723] S. Wickner, *Proc. Natl. Acad. Sci. USA* **73** (1976) 3511.

[6.724] W. Wickner, A. Kornberg, *Proc. Natl. Acad. Sci. USA* **70** (1973) 3679.

[6.725] R. G. Wiegand, G. N. Godson, C. M. Radding, *J. Biol. Chem.* **250** (1975) 8848.

[6.726] M. Wigler, D. Levy, M. Perucho, *Cell* **24** (1981) 33.

[6.727] K. W. Wilcox, M. Orlosky, E. A. Friedman, H. O. Smith, *Fed. Proc. Fed. Am. Soc. Exp. Biol.* **34** (1975) 515.

[6.728] F. R. Williams, T. Godefroy, E. Mery, J. Yon, M. Grunberg- Manago, *Biochem. Biophys. Res. Commun.* **19** (1965) 25.

[6.729] G. G. Wilson, N. E. Murray, *J. Mol. Biol.* **132** (1979) 471.

[6.730] G. Winter, G. G. Brownlee, *Nucleic Acids Res.* **5** (1978) 3129.

[6.731] M. Winters, M. Edmonds, *J. Biol. Chem.* **248** (1973) 4756.

[6.732] C. R. Woese, G. E. Fox, *Proc. Natl. Acad. Sci. USA* **74** (1977) 5088.

[6.733] M. Wright, G. Buttin, J. Hurwitz, *J. Biol. Chem.* **246** (1971) 6543.

[6.734] D. M. Yaiko, M. C. Valentine, B. Weiss, *J. Mol. Biol.* **85** (1974) 323.

[6.735] Y. Yamada, H. Ishikura, *Biochim. Biophys. Acta* **402** (1975) 285.

[6.736] M. Yamaguchi, A. Matsukage, T. Takahashi, *J. Biol. Chem.* **255** (1980) 7002.

[6.737] M. Yamaguchi, K. Tanabe, Y. N. Taguchi, M. Nishizawa, T. Takahashi, A. Matsukage, *J. Biol. Chem.* **255** (1980) 9942.

[6.738] J. Yarranton, R. H. Das, M. L. Gefter, *J. Biol. Chem.* **254** (1979) 12002

[6.739] J. Yarranton, M. L. Gefter, *Proc. Natl. Acad. Sci. USA* **76** (1979) 1658.

[6.740] O. J. Yoo, P. Dwyer-Hallquist, K. L. Agarwal, *Nucleic Acids Res.* **10** (1982) 6511.

[6.741] H. Youssoufian, C. Mulder, *J. Mol. Biol.* **150** (1981) 133.

[6.742] K. Zechel, K. Weber, *Eur. J. Biochem.* **77** (1977) 133.

[6.743] W. Zillig, P. Palm, A. Heil in R. Losick, M. Chamberlin (eds.): *RNA Polymerase,* Cold Spring Harbor Laboratory, Cold Spring Harbor, New York 1976, p. 101.

[6.744] S. B. Zimmermann, B. H. Pheiffer in P. D. Boyer (ed.): *The Enzymes,* vol. 14, Academic Press, New York 1982, p. 315.

[6.745] K. Zinn, D. Dimaio, T. Maniatis, *Cell* **34** (1983) 865.

References for Chapter 7

[7.1]  K. Aunstrup: "Production, Isolation and Economics of Extracellular Enzymes," in *Applied Biochemistry and Bioengineering*, Academic Press, New York 1979.
[7.2]  K. Aunstrup: *Enzyme Technology,* "Problems and possibilities," *Proc. Third Eur. Congress on Biotechnology,* München **4** (1984) 143.

References for Chapter 8

[8.1]  W. H. B. Denner, R. D. Farrow, J. R. Reichelt in T. Godfrey, J. R. Reichelt (eds.): *Industrial Enzymology,* The Nature Press, New York 1983, pp. 10, 136, and 157–169.

# 10. Index

Tubular bowl centrifuge  49
Turnover number  14
(+)-(4R)-Twistanone  [*13537-95-6*]  146, 147
L-Tyrosine  [*60-18-4*]
  from phenol, pyruvate, and ammonia  140
L-Tyrosine phenol lyase (E.C. 4.1.99.2)  [*9059-31-8*]
  140

UDP– glucose epimerase  149
UDP– glucose pyrophosphorylase  149
Ukidan  181
Ultrafiltration
  for enzyme concentration  52
Ultrasound
  for cell disruption  45, 46
Ultrogele  57
Unhairing
  in leather production, enzymes in  116
Units
  definition of, for enzyme activity  **20**
  definition of, for restriction endonucleases  28
  in enzymology  **77**
  international, of enzyme activity  20
Urate oxidase (E.C. 1.7.3.3)  [*9002-12-4*]  159
Urea  [*57-13-6*]
  enzymatic analysis  163, 178
Urease (E.C. 3.5.1.5)  [*9002-13-5*]  159, 163, 178
  covalent chromatography  59
  inhibition by ATP  18
  from jack bean meal  2
Uricase (E.C. 1.7.3.3)  [*9002-12-4*]  159
Urokinase (E.C. 3.4.21.31)  [*9039-53-6*]  161
  in thrombolytic therapy  181, 182
Ustilago maydis nuclease, *see Nuclease, DNA*
UV absorption
  for protein determination  29

Vaccina virus DNA polymerase, *see Polymerase, DNA*
L-Valine  [*72-18-4*]
  enzymatic production  131, 132, 135
  enzymatic synthesis  136
  production with immobilized aminoacylase  74
Vegetable processing
  enzymes in  126
Velyn  182
Venacil  180

Viscosimetry
  for enzyme analysis  25
Vitamin C  [*50-81-7*]
  *see also Ascorbic acid*
  enzymatic analysis  171

Washing performance
  of detergent enzymes  112
Waste disposal
  in enzyme production  62
Waste water
  analysis of urea in  178
Whey
  lactose cleavage in  123
  recycling of rennet enzymes from  123
Wine
  enzymatic detection of sugar in  168
  enzymatic glycerol analysis in  175
Wine production
  enzymes in  127
Woodward's reagent K
  for enzyme immobilization  68
Workplace safety  253
Wound debridement  180
Wound healing
  enzymes in  180

Xanthine oxidase (E.C. 1.1.3.22)  [*9002-17-9*]  159
  in milk  120
Xanthomonas oryzae phage XP12  207
*Xba*I, *see Restriction endonuclease*
*Xho*I, *see Restriction endonuclease*
*Xho*II, *see Restriction endonuclease*
*Xma*I, *see Restriction endonuclease*
*Xmm*I, *see Restriction endonuclease*
Xylanases  83
Xylitol  [*87-99-0*]
  enzymatic analysis  175
Xylose isomerase (E.C. 5.3.1.5)  [*9023-82-9*]  103

Yeast invertase (E.C. 3.2.1.26)  [*9001-57-4*]
  in sucrose inversion  130
Yogurt
  galactose determination in  168

Zone electrophoresis
  for enzyme purification  53